COMPUTER TECHNIQUES
in
POWER SYSTEM ANALYSIS

Third Edition

COMPUTER TECHNIQUES *in* POWER SYSTEM ANALYSIS

Third Edition

M A Pai

Department of Electrical and Computer Engineering
University of Illinois at Urbana Champaign
USA

Dheeman Chatterjee

Department of Electrical Engineering
Indian Institute of Technology Kharagpur
India

McGraw Hill Education (India) Private Limited

NEW DELHI

McGraw Hill Education Offices

New Delhi New York St Louis San Francisco Auckland Bogotá Caracas
Kuala Lumpur Lisbon London Madrid Mexico City Milan Montreal
San Juan Santiago Singapore Sydney Tokyo Toronto

McGraw Hill Education (India) Private Limited

Published by McGraw Hill Education (India) Private Limited,
P-24, Green Park Extension, New Delhi 110 016.

Computer Techniques in Power System Analysis, 3e

ISBN 13: 978-93-392-1367-1
ISBN 10: 93-392-1367-X

Managing Director: *Kaushik Bellani*

Head—Higher Education (Publishing and Marketing): *Vibha Mahajan*

Senior Publishing Manager (SEM & Tech. Ed.): *Shalini Jha*
Editorial Executive: *Koyel Ghosh*
Manager—Production Systems: *Satinder S Baveja*
Assistant Manager—Editorial Services: *Sohini Mukherjee*
Senior Manager—Production: *P L Pandita*

Assistant General Manager—Higher Education (Marketing): *Vijay Sarathi*
Assistant Product Manager: *Tina Jajoriya*
Senior Graphic Designer—Cover: *Meenu Raghav*

General Manager—Production: *Rajender P Ghansela*
Manager—Production: *Reji Kumar*

Typeset at Script Makers, 19, A1-B, DDA Market, Paschim Vihar, New Delhi 110 063 and printed at ****

Cover Designer:

CONTENTS

PREFACE

Background

The first edition of the book was the result of teaching an elective for the final-year undergraduate class at IIT Kanpur. The content of this book has also been used for several short courses for industry, to convey the essential system concepts involved in power network analysis to EE and non-EE audiences. Portions of the book have been class tested at the University of Illinois in a senior-level elective course on "Operation and Control of Power Systems."

Increase in the demand for electricity is a constant feature across the globe. This necessitates newer sources of energy and larger and more complex transmission and distribution networks. In the last few decades, there have been numerous grid failures affecting the lives of millions of people. Hence, a detailed study is required for running the existing power systems and planning their extension. Academicians as well as practicing engineers are switching over to a study of the numerous problems associated with planning and operation of large interconnected power systems. A power system analysis and operation course deals with the functioning principles of the power system—component-wise and as a whole. However, computer-based simulation is almost mandatorily associated with all these analyses. An effort has been made in this book to bridge the gap between the physics of the power system and the simulation. If a systematic solution of the problems of power systems is to be attempted, a necessary prerequisite would be firm theoretical and conceptual clarity in modern power system analysis. This book is an effort in this direction. It seeks to build on the courses the student has already had in the area of circuit and system theory. An earlier exposition to a course in power system analysis is preferred, though not essential.

Target Audience

The book can be used for final-year classes in electrical engineering or a first course at the postgraduate level. The material can be comfortably covered in one semester. The set of exercises at the end of the book is meant to test the students' understanding of the subject as well as to give them computational experience.

The Revision

The revision was undertaken with Dr. Dheeman Chatterjee as co-author in response to the demands of the changing power sector scenario. Development of power electronics technology suitable for use in the voltage and power levels of transmission and distribution has provided power system operators and planners a new tool for better management of the system. Application of power electronics adds flexibility and capability of fast control and economy. For example, high-voltage direct current (HVDC) transmission comes out to be cheaper in spite of the large cost of terminal equipment when used for carrying power over a long distance. Flexible AC transmission system (FACTS) helps utilize the transmission infrastructure better by enabling operation closer to the limits.

The depleting sources of fossil fuel and their environmental impacts are forcing decision-makers to look for alternate renewable energy sources. The share of wind power, the cheapest renewable source, is increasing exponentially in different parts of the globe including the US and India. The variable nature of the source poses questions on possible impacts on the power system. Hence, future power system analysis and planning must include FACTS, HVDC and renewable sources like wind. The present revision has taken up these as the need of the hour. The modeling of HVDC and FACTS has been included in Chapter 2 along with the modeling of other power system components. Subsequently, the techniques for inclusion of FACTS and HVDC in load flow and transient stability analysis have been discussed in chapters 6 and 8 respectively. A new chapter (Chapter 9) on wind energy system has been included. This includes a brief discussion of the methods for inclusion of the wind turbine and induction generator (both singly excited and doubly fed) in power system analysis programes like load flow and stability.

One important aspect of power system operation is optimal power flow considering the economic load dispatch. A brief introduction to that has been provided as a new chapter (Chapter 7) to make the content more complete.

A summary of the new features is given below:

- New chapter on Economic Load Dispatch
- New chapter on Renewable Sources—Wind Energy Systems
- Comprehensive updates on HVDC, FACTS and WP
- Refreshed solved examples and illustrations

Salient Features

- Detailed study of the nature of problems requiring computational techniques in power system planning and operation
- Clear and rigorous description of bus impedance algorithm, short-circuit studies, power flow studies, etc.
- Systematic presentation of load flow, transient stability studies of multi-machine systems with emphasis on numerical simulation
- Test data of IEEE standard systems for ready reference
- Rich pedagogy includes
    - Illustrations: **134**
    - Unsolved numerical problems: **49**
    - Solved examples: **33**

Organization

There are nine chapters in the book. A brief chapter-wise overview is as follows.

Chapter 1 gives an overview of the nature of problems arising during the planning and operation of power systems, which require computational techniques to solve.

Chapter 2 provides the terminal relations of the components of modern power systems including generators, transformers, transmission lines (HVAC and HVDC) and FACTS controllers.

Chapter 3 is devoted to an elementary exposition to linear graph theory and its use in deriving the multi-terminal representation of the transmission network.

Chapter 4 contains the discussion of the algorithm for developing the bus impedance matrix of the transmission network based on graph theory.

Chapter 5 describes short circuit analysis using the network representations where both phase impedance form and symmetrical component formulations are presented.

Chapter 6 has a systematic presentation of the Gauss-Seidel method, the exact Newton-Raphson method and its various simplifications leading up to the DC load flow. The method for inclusion of FACTS and HVDC in the load-flow programme is also presented.

Chapter 7 has a brief introduction of economic load dispatch and optimal power flow.

Chapter 8 deals with the transient stability studies of multi-machine systems with emphasis on numerical simulation. Inclusion of FACTS controllers in stability studies is discussed.

Chapter 9 includes a brief discussion of the methods for inclusion of wind generation in power system analysis programmes like load flow and stability.

Acknowledgements

The initial idea of writing the book started in 1970-71 when the first author conducted a series of two short courses for power engineers at the Memorial University of Newfoundland, Canada. The discussions with Professor H K Kesavan at the University of Waterloo were very helpful in clarifying many of the concepts of graph theory as applied to power systems. The bus impedance algorithm, as developed in Chapter 4 with a firm topological basis, is based on one of the chapters in the PhD dissertation of Dr R Jegatheesan at IIT Kanpur, done under the supervision of the first author. The load flow chapter owes a lot to an excellent review article by B Stott, appearing in the July 1974 special issue of *Proceeding of IEEE*. In the working and checking of various numerical examples, the first author would like to thank his students Brijendra Sharma, R Radhakrishna, and N Armugam. The first author wishes to acknowledge gratefully, the support received from IIT Kanpur.

For the second edition, the first author would like to thank Terry Peterson for the excellent typing job as well as doing the drawings in the book. The first author would also like to thank the University of Illinois and the Power Affiliates Program for the support in completing this project.

For the third edition, the authors would like to express immense sense of gratitude towards Dr Tony B Nguyen for his help in writing Chapter 7 on Optimal Power Flow. In addition to his contributions towards the content and the examples, his suggestions have helped the correctness and logical flow of the whole chapter including the economic load dispatch part. The authors would also like to thank Mr Arghya Mitra of IIT Kharagpur for his help in the computer solutions of the examples in Chapter 9 (Wind Energy System).

M A Pai
University of Illinois, USA

Dheeman Chatterjee
IIT Kharagpur
14 April, 2014 India

Publisher's Note

McGraw-Hill Education (India) invites suggestions and comments from you, all of which can be sent to *info.india@mheducation.com* (kindly mention the title and author name in the subject line).

Piracy-related issues may also be reported.

1 INTRODUCTION

1.1 CHANGING SCENARIO OF POWER AND ENERGY SYSTEMS

The power industry is undergoing vast changes on a global basis. Perhaps the most significant change is the restructuring of the industry. The old concept of a vertically integrated utility (VIU) owning generation, transmission, and distribution has given way to a system where the generation part is decoupled from the rest of the system to form separate entities and they, along with other independent power producers (IPPs), will operate in a competitive environment to supply power to the end users. The transmission network, however, is not amenable to restructuring. While generation represents nearly half the cost, transmission represents about 8% of the cost. Competitive markets mean competitive generation. The transmission network does not lend itself to competition. It has to be controlled by the owners, namely, the different transmission companies. In India, however, transmission is the sole responsibility of the Power Grid Corporation of India Limited (PGCIL). The operation and control is the responsibility of the Power System Operation Corporation (POSOCO) Limited. The control of transmission is important since it determines which generators get access to the load. Hence, it is called open access transmission. At the same time, the transmission entity, separate from the generating companies, should not have a material interest in deciding which generators are going to be successful in selling power either in terms of the price or the customer. In addition to the transmission companies, at lower voltages, we may have distribution companies (called DISCOs).

Since the transmission system obeying the laws of physics controls the flow of power, it was recognized early that power industry deregulation would not occur at the same speed as other industries. Historically, the UK. power industry was the first to get deregulated. However, for them it was easier since the whole of UK. was a VIU prior to deregulation. They do not have an Independent System Operator (ISO) but a for-profit, regulated independent transmission company (ITC). The UK. model is therefore more appropriate for the developing countries. The Nordic countries in Europe (Norway, Sweden, and Finland) have formed a pool called the Nord Pool, and deregulation has worked very well there. In the US, the power industry is fully deregulated. In India, the power sector reforms are being implemented, first at the state level with DISCOs throughout the country and then at the All-India level. At the All-India level, the transmission is controlled by the Power Grid Corporation of India (PGCIL). The bulk control of power is vested in the National Load Dispatch Control

Centre (NLDC) with the Regional Load Dispatch Centres (RLDC). The entire country is now a single grid with the common frequency of 50 Hz. Power can be dispatched from one part of the country to another. GPS based phasor measurement units (PMU) at various substations enable the operators to have real time information about the system variables—referred to as a situational awareness of the system.

Another major change that has picked momentum in the past few years is the introduction of a larger share of renewable power generation. Till today, the major share of electrical power comes from different forms of fossil fuels. The damaging effect of the use of fossil fuel on the environment is well known. The reserves of coal and petroleum are also limited. Large hydro power and nuclear power are two major alternatives. However, setting up of a large hydro generating facility involves the construction of big dams which requires displacement of large populations. In addition, it has other environmental impacts also. With nuclear power, there is always a perception of risk whose enormity gives rise to resistance from the civilians. So, decision makers across the globe are looking out for alternate renewable sources to augment power generation.

The two major sources which presently provide the bulk of the renewable power generation are wind and solar power. From the Indian perspective, India has an installed capacity of 28.2 GW of renewable sources which is just above 10% of the total installed capacity of 227 GW [36]. Out of this, about 18.5 GW comes from wind and 1.5 GW from solar [37]. Also there is a much larger potential for both of these sources and efforts are there to increase the installed capacity of the renewables at a fast rate. Now, the main problem of these renewable sources is that they are variable in nature. They vary over the year and also over the day. Hence, there is a challenge of integrating these renewable energy sources (RES) into the grid. Therefore, it becomes important to include the effects of these sources in different power system studies, e.g., load flow and stability studies. The methods of modeling wind power for these power system studies will be discussed in one of the chapters.

1.2 PHYSICAL SYSTEM, COMPUTATION, AND CONTROLS

The physical system consists of the generators, the transmission network, and the distribution network. The transmission network may be high voltage AC or may be high voltage DC (HVDC). Also, there are different controllers which include the exciter and the turbine governor controls, the power system stabilizers (PSS) to damp out oscillations, and the tap changing transformers. Another new type of device which helps to control the transmission and distribution system variables is the Flexible AC Transmission System (FACTS) controllers. These are used for rerouting as well as for enhancing the power transfer capability of transmission lines, voltage support, etc. With the interconnection of all these components, the power system becomes a highly complex one.

1.2.1 Digital Computers in Power System Simulation

Every engineering discipline in the last five decades has felt the impact of computer-aided analysis and design. The availability of faster and larger digital computers has opened up an area of intense and challenging activity for electrical engineers. The computer has been increasingly used for on-line control and optimization of complex control processes such as in-space vehicles, industrial systems, and manufacturing systems.

The area of power systems was no exception to feeling the impact of digital computers and the problems faced were no less challenging. Indeed, they are more complex both in terms of the dimension (physically and mathematically) and the kinds of inherent nonlinearities that are present. The size of power systems has grown steadily, and with the highly interconnected nature of the power systems, economic operation becomes a challenge for the operator. Hence, economic load dispatch, frequency stability, transient stability, protective relaying, and coordination become important issues. Deregulation poses even more challenging problems such as congestion, interaction with market dynamics, etc. The solutions for many of these problems, which have generally been on an approximate and sometimes heuristic basis, need an integrated approach. The big North East power failure in the USA in 1965, the blackout of 2003 in the USA as well as in Italy, and the 2012 grid failure in India, place a fresh emphasis on the need for adequate modeling as well as simulation.

Simulation in the area of power systems has had a long history. In the early part of the last century, even when power systems were small (with a single power station supplying power to a particular area), the problems were complex enough to make the power engineers recognize the need for both modeling and simulation. The pioneering research in synchronous machine theory and transmission lines (which was incidentally needed for communication engineering as well) was the outgrowth of such strong motivation. The tool for simulation was first in the form of a DC network analyzer for short-circuit and load flow studies, and then an AC network analyzer to accommodate transient stability studies. The impact of the digital computer on simulation was first felt in the late 1950s and the early 1960s, and saw a steady but rapid rise in their use for many of the studies made earlier with the network analyzers. Now the computer as well as communication aspects constitute a vital backbone of the vast power system.

The studies done at the planning level of any utility includes short-circuit, load flow, and transient stability, constituting the bulk of the computer studies. At the operation level, computers are used for automatic generation control (load frequency control and economic load dispatch), state estimation, on-line load flow, security monitoring, security assessment, and preventive control, etc. Thus there exists a continued need for emphasis on computer-aided power system analysis and operation. The analytical aspects underlying these studies lie at the interface of power and other fields such as the circuit and system theory, control theory, and computer science. A power system can be viewed as an interconnection of three main subsystems:

1. Generator system
2. Transmission system
3. Loads

Each of these subsystems can in turn be viewed as an interconnection of subcomponents. For example, the generator system is comprised of the synchronous machine, the voltage regulator, the power system stabilizer (PSS), and the prime mover (hydro or steam) with its governing mechanism. The prime mover is supplied by any of the energy sources such as the fossil fuel, nuclear, or hydroelectric. The transmission system consists of the transmission lines, transformers (tap changing and phase shifting), protective relay apparatus and circuit breakers, static capacitors, and shunt reactors, etc. With increased transmission distance, it becomes economic to replace the high voltage AC lines with the high voltage DC (HVDC) transmission lines. Also, HVDC being an asynchronous link, two areas of a power system connected by the HVDC do not have the problem of maintaining synchronism. The restructuring and deregulation of the electrical energy industry in different parts of the world has brought in more competition among the utilities. Hence, in the pursuit of higher operational efficiency, the existing infrastructure is being pushed to their limits. The flexible AC transmission system (FACTS) is one tool which helps in achieving this goal without endangering system reliability. Lastly, the load is modeled either as constant power, voltage dependent, current dependent, or in its simplest form, as static impedance. A power system is thus a complex system in the real sense of the word. It was common, prior to restructuring, for a power network belonging to a particular utility, to be connected to the neighboring utilities through transmission lines called tie lines. Such an arrangement is termed a power pool or regional grid depending on the structure and functioning of the energy supply in a given economy. With deregulation, generation is now decoupled from transmission in terms of ownership and coordination. Tie-lines will still remain, but any generating company may supply any load through the open access transmission network. There is an entity called the independent system operator (ISO) who will ensure that there is no congestion in the network. At the higher level, generators take part in the bidding process and the market design allows the proper regulation of prices through the ISO. In India, the national load dispatch centre (NLDC) acts as the ISO.

There are a number of decisions to be taken into account in a power system both at the operational (day-to-day basis) and at the planning level (on a long-range basis). Any such decision process must be based on analysis with very heavy emphasis on the modeling and simulation techniques. The kinds of studies made by utilities are so numerous that it is difficult to list all of them. A representative list of the studies is given below:

1. Short-circuit studies
2. Load flow studies, also called power flow studies
3. Economic Load Dispatch (ELD) and Optimal Power Flow (OPF)
4. Transient stability (for large disturbances)
5. Small signal stability (for small disturbances)
6. EHV transient simulation
7. Relay coordination studies
8. Load forecasting
9. Unit commitment and maintenance scheduling
10. Economic allocation of generation
11. Economic hydrothermal scheduling
12. Spinning reserve determination
13. Steady-state contingency analysis (security assessment)

14. Dynamic security analysis
15. Visualization of the power system
16. Integration of renewable sources (like wind and solar power) in the grid

1.2.2 Control Viewpoint

A power system has a number of control loops. For frequency control, there is a fine tuned Automatic Generation Control (AGC) system. For voltage control, the excitation system is the primary device. The power system stabilizer controls the power system oscillations. In addition, some power electronics based controllers have been introduced in the transmission and distribution system. These are connected either in series with a transmission line to control the power flow and to damp oscillations (e.g. thyristor controlled series capacitor, unified power flow controller), or in shunt at a bus to control the reactive power injection and the bus voltage (e.g. static VAR controller, static synchronous compensator). These devices are part of the Flexible AC Transmission System (FACTS) devices. Control of HVDC lines is another important issue.

1.3 REPRESENTATION OF POWER SYSTEM COMPONENTS

The electric supply systems are invariably three-phase systems. The distribution of loads on the network is such as to ensure an almost balanced operation on the three phases under steady-state conditions. Therefore, the voltage and current relations in one phase are sufficient to complete the analysis for most purposes. When a disturbance (fault) occurs, this may no longer be true unless it is a symmetrical three-phase fault, which affects all the three phases equally. However, not all abnormal conditions belong to this category. For example, if one phase gets shorted to the ground, it produces unbalanced currents and voltages in the three phases throughout the system. The case of unbalanced abnormal operation can be handled very conveniently through linear transformations on the mathematical model representing the three-phase system. A widely used transformation is the well-known symmetrical component transformation. In either case, i.e., balanced or unbalanced operation, a single-line diagram of the network is still the primary basis for analysis.

The single-line diagram pertains to the electrical portion of the transmission network only and is actually abstracted from a full system diagram that includes comprehensive information about the transmission system. This information includes types of connections of three-phase transformers on the primary/secondary side, its rating, voltage ratio, circuit breakers and their type (e.g., air-blast, oil breaker, etc.), details of transmission line (size, length, etc.), location of isolators, arrangement of the bus bars in a given generating station, and many other details. For a power system simulation, many of these details are not necessary and hence the single-line diagram contains only the salient features pertaining to the studies generally undertaken. Fig. 1.1 shows, for example, the single-line diagram of a 10 machine, 39-bus system.

All the parameters in a single-line diagram are given in terms of per unit quantities. It is assumed that the reader is familiar with this notation. Any text on power system analysis may be consulted for this purpose. Throughout this book, we shall assume that the system is given in the form of a single-line diagram with all the quantities specified on a per unit (p.u.) basis. One of the significant advantages of such a specification is that all the bus voltages have their nominal voltage magnitudes around unity. The transmission line information is supplied in terms of R, X, and B, the nominal π parameters. The static capacitors, shown explicitly by their symbols, are used to improve the voltage levels at certain buses. Their admittance is specified. If there are static shunt reactors or series capacitors, their p.u. impedance values must be specified.

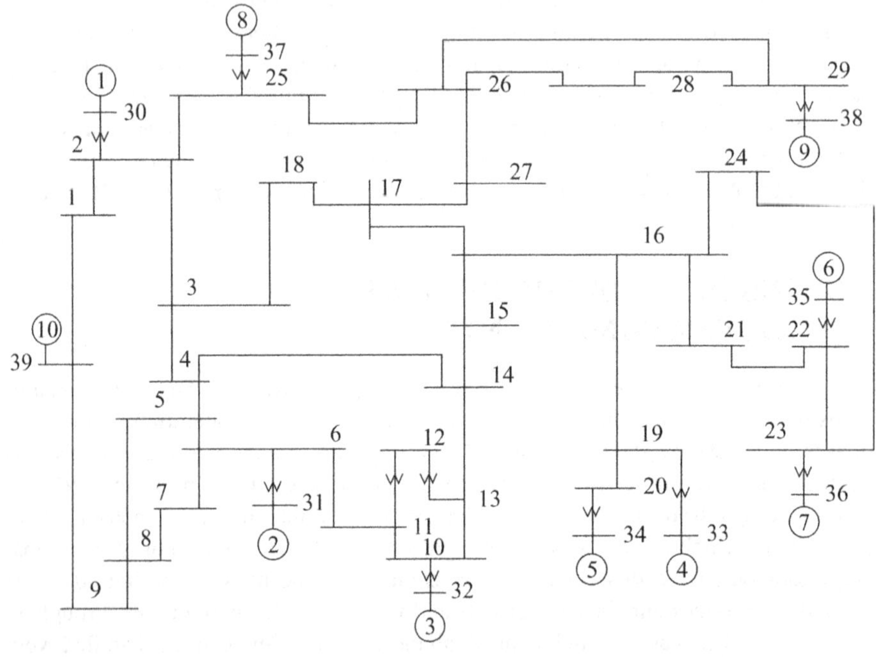

Fig. 1.1 *Single-line diagram of a 10 machine 39 bus system*

The transformers having nominal turns ratio do not appear in a single-line diagram because of the use of the p.u. system. However, there are certain transformers that have off-nominal turns ratios for the purpose of voltage regulation at the neighboring buses. Physically, this is achieved through taps that can be changed in discrete steps automatically or manually depending on voltage deviations at the nearest bus. Tap changing transformers are shown by the appropriate symbol.

A typical power system will have a large number of buses, but generators will be connected at only a few buses. At a given station there may be several generators feeding the bus, but in a single-line diagram, it is schematically represented by means of a small circle with the symbol '~' inside it. Unlike generators, loads will exist at most of the buses. These are indicated by means of arrows. Loads are specified by the real power P and the reactive power Q. Since P = Real part of VI^* and Q = Imaginary part of VI^*, where V and I are both in phasor notation, the load specification is always a product relationship between the voltage and the current. It thus constitutes a nonlinear component.

1.4 NATURE AND SCOPE OF POWER SYSTEM STUDIES

1.4.1 Load Flow Studies

The principal objective of a utility is to deliver power to the consumer satisfactorily (i.e., with voltage and frequency within limits) and in a manner most economical to the company from the transmission and distribution point of view. Given a certain load demand at the buses, what is the most effective manner in which generation can be scheduled at various generating buses so that voltages at the buses are within acceptable limits? This is the crux of the load flow study. In the restructured scenario it is the responsibility of the ISO. At any bus there are four quantities that are of interest, P, Q, $|V|$, and θ, which is the angle associated with the phasor V. For the sake of simplicity let us assume that at the generator buses both P and Q are specified, as in the case of load buses. Then the problem is one of finding $|V|$ and θ at the buses given at net P and net Q at all the buses. At each bus, the conservation of real and reactive power equations are written and we get a set of 2 N on linear equations for an N bus system. Mathematically, this may be correct. However, from a physical viewpoint, there is a basic flaw. By specifying net P and Q at all the buses, we have fixed the transmission losses a-priori (transmission losses = total generation – total load). This is unrealistic, since a feasible solution may not exist. Because of this, we must have a bus where P and Q are not fixed but which takes up the transmission line losses. In a system, this is generally the generating bus having maximum capacity. Such a bus is called, in power system terminology, a *slack* bus or a *swing* bus. With one of the buses designated as the slack bus, mathematically we will now have $(2N - 2)$ nonlinear equations. If the solution of these nonlinear equations results in a satisfactory voltage profile, we have a feasible load flow. Now we relax the assumption that net P and net Q are specified at all the buses. At the generator buses, through appropriate control of excitation and voltage regulating devices, it is desirable to fix P and $|V|$ and let Q vary within certain limits. This will result in a better voltage profile. Such a bus is called a voltage regulated bus or a P-V bus. If there are M generator buses, we can formulate the problem with $(2N - 2 - 2M)$ equations. Such a formulation will result in converged load flow solution in most cases. In addition to having control over Q at the generator bus, it is also possible to control the taps on the off-nominal transformers. With these control parameters, it is found that a larger set of feasible voltage profiles can be achieved. Thus, it is clear that there is no unique load flow solution as such but a large number of alternative choices are possible for a different set of control parameters. A unique choice can be made by defining a cost or objective function such as minimizing the fuel cost, or transmission losses, or both. Such a formulation is called the optimal power flow (OPF) problem. Load flow studies are done by the operation as well as by the planning engineers. Some of the information obtained from load flow studies is listed as follows:

1. Flow of MW and MWAr in the different lines of the network
2. Bus voltages, both magnitude and phase angles
3. Real and reactive power generation
4. The effect of rearranging circuits and the effect of new circuits on loading
5. The effect of temporary loss of generation or loss of transmission lines
6. Optimum rating and tap range of transformers
7. Improvement from change of conductor size and system voltage

1.4.2 Economic Load Dispatch and Optimal Power Flow

With load flow studies, for a given load, we specify generation at all except the slack bus and get a solution. But from an economic point, the total fuel cost may not be at the minimum value. This is accomplished by allocating the generation economically among the different generators depending on their fuel cost curves. One can also minimize the transmission losses. At a higher level, it is possible to do a more sophisticated optimization called the *optimal power flow* that optimally computes both the real and the reactive power generation taking into account the constraints in the transmission network. In the restructured power system, this is done since it is possible to get prices at any desired node in the network. This will be discussed briefly in one of the chapters.

1.4.3 Short-circuit Studies

Short-circuit studies are very important from the operating and planning points of view. Abnormal conditions can arise in a power system owing to short-circuits between two phases, or a line snapping and making contact with the ground, or a lightning strike hitting a particular transmission line. These give rise to heavy currents in the system. Before these currents can do any damage to the equipment such as the expensive generators and transformers, the faulty part of the transmission system must be isolated. This is done by the circuit breakers actuated through sensing relays. Hence, the determination of the interrupting duties of circuit breakers as well as relay settings is needed. This information can be easily calculated if we know the currents in the system upon the occurrence of a fault. Such investigations are termed as short-circuit studies. Since the currents during abnormal operations such as faults, are much greater than the currents during normal operation due to loads, it is customary to neglect the loads for a short-circuit study. Consequently, the short-circuit study becomes essentially a linear network problem since the generator will be represented by means of a voltage source in series with a reactance (generally the transient or sub-transient reactance). Short-circuit studies are essential to the protection engineer who is concerned with ensuring that the system is protected adequately by primary as well as by back-up means.

1.4.4 Transient Stability

The need to preserve the stability of the system under abnormal conditions, such as under-faulted conditions, is of paramount importance. Under normal conditions, all the generators run at close to synchronous speeds and the system is said to be in "synchronism". When faults occur, especially near the generators (say three-phase to ground fault at the generator bus), there is momentarily an excess of mechanical power supplied through the prime movers and the electrical output of the generator is zero. This excess of mechanical power over the electrical power tends to accelerate the rotor, and if the fault is not cleared quickly, the acceleration of the generator may result in the loss of synchronism in the system. The power balance at other generators in the system also gets disturbed, and consequently, their rotor angles vary as a function of time. This manifests itself in a series of chain reactions with power flows changing drastically in the network within a few seconds. This may result in circuit breakers tripping remote from the actual fault. In order to prevent such a loss of stability and its consequent undesirable effects, it is necessary to clear the faults quickly. Since in this book attention is focused on the loss of synchronism only, it is natural to monitor the variations of the rotor angles as a function of time. The length of time interval over which a study is made depends on the objective. A 1-5 second interval is usually sufficient for getting an approximate idea regarding transient stability and this takes into account the dynamic model of the machine, the excitation system, and the governor. Beyond this time interval, the dynamics associated with the turbine and load frequency control can be considered. Fig. 1.2 gives an idea of the time scales and the duration of principal effects that need to be considered.

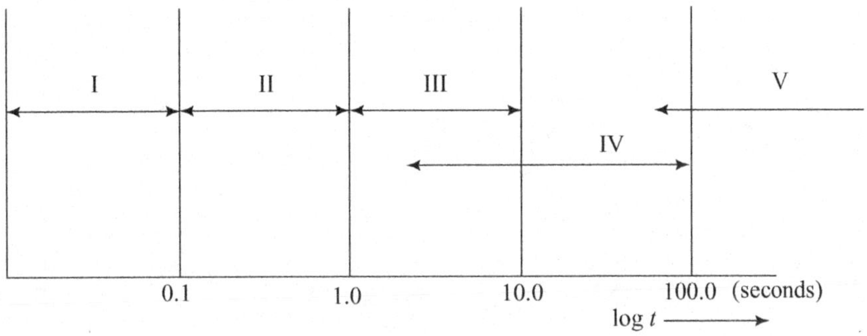

Region	Time (in seconds)	Phenomena
I	$0 < t < 0.1$	Electromagnetic transient and switching surge
II	$0.1 < t < 5$	Synchronous machine dynamics and its controls
III	$5 < t < 10$	Turbine dynamics
IV	$5 < t < 100$	Load frequency control
V	$t > 60$	Boiler dynamics

Fig. 1.2 *Time scale and the duration of principal effects*

1.5 CONCLUSION

In this chapter, we have given an overview of the rapidly changing power system structure. The important role of the computer has been stressed as well as the various studies that are performed off-line at the planning level as well as the on-line studies.

2 POWER SYSTEM COMPONENTS

2.1 INTRODUCTION

As mentioned in Chapter 1, the power system consists of three main parts, namely (a) generation, (b) transmission, and (c) distribution. Generation involves the synchronous machines along with the exciter, the governor, and the turbine. The transmission system includes two main components: (i) the transformers, and (ii) the transmission lines. The distribution system includes the distribution lines, the transformers, and the different types of loads, namely constant impedance, constant current, and constant power. For simulating/analyzing a power system, models of all these components are required. In this chapter, suitable models of all these components will be discussed.

High Voltage DC (HVDC) transmission is being used in power system networks for a long time now. Its advantages over AC transmission for transmitting power over very long distances are well accepted. HVDC links are established in many cases to achieve asynchronous operation of the two sections of the system. With the advancement in power electronics, the HVDC and the flexible AC transmission system (FACTS) have become integral parts of modern day power systems. Hence, the analysis of the present day power system requires the models of HVDC and FACTS, which are also discussed here. It is not proposed to discuss these in detail, but to briefly deal with their modeling so far as they are pertinent to our studies. The complexity of the model depends to a great extent on the type of study undertaken and the accuracy desired.

However, before discussing the modeling, a brief introduction of the per unit system is required. The voltage levels at the generation, the transmission and the distribution differ by orders of magnitude. To avoid difficulty in calculation with such different voltage levels, a per unit system is used which brings the different components of the power system to the same platform. To start with, the method of calculation of the per unit values of different variables is discussed here.

2.2 PER UNIT SYSTEM

The first step in introducing the per unit system is to select a base value for the system variables. As the voltage, current, power and impedance of a particular part of the system are related, choosing the base quantity for any two of these is sufficient. Usually, the base quantities of the system voltage and the apparent power are chosen first. Let these be denoted by V_{base} and VA_{base} respectively. Then, the base for the current (I_{base}) and the impedance (z_{base}) can be calculated as:

$$I_{\text{base}} = \frac{VA_{\text{base}}}{V_{\text{base}}} \tag{2.1}$$

$$z_{\text{base}} = \frac{V_{\text{base}}}{I_{\text{base}}} = \frac{(V_{\text{base}})^2}{VA_{\text{base}}} \tag{2.2}$$

In case of large power systems, usually the base chosen for the voltage and the power are in kV (kV_{base}) and MVA (MVA_{base}) respectively. So in that case, the base for the impedance can be calculated as:

$$z_{\text{base}} = \frac{(kV_{\text{base}})^2}{MVA_{\text{base}}} \tag{2.3}$$

The choice of the base quantity can be totally arbitrary. However, it should be kept in mind that if the phase (line to neutral) voltage is chosen as the voltage base and the per phase power is chosen as the power base, then Eqs. (2.1)–(2.3) can be used. However, if the line to line voltage is chosen as the voltage base and the three-phase power is chosen as the power base, Eq. (2.1) should be modified to:

$$I_{\text{base}} = \frac{VA_{\text{base}}}{\sqrt{3}\,V_{\text{base}}} \tag{2.4}$$

The z_{base} can still be found using Eq. (2.3).

Usually, the rated voltage and power of the particular power system component is chosen as the kV_{base} and the MVA_{base} respectively. The corresponding base values of impedance and current are calculated using Eqs. (2.3) and (2.4). Then, the per unit values of the variables are defined as the ratio of the actual value to the base value. For example, the per unit impedance is given as:

$$z_{pu} = \frac{z_{\text{actual}}\,(\Omega)}{z_{\text{base}}} = z_{\text{actual}}\,(\Omega)\,\frac{MVA_{\text{base}}}{(kV_{\text{base}})^2} \tag{2.5}$$

In case the base values are changed, the per unit values of the variables in the new base can be computed using the old and the new base values, and the per unit values in the old base as:

$$(pu)_{\text{new}} = (pu)_{\text{old}} \cdot \frac{(\text{base})_{\text{old}}}{(\text{base})_{\text{new}}} \tag{2.6}$$

For the analysis of the whole system, one common MVA base is considered for the whole system whereas the kV bases of the respective components are retained. Then, the per unit values of the parameters and variables of the different components with respect to these common bases are obtained using Eq. (2.6).

For majority of the studies, the power system is considered to be balanced and hence calculation on a per phase basis is sufficient. In these cases, the system can be represented by a single phase which is called the single line representation. The single line representation of a 10 machine 39 bus system has been shown in Chapter 1, Fig. 1.1.

2.3 SYNCHRONOUS GENERATOR REPRESENTATION

Models of synchronous machines are required for dynamic studies, i.e., the study of the response of the system when it is subjected to some disturbance. Depending on the magnitude of the disturbance, dynamic studies can be small signal (disturbances like small changes in the load) or transient (large disturbances like a fault). In this book, we will not deal with the small signal studies. The detailed model of the synchronous machine is rarely used in transient stability studies. Instead, two different reduced order models of synchronous machines are mostly used, which are: (a) the Flux-decay model, and (b) the Classical model. These representations are valid for dynamic phenomena in the range of 1–5 s.

The notations used for the different system variables and parameters while describing the models are as given below.

δ = the angular position of the rotor with respect to a synchronously rotating reference frame;
ω_r = the rotor speed;
ω_s = the synchronous speed;
H = the inertia constant in sec;
k_D = the damping coefficient;
P_m = the mechanical power in p.u.;
P_e = the electrical power in p.u.;
x_d and x_q = the d-axis and the q-axis synchronous reactance respectively in p.u.;
x'_d = the d-axis transient reactance in p.u.;
E'_q = the q-axis component of the voltage behind the transient reactance in p.u.;
T'_{do} = the d-axis open circuit time constant in p.u.;
I_d and I_q = the d-axis and the q-axis components of the stator current in p.u.;
E_{fd} = the exciter voltage in p.u.;
K_A and T_A = the gain and the time constant of the exciter respectively;
$V\angle\theta$ = the terminal voltage of the machine, where V is in p.u. and θ is in degrees;

The terminal voltage of the machine is the same as the voltage of the bus at which the machine is connected with the system. Here all the reactances, voltage magnitudes and the powers are in per unit. E'_q is actually a voltage proportional to the total flux linkage of the generator field winding. Hence, T'_{do} is the field circuit time constant.

Flux-decay Model with Fast Exciter

The assumptions considered while using this model are listed below.

(i) The machine is operating under balanced three-phase positive-sequence conditions.
(ii) Synchronous machine damper winding time constants are much less compared to that of the field winding.
(iii) The stator resistance is neglected.
(iv) Turbine-governor dynamics is not considered.

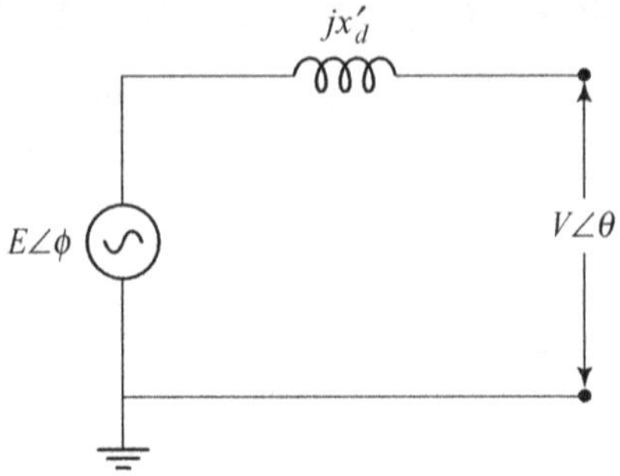

Fig. 2.1 *Flux-decay model of generator*

The generators are represented by the flux-decay model as shown in Fig. 2.1. In Fig. 2.1, the generator internal voltage and its angle, $E\angle\phi$, is given as:

$$E\angle\phi=\left[\left(1-\frac{x'_d}{x_q}\right)V\sin(\delta-\theta)+jE'_q\right]e^{j(\delta-\pi/2)} \tag{2.7}$$

The machine dynamics are described by the following equations:

$$\frac{d\delta}{dt}=\omega_r-\omega_s \tag{2.8}$$

$$\frac{2H}{\omega_s}\frac{d\omega_r}{dt}=P_m-P_e-\frac{k_D}{\omega_s}(\omega_r-\omega_s) \tag{2.9}$$

$$T'_{do}\frac{dE'_q}{dt}=-E'_q-(x_d-x'_d)I_d+E_{fd} \tag{2.10}$$

The electrical power output, P_e is given as:

$$P_e=E'_qI_q-\left(x_q-x'_d\right)I_dI_q \tag{2.11}$$

Neglecting the dynamics, the stator winding is represented by the following algebraic equations:

$$V\sin(\delta-\theta)-x_qI_q=0 \tag{2.12}$$

$$V\cos(\delta-\theta)+x_dI_d-E'_q=0 \tag{2.13}$$

Replacing the rotor speed (ω_r) with the per unit speed deviation ($\Delta\omega_r$) in Eqs. (2.8) and (2.9) using the relations $\Delta\omega_r=(\omega_r-\omega_s)/\omega_s$ and $\frac{d\Delta\omega_r}{dt}=\frac{1}{\omega_s}\frac{d\omega_r}{dt}$, we get the modified equations:

$$\frac{d\delta}{dt}=\omega_s\Delta\omega_r \tag{2.14}$$

$$2H\frac{d\Delta\omega_r}{dt} = P_m - P_e - k_D\Delta\omega_r \tag{2.15}$$

Using Eqs. (2.12) – (2.13), I_d and I_q are eliminated from Eq. (2.10) and Eq. (2.11) as:

$$T'_{do}\frac{dE'_q}{dt} = -\frac{x_d}{x'_d}E'_q + \left(\frac{x_d}{x'_d} - 1\right)V\cos(\delta-\theta) + E_{fd} \tag{2.16}$$

$$P_e = E'_q V\sin(\delta-\theta)/x'_d + 0.5\left(1/x_q - 1/x'_d\right)V^2\sin 2(\delta-\theta) \tag{2.17}$$

If the power output at the generator terminal ($P_g + jQ_g$) and the terminal voltage $V\angle\theta$ is known, the internal emf $E\angle\phi$ (classical model) can be obtained as:

$$I_G\angle\gamma_G = \left(\frac{P_G + jQ_G}{V\angle\theta}\right)^* \tag{2.18}$$

$$E\angle\phi = V\angle\theta + jx'_d\left(I_G\angle\gamma_G\right) \tag{2.19}$$

It can be shown that the rotor angle δ can be obtained as the angle of $\left[V\angle\theta + jx'_q\left(I_G\angle\gamma_G\right)\right]$. Also it is to be noted that $P_g = P_e$ (as the stator resistance is neglected).

A simplified static exciter model with one gain and one time constant is considered (Fig. 2.2).

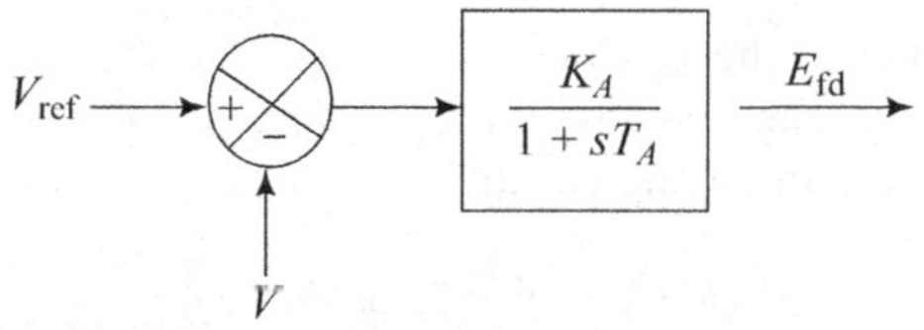

Fig. 2.2 *Static exciter model*

The dynamics of the exciter can be described as:

$$T_A\frac{dE_{fd}}{dt} = -E_{fd} + (V_{ref} - V)K_A \tag{2.20}$$

Classical Model

This is a model which is obtained by further simplification of the flux-decay model. The additional assumptions used are the following:

(i) The dynamics of the exciter are not considered in this model.

(ii) Considering the small period of analysis, the field time constant can be considered as very large, i.e., $T'_{do} = \infty$

(iii) Also it is assumed that $x_q = x'_d$

With these assumptions, E'_q becomes constant and equal to its initial value E'_{q0}. The generator model now consists of a voltage of constant magnitude E'_{q0} behind a transient reactance x'_d, as shown in Fig. 2.3. The angle of the voltage is the generator rotor angle δ.

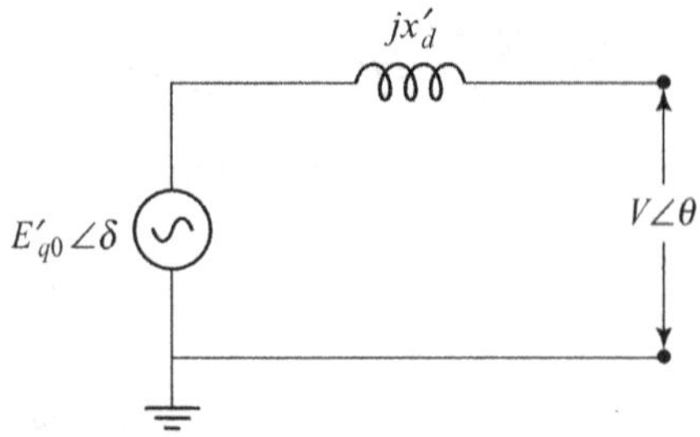

Fig. 2.3 *Classical model (constant voltage behind transient reactance)*

The generator equations then become:

$$\frac{d\delta}{dt} = \omega_s \Delta\omega_r \tag{2.21}$$

$$2H\frac{d\Delta\omega_r}{dt} = P_m - P_e - k_D\Delta\omega_r \tag{2.22}$$

Where the electrical power output P_e is given as:

$$P_e = \frac{E'_{q0}V\sin(\delta-\theta)}{x'_d} \tag{2.23}$$

If the power output of the generator at its terminal $(P_g + jQ_g)$ and the terminal voltage $V\angle\theta$ is known, the internal emf $E'_{q0}\angle\delta$ (classical model) can be obtained using Eq. (2.18) and the following equation:

$$E'_{q0}\angle\delta = V\angle\theta + jx'_d\left(I_G\angle\gamma_G\right) \tag{2.24}$$

2.4 TRANSMISSION LINES

In all the studies to be considered in this book, the transmission network can be assumed to be in sinusoidal steady-state since the time constants associated with it are much smaller compared to the time constants of the synchronous machine.

2.4.1 Uniform Long Line

A transmission line has its resistance, inductance, capacitance, and leakage conductance uniformly distributed along its length and they can be calculated on a per phase, per unit length basis using the parameters of the line. In all practical cases, the leakage conductance to ground is negligibly small. We are essentially interested in the relation between the voltage and the current at the two ends of the line—one of them called the *sending end* and the other, the *receiving end* of the line. The distances are measured from the receiving end. There is a progressive change in both the voltage and the current along the line because of the distributed nature of the series impedance and the shunt admittance to neutral. Refer to Fig. 2.4, where, on a per phase to neutral basis, element of length dx at a distance x from the receiving end is concerned.

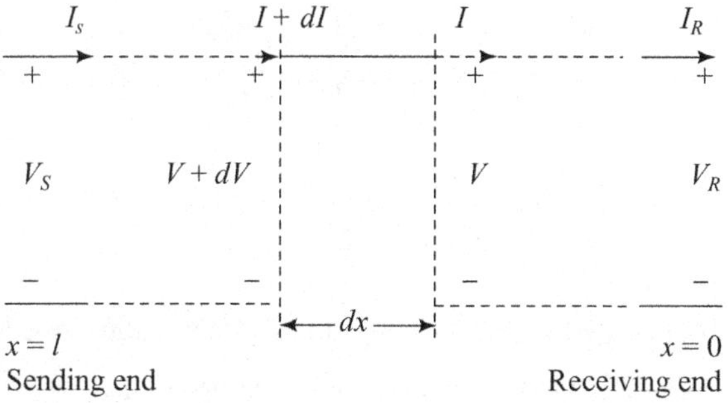

Fig. 2.4 *Voltage-current relations in a small element length of a transmission line*

For this line element, we can write the following relations:

$$dV = Izdx \qquad \Rightarrow \frac{dV}{dx} = Iz \tag{2.25}$$

$$dI = Vydx \qquad \Rightarrow \frac{dI}{dx} = Vy \tag{2.26}$$

where z = series impedance per phase per unit length, and y = shunt admittance per phase per unit length.

Differentiation of Eq. (2.1) and Eq. (2.2) with respect to x gives

$$\frac{d^2V}{dx^2} = z\frac{dI}{dx} \tag{2.27}$$

$$\frac{d^2I}{dx^2} = y\frac{dV}{dx} \tag{2.28}$$

Substituting Eqs. (2.25) and (2.26) in Eqs. (2.27) and (2.28) respectively, we get

$$\frac{d^2V}{dx^2} = zyV \tag{2.29}$$

$$\frac{d^2I}{dx^2} = zyI \tag{2.30}$$

The solution of Eq. (2.29) is of the form

$$V = A_1 \exp\left(\sqrt{zy}x\right) + A_2 \exp\left(-\sqrt{zy}x\right) \tag{2.31}$$

Substitution of the first derivative of Eq. (2.31) with respect to x into Eq. (2.25) results in an expression for the current I as

$$I = \frac{1}{\sqrt{(z/y)}} A_1 \exp\left(\sqrt{zy}x\right) - \frac{1}{\sqrt{z/y}} A_2 \exp\left(-\sqrt{zy}x\right) \tag{2.32}$$

We know that $V = V_R$ and $I = I_R$ at $x = 0$. Using this in Eqs. (2.7) and (2.8), we get

$$A_1 = \frac{V_R + \sqrt{(z/y)}\, I_R}{2} \tag{2.33}$$

$$A_2 = \frac{V_R - \sqrt{(z/y)}\, I_R}{2} \tag{2.34}$$

Here we introduce two new terms—(i) the characteristic impedance of the line defined by $Z_c = \sqrt{(z/y)}$ and (ii) the propagation constant $\gamma = \sqrt{zy}$. The expressions of the voltage and the current at a distance x from the receiving end can be written using Eqs. (2.33) and (2.34) in Eqs. (2.31) and (2.32) respectively, as

$$V(x) = \frac{V_R + I_R Z_c}{2} \exp(\gamma x) + \frac{V_R - I_R Z_c}{2} \exp(-\gamma x) \tag{2.35}$$

$$I(x) = \frac{(V_R/Z_c) + I_R}{2} \exp(\gamma x) - \frac{(V_R/Z_c) - I_R}{2} \exp(-\gamma x) \tag{2.36}$$

Equation (2.35) can be rewritten as:

$$V(x) = V_R \frac{1}{2}(\exp(\gamma x) + \exp(-\gamma x)) + I_R Z_c \frac{1}{2}(\exp(\gamma x) - \exp(-\gamma x))$$

$$= V_R \cosh \gamma x + I_R Z_c \sinh \gamma x \tag{2.37}$$

Similarly, we can rewrite the expression for current as

$$I(x) = I_R \cosh \gamma x + \frac{V_R}{Z_c} \sinh \gamma x \tag{2.38}$$

The voltage and current at the sending end of the line are obtained by setting $x = \ell$ in Eqs. (2.37) and (2.38) as

$$V_s = V_R \cosh \gamma \ell + I_R Z_c \sinh \gamma \ell \tag{2.39}$$

$$I_s = \frac{V_R}{Z_c} \sinh \gamma \ell + I_R \cosh \gamma \ell \tag{2.40}$$

2.4.2 Equivalent Network of Long Line

Suppose we wish to represent Eqs. (2.39) and (2.40) as an equivalent circuit of the form shown in Fig. 2.5 (called the π-equivalent). Then from Fig. 2.5,

$$V_S = V_R + (I_R + V_R Y_{\pi 2}) Z_\pi = [1 + Z_\pi Y_{\pi 2}] V_R + Z_\pi I_R \tag{2.41}$$

$$\begin{aligned} I_S &= (I_R + V_R Y_{\pi 2}) + V_S Y_{\pi 1} = (I_R + V_R Y_{\pi 2}) + [(1 + Z_\pi Y_{\pi 2}) V_R + Z_\pi I_R] Y_{\pi 1} \\ &= [(Y_{\pi 1} + Y_{\pi 2}) + Z_\pi Y_{\pi 1} Y_{\pi 2}] V_R + (1 + Z_\pi Y_{\pi 1}) I_R \end{aligned} \tag{2.42}$$

Figure 2.5 being the equivalent circuit representation, Eqs. (2.41) and (2.42) should be identical to Eqs. (2.39) and (2.40) respectively.

For this to happen, we must have

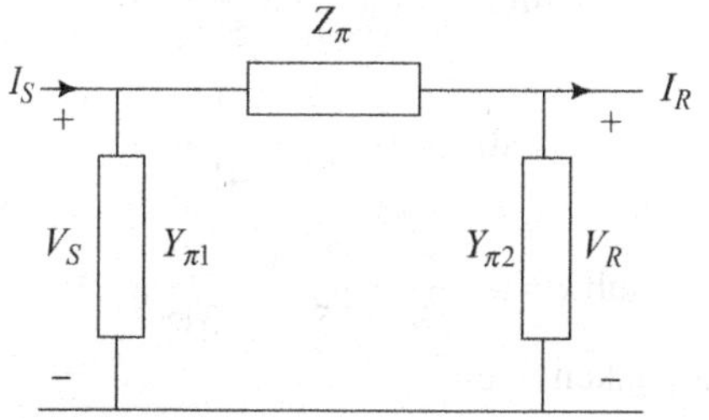

Fig. 2.5 *π-equivalent of the transmission line*

$$Z_\pi = Z_c \sinh \gamma \ell \tag{2.43}$$

$$Y_{\pi 1} = Y_{\pi 2} = Y_\pi \tag{2.44}$$

$$(1 + Z_\pi Y_\pi) = \cosh \gamma \ell \tag{2.45}$$

Hence,

$$Y_\pi = \frac{\cosh \gamma \ell - 1}{Z_c \sinh \gamma \ell} = \frac{1}{Z_c} \tanh \frac{\gamma \ell}{2} \tag{2.46}$$

Substitution of Z_c and γ by their respective expressions in Eqs. (2.43) and (2.46), and subsequent rearrangement results in the following:

$$Z_\pi = Z_c \gamma \ell \frac{\sinh \gamma \ell}{\gamma \ell} = z\ell \frac{\sinh \gamma \ell}{\gamma \ell} \tag{2.47}$$

$$Y_\pi = \frac{(\gamma \ell/2)}{Z_c} \frac{\tanh(\gamma \ell/2)}{\gamma \ell/2} = \frac{y\ell}{2} \frac{\tanh(\gamma \ell/2)}{\gamma \ell/2} \tag{2.48}$$

Thus the final form of equivalent π representation becomes as shown in Fig. 2.6.

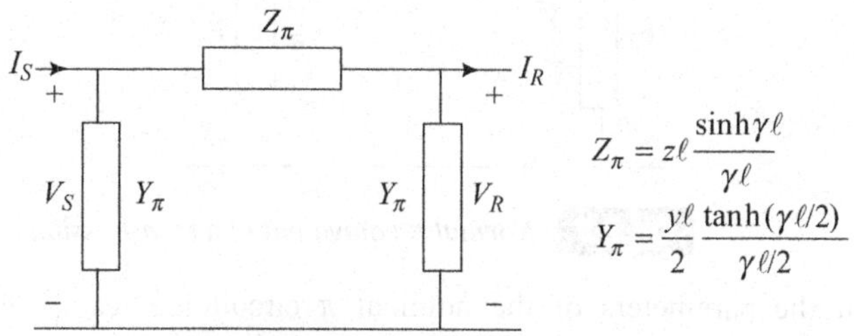

Fig. 2.6 *π-equivalent of the transmission line*

The terms Z_π and Y_π can be calculated to any degree of accuracy desired from the tables for hyperbolic functions. In most of the cases, power series expansion with calculation up to two or three terms will give acceptable accuracy.

$$\sinh x = x + \frac{x^3}{3!} + \frac{x^5}{5!} + \cdots \quad (2.49a)$$

$$\cosh x = 1 + \frac{x^2}{2!} + \frac{x^4}{4!} + \cdots \quad (2.49b)$$

$$\tanh x = x - \frac{x^3}{3} + \frac{2}{15}x^5 - \frac{17}{315}x^7 + \cdots \quad (2.49c)$$

If the first two terms are taken, then

$$Z_\pi \approx z\ell\left[1 + \frac{(\gamma\ell)^2}{6}\right] \quad (2.50a)$$

$$Y_\pi \approx \frac{y\ell}{2}\left[1 - \frac{1}{3}\left(\frac{\gamma\ell}{2}\right)^2\right] = \frac{y\ell}{2}\left[1 - \frac{(\gamma\ell)^2}{12}\right] \quad (2.50b)$$

2.4.3 Equivalent Circuit for Medium Lines

In the case of medium lines (length between 80 km and 240 km), it can be assumed that $\gamma\ell << 1$. In this case, we may neglect the second order terms in Eq. (2.27).

Hence,

$$Z_\pi = z\ell = Z \text{ (total series impedance)}$$

$$Y_\pi = \frac{y\ell}{2} = \frac{Y}{2} \text{ (half shunt admittance)}$$

The π circuit obtained under such an approximation is called nominal π circuit (Fig. 2.7).

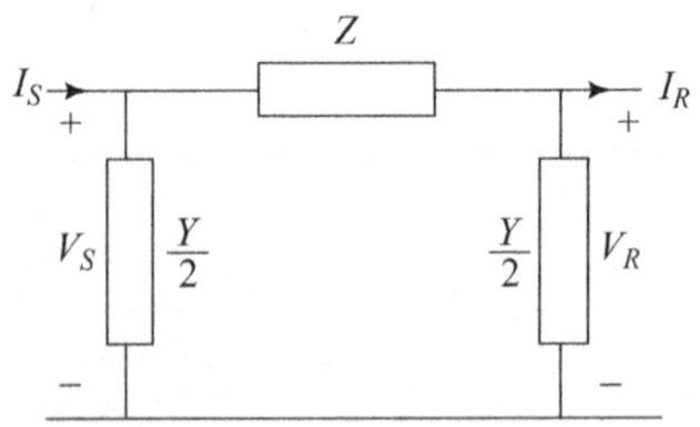

Fig. 2.7 *Normal π-equivalent of a transmission line*

Note that the parameters of the nominal π circuit are readily expressed in terms of the total series impedance Z and the total shunt admittance Y (also called *line charging*). π-representation is the preferred model in system studies.

In the case of a very short line (80 km or less), the shunt admittance is neglected and the nominal π-representation reduces to that of a simple series impedance Z (Fig. 2.8).

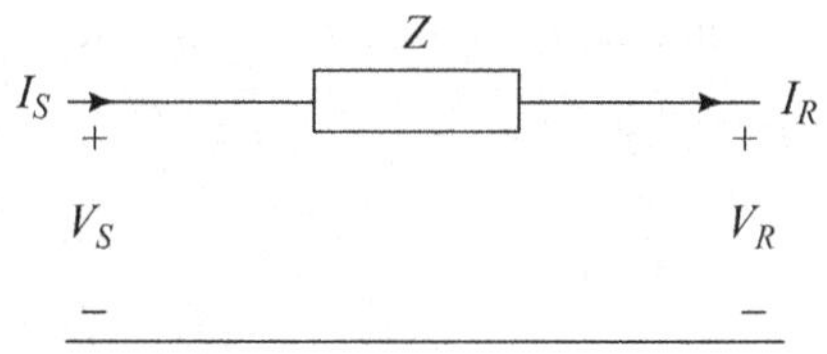

Fig. 2.8 *Representation of a short transmission line*

2.5 TRANSFORMERS

2.5.1 Two-winding Transformer

The equivalent circuit of a two-winding (Fig. 2.9) transformer is shown in Fig. 2.10, where all the quantities are referred to side 1. This means

$$\left(r_2' + jx_{l/2}'\right) = \left(\frac{N_1}{N_2}\right)^2 \left(r_2 + jx_{l/2}\right) \tag{2.51}$$

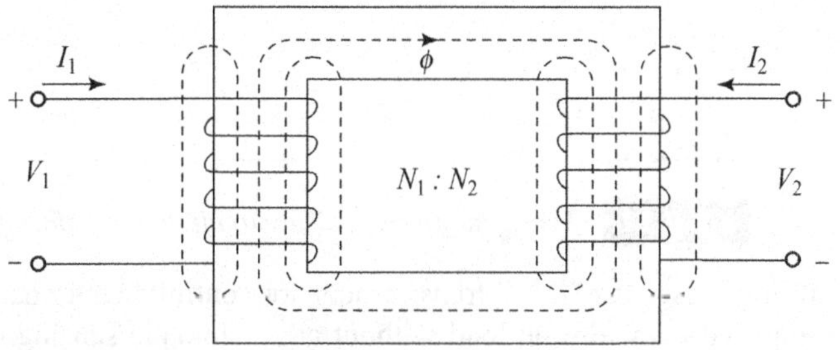

Fig. 2.9 *A two-winding transformer*

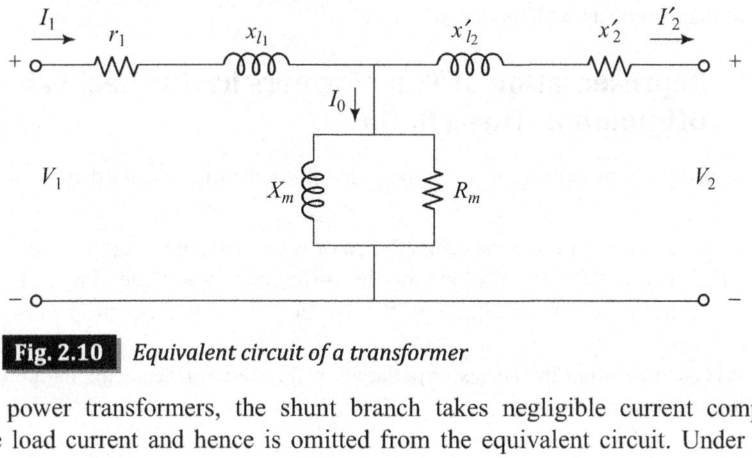

Fig. 2.10 *Equivalent circuit of a transformer*

In power transformers, the shunt branch takes negligible current compared to the load current and hence is omitted from the equivalent circuit. Under these circumstances, the equivalent circuit reduces to a simple series impedance (Fig. 2.8), where

$$R = r_1 + \left(\frac{N_1}{N_2}\right)^2 r_2 \quad \text{and} \quad X = x_{l_1} + \left(\frac{N_1}{N_2}\right)^2 x_{l_2} \tag{2.52}$$

2.5.2 Tap-changing Transformers

It is often required to adjust the turns ratio of the transformers to keep the voltage levels at the neighboring load buses within acceptable limits. For this purpose, transformers with a tap-changing mechanism are used. Taps are generally located on the HV side to provide smoother control. When the turns ratio N is equal to the ratio of rated voltages, we say the ratio is *nominal*. Hence in the single-line diagram, the transformer is represented by its leakage impedance ($R + jX$) as given by Eq. (2.52). However, when the ratios are not equal, we say the particular transformer has an *off-nominal turns ratio*. Such transformers are often designated as on-load tap changers (OLTC) or underload tap changers (ULTC).

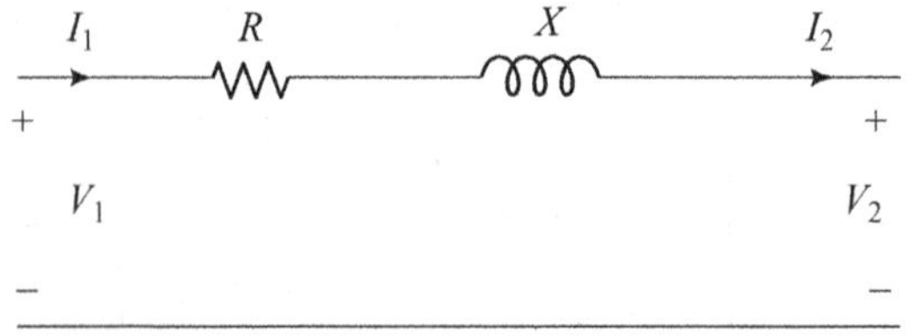

Fig. 2.11 *Approximate equivalent circuit of a transformer*

Though in most cases the OLTC transformers are controlled automatically, taps can also be adjusted manually on-load without any automatic sensing device, but based on both experience and prior load flow calculations. The tap ratio can either be a real number or a complex number. In the latter case, the transformers are called phase-shifting transformers, because the voltages on the two sides of the transformer differ in magnitude as well as phase.

2.5.3 Representation of Transformers having Real Valued Off-nominal Turns Ratio

The following considerations arise in the case of transformers with off-nominal turns ratio.

1. The p.u. value of the series impedance of the transformer is put in series with an ideal transformer to allow for the difference in voltage (Fig. 2.12). The off-nominal ratio represented by the symbol 'a' would assume a value close to unity.
2. It is assumed that the series impedance of the transformer is unchanged when the tap position is varied.

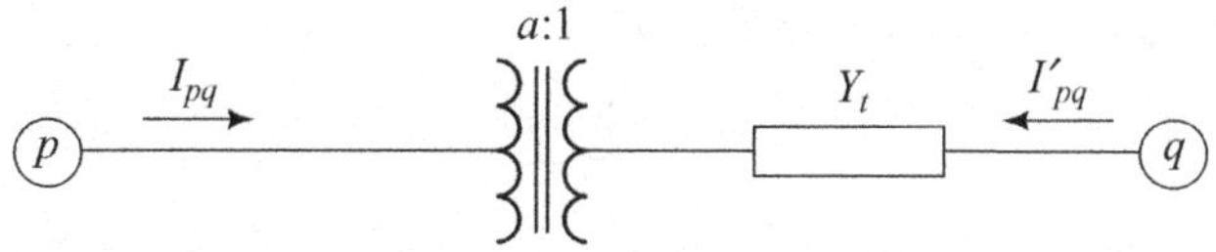

Fig. 2.12 *Representation of transformers with off-nominal turns ratio*

With the transformer ratio being normalized as a: 1, the non-unity side is called the tap side. Thus, the series admittance of the transformer is connected to the unity side. The equivalent circuit is now derived from elementary considerations.
At node **p**:

$$I_{pq} = (V_p - aV_q)\frac{Y_t}{a^2} = \frac{V_p Y_t}{a^2} - \frac{V_q Y_t}{a} \tag{2.53}$$

Similarly, at node **q**:

$$I'_{pq} = \left(V_q - \frac{V_p}{a}\right)Y_t \tag{2.54}$$

Equations (2.53) and (2.54) can be represented by means of an equivalent circuit as shown in Fig. 2.13.

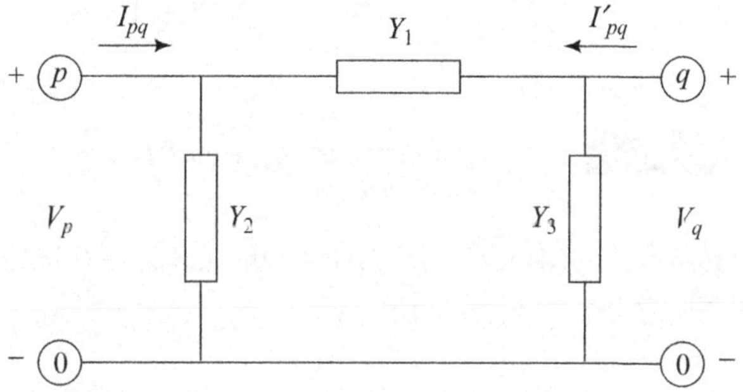

Fig. 2.13 *Equivalent circuit of off-nominal transformer*

Writing node equation in Fig. 2.13, we get

$$I_{pq} = V_p Y_2 + (V_p - V_q)Y_1 \tag{2.55}$$

$$I'_{pq} = V_q Y_3 + (V_q - V_p)Y_1 \tag{2.56}$$

Comparing Eqs. (2.55) and (2.31) with Eqs. (2.28) and (2.29) respectively, we get

$$Y_1 + Y_2 = \frac{Y_t}{a^2} \tag{2.57}$$

$$Y_1 = \frac{Y_t}{a} \tag{2.58}$$

and

$$Y_1 + Y_3 = Y_t \tag{2.59}$$

Hence from Eq. (2.57) and Eq. (2.58), we get

$$Y_2 = \frac{Y_t}{a^2} - \frac{Y_t}{a} \tag{2.60}$$

Also from Eq. (2.59),

$$Y_3 = Y_t - \frac{Y_t}{a} \tag{2.61}$$

The final equivalent circuit is shown in Fig. 2.14. Note that all the admittances in the equivalent circuit are functions of the turns ratio 'a'. Furthermore, the signs associated with Y_2 and Y_3 are always opposite. For example, if Y_t represents an inductance, then for $a > 1$, Y_2 is an inductance and Y_3 is a capacitance. For $a > 1$, Y_2 is a capacitance and Y_3 is an inductance. As 'a' approaches unity, both the shunt branches in the π-equivalent become infinite impedances and the series admittance approaches the value Y_t.

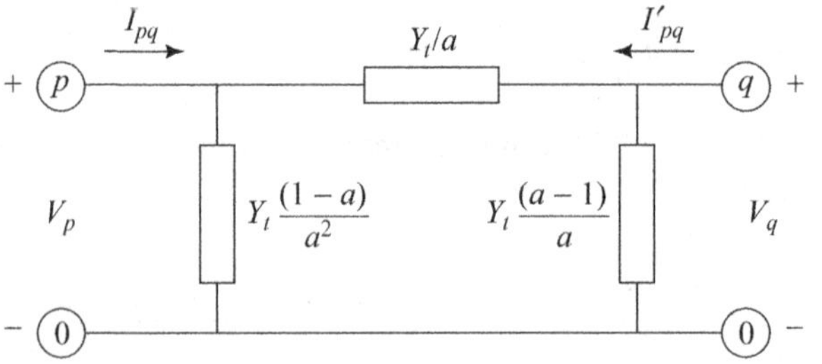

Fig. 2.14 *Final equivalent circuit of transformer with off-nominal turns ratio*

If the transformer impedance is referred to the tap side (Fig. 2.15), then the π-equivalent circuit is given in Fig. 2.16.

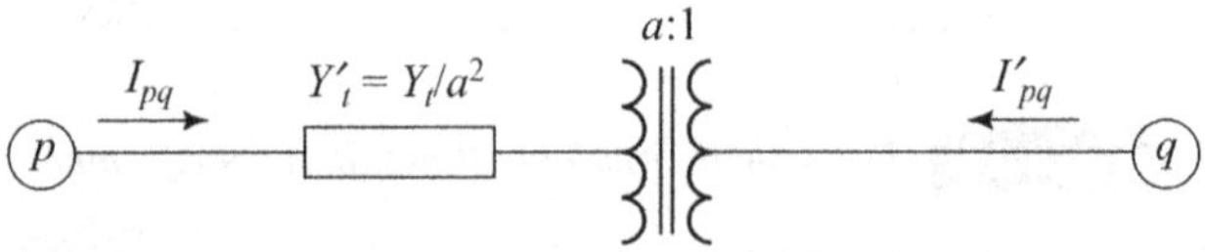

Fig. 2.15 *Off-nominal transformer with impedance on tap side*

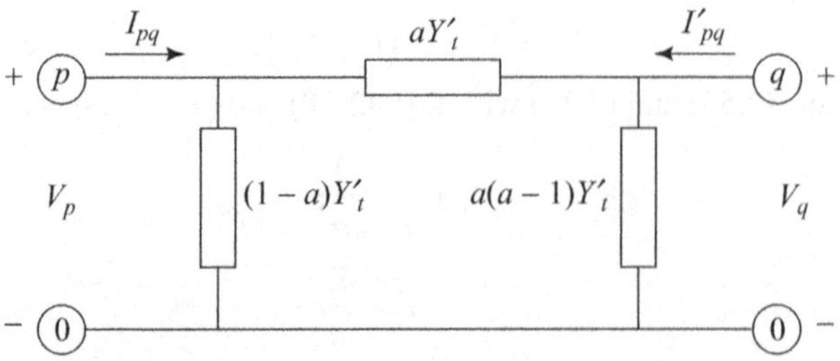

Fig. 2.16 *π-equivalent for the off-nominal transformer with impedance on tap side*

2.5.4 Phase Shifting Transformers

In interconnected power systems involving loop circuits or parallel lines, the real power flowing through a line can be controlled by having a phase-shifting transformer, i.e., a transformer whose turns ratio is a complex number having a certain magnitude and phase angle. The amount of the phase angle depends on the tap position.

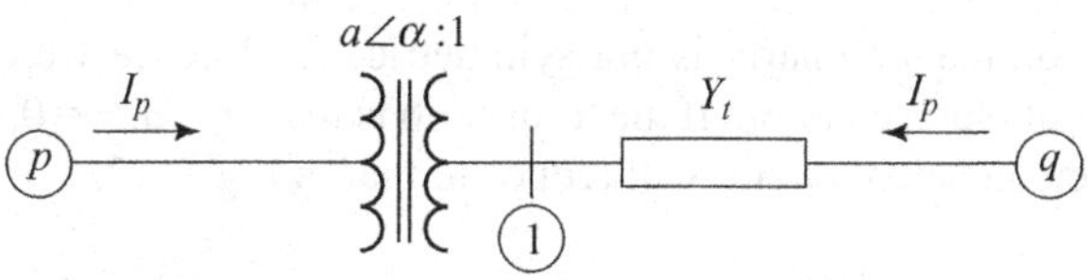

Fig. 2.17 *Phase-shifting transformer representation*

The derivation of the equivalent circuit for a phase-shifting transformer is somewhat more complex, but we can have a two-port description. The phase shifting transformer can be represented as in Fig. 2.17. The phase angle step may not be equal at different positions of the tap. However, for power flow and transient stability studies, equal step size is usually assumed. By definition,

$$\frac{V_p}{V_r} = a\angle\alpha = a(\cos\alpha + j\sin\alpha) = a_s + jb_s \tag{2.62}$$

where α is the phase shift from the bus r to p. It is positive when V_p leads V_r. Since the transformer is ideal, we have,

$$V_p I_p^* = -V_r I_q^* \tag{2.63}$$

Hence,

$$I_p = \frac{-V_r^*}{V_p^*} I_q = -\frac{1}{(a_s - jb_s)} I_q = \frac{Y_t(V_r - V_q)}{(a_s - jb_s)} \tag{2.64}$$

Substituting for V_r from Eq. (2.62)

$$I_p = \frac{Y_t}{(a_s - jb_s)}\left(\frac{V_p}{(a_s + jb_s)} - V_q\right) = Y_t\left(\frac{V_p}{a_s^2 + b_s^2} - \frac{V_q}{a_s - jb_s}\right) \tag{2.65}$$

From Eq. (2.64),

$$I_q = -(a_s - jb_s)I_p \tag{2.66}$$

Substituting for I_p from Eq. (2.65),

$$I_q = \frac{-Y_t V_p}{(a_s + jb_s)} + Y_t V_q \tag{2.67}$$

Combining Eqs. (2.65) and (2.67), we get

$$\begin{bmatrix} I_p \\ I_q \end{bmatrix} = \begin{bmatrix} \dfrac{Y_t}{a_s^2 + b_s^2} & \dfrac{-Y_t}{a_s - jb_s} \\ \dfrac{-Y_t}{a_s + jb_s} & Y_t \end{bmatrix} \begin{bmatrix} V_p \\ V_q \end{bmatrix} \tag{2.68}$$

Note that the admittance matrix is not symmetrical and hence we cannot have an equivalent circuit representation. If the turns ratio is real, $a_s = a$, $b = 0$, i.e., there is no phase shift, the equivalent circuit will reduce to that in Fig. 2.12.

2.5.5 Three-winding Transformers

Three-winding transformers are used in situations where voltages at two different levels are to be supplied from the primary. The other two windings are called the secondary and the tertiary windings (Fig. 2.18). Tertiary windings are also sometimes used for other purposes such as connecting shunt capacitors or for suppression of third harmonics. The equivalent circuit for the three-winding transformer will now be derived. We shall designate 1-1′ as the primary winding p, 2-2′ as the secondary winding s and 3-3′ as the tertiary winding t. We neglect the exciting or the magnetizing current.

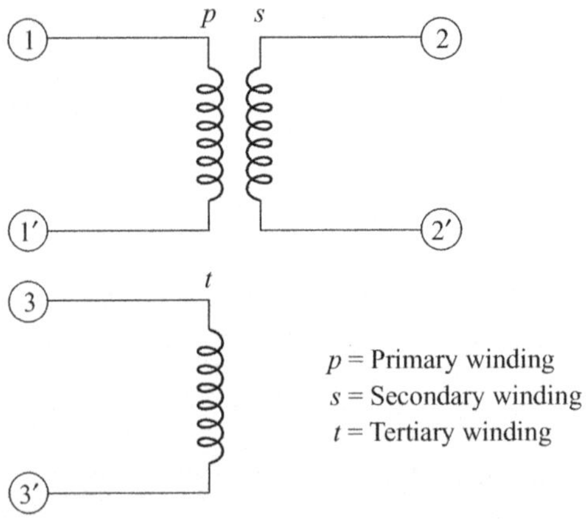

Fig. 2.18 *Three-winding transformer*

The following parameters can be measured experimentally:

Z_{ps} = Impedance of primary with secondary short circuited and tertiary open.
Z_{pt} = Impedance of primary with tertiary short circuited and secondary open.
Z'_{st} = Impedance of secondary with tertiary short circuited and primary open.

Suppose Z'_{st} were referred to the primary side, then,

$$Z_{st} = \left(N_p / N_s\right)^2 Z'_{st} \tag{2.69}$$

Let the equivalent circuit of the transformer be sought in the form shown in Fig. 2.19 with Z_p, Z_s, and Z_t referred to the primary side. Then, from an interpretation of the short circuit measurements,

$$Z_{ps} = Z_p + Z_s \tag{2.70}$$

$$Z_{pt} = Z_p + Z_t \tag{2.71}$$

$$Z_{st} = Z_s + Z_t \tag{2.72}$$

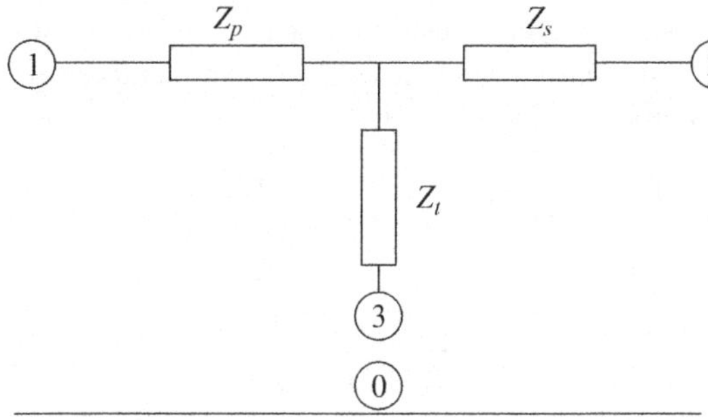

Fig. 2.19 *Equivalent circuit of the tree-winding transformer*

Subtracting Eq. (2.72) from Eq. (2.71), we get

$$Z_{pt} - Z_{st} = Z_p - Z_s \tag{2.73}$$

From Eqs. (2.70) and (2.73), we get

$$Z_p = \frac{1}{2}\left(Z_{ps} + Z_{pt} - Z_{st}\right) \tag{2.74}$$

Similarly,

$$Z_s = \frac{1}{2}\left(Z_{ps} + Z_{st} - Z_{pt}\right) \tag{2.75}$$

$$Z_t = \frac{1}{2}\left(Z_{pt} + Z_{st} - Z_{ps}\right) \tag{2.76}$$

Ground bus 0 is shown isolated because we are neglecting the shunt impedances of the transformers. Terminal 1 is connected to the supply and terminals 2 and 3 to the load. If the tertiary winding is used for the suppression of harmonics, then terminal 3 is floating.

2.6 LOAD REPRESENTATION

In the representation of loads for load flow and stability studies, it is important to know the variation of the real and reactive power with the variation of voltage. At a typical bus, the load may consist of the following:

1. Induction motors, 50–70%
2. Heating and lighting, 20–30%
3. Synchronous motors, 5–10%

Although it would be accurate to consider the *P-V* and *Q-V* characteristics of each of these loads for simulation, the analytic treatment would be very complicated. For analytic purposes, there are mainly three ways of representing the load:

Constant Impedance Representation This is often used for the representation of loads in stability studies when there are convergence problems or when simple models are used for synchronous machines. As the name suggests, the loads are considered as fixed value impedances. If the active and reactive power consumed by the load is $(P_{LO} + jQ_{LO})$, the nominal voltage at the load terminal is V_0, and the nominal current drawn by the load is I_0, then,

$$(P_{LO} + jQ_{LO}) = V_0 I_0^* \Rightarrow I_0 = \frac{(P_{LO} - jQ_{LO})}{V_0^*} \tag{2.77}$$

Therefore,

$$Z_L = \frac{V_0}{I_0} = \frac{|V_0|^2}{(P_L - jQ_L)} \tag{2.78}$$

where Z_L is the value of the load impedance.

As the impedance is considered constant, the load power changes with any change in the voltage at the load terminal. For a voltage value V, the power $(P_L + jQ_L)$ can be computed as:

$$(P_L + jQ_L) = (P_{LO} + jQ_{LO})\left(\frac{V_0}{V_0}\right)^2 \tag{2.79}$$

Constant Current Representation In this case, the load current I_0 is computed as:

$$I_0 = \frac{(P_{LO} - jQ_{LO})}{V_0^*} = |I_0|\angle(\theta - \phi) \tag{2.80}$$

where $V_0 = |V_0|\angle\theta$ and $\phi = \tan^{-1}\left(\frac{Q_{LO}}{P_{LO}}\right)$. Here ϕ is the power factor angle. The magnitude of I_0 is held constant. Therefore, the load power varies with the variation of load terminal voltage as:

$$(P_L + jQ_L) = (P_{LO} + jQ_{LO})\left(\frac{V_0}{V_0}\right) \tag{2.81}$$

Constant Power Representation Here, both the specified MW and MVAr are assumed to be constant. That means there is no variation of the power with the load terminal voltage. This is the representation used in load flow studies.

In some cases, the load power at a bus of the power system is considered as a mix of all the three types. Then the total load power is represented as:

$$(P_L + jQ_L) = (P_{LO} + jQ_{LO})\left[a_1\left(\frac{V_0}{V_0}\right)^2 + a_2\left(\frac{V_0}{V_0}\right) + a_3\right] \tag{2.82}$$

Here $a_1 - a_3$ are the proportions of each type of load.

2.7 REPRESENTATION OF FACTS CONTROLLERS [23]

FACTS controllers are mainly of two types, namely the thyristor controlled variable impedance type and the switching converter type. A brief description of each type, their further classification, and the power system representations of the most commonly used FACTS controllers are given below.

2.7.1 Thyristor Controlled Variable Impedance Type

This type of FACTS controllers use line commutated thyristors as the switching device. A combination of a number of inductors and capacitors are used. Some or all of these circuit elements are brought in or taken out of the circuit (transmission system) for a specified duration in every cycle of the AC supply with the use of the thyristors. This results in a change in the effective reactance of the combination as seen by the transmission system. Thus, a variation in the reactive power exchange between the FACTS controller and the power system is affected. However, this reactive power is proportional to the square of the bus voltage or the line current.

Depending on the connection of the controller to the system, they can be (i) shunt type or (ii) series type. The shunt connected variable impedance type FACTS controller is termed as static VAR compensator (SVC). SVC can have different configurations like—fixed capacitor thyristor controlled reactor (FC-TCR), and thyristor switched capacitor thyristor controlled reactor (TSC-TCR). Similarly, the configurations of the series connected variable impedance type FACTS controllers are—thyristor switched series capacitor (TSSC), and thyristor controlled series capacitor (TCSC).

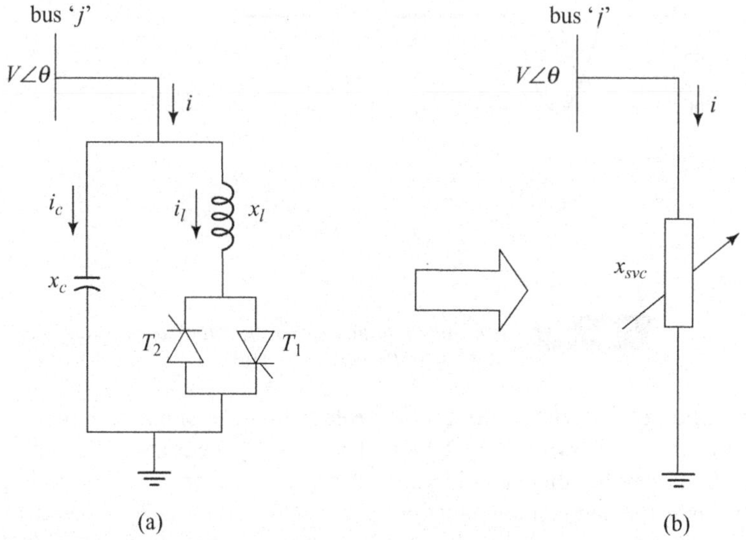

Fig. 2.20 *(a) FC-TCR, (b) Equivalent representation*

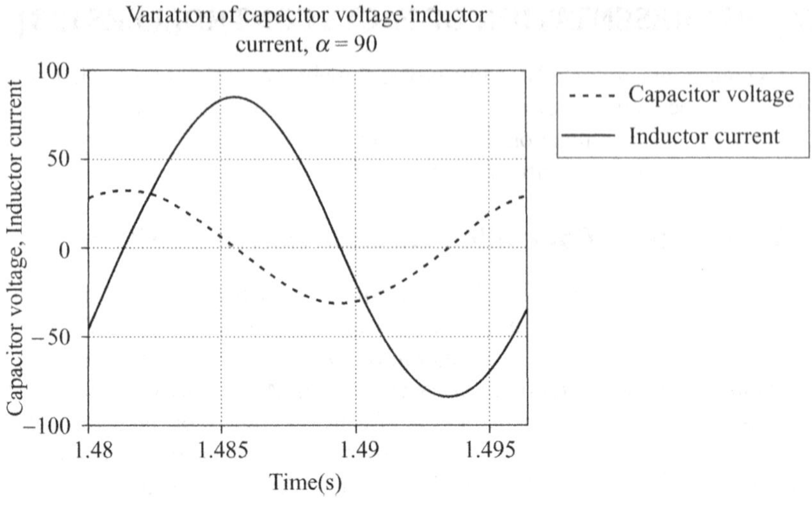

(a)

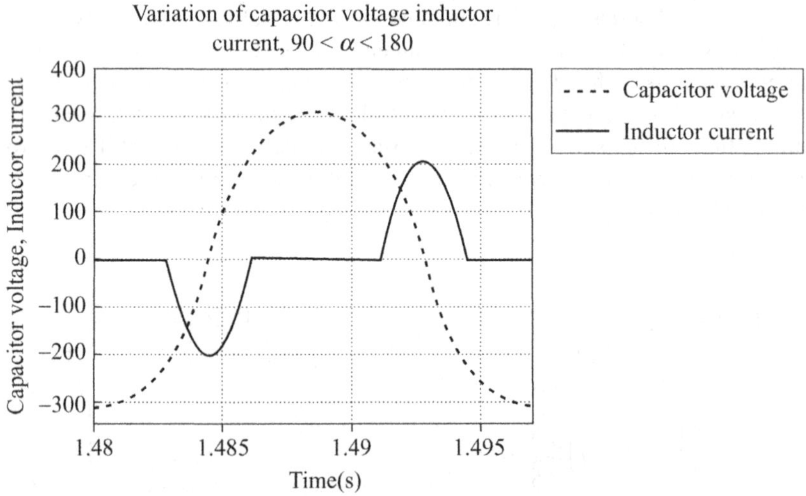

(b)

Fig. 2.21 *Variation of inductor current, capacitor voltage of FC-TCR for (a) $\alpha = 90°$, (b) $90° < \alpha < 180°$*

Fixed Capacitor-thyristor Controlled Reactor The single-line representation of the FC-TCR is shown in Fig. 2.20(a). The FC-TCR consists of an inductor with a pair of anti-parallel thyristors in series. A fixed capacitor is connected in parallel with the inductor-thyristor combination. The values of the reactances of the inductor and the capacitor are x_l and x_c respectively. This combination is connected to the power system bus 'j' in shunt, so that the voltage across the capacitor is the phase value of the bus voltage $V\angle\theta$. The same voltage is applied across the inductor when the thyristor is ON. The anti-parallel thyristors are used for the two half cycles of the supply voltage. The thyristor T_1 is switched ON in the +ve half cycle and the current starts flowing through the inductor. T_1 remains ON till the current reaches its natural zero. Switching of the thyristor T_2 in the –ve half cycle results in the same sequence of events and allows the flow of the –ve half cycle of the inductor current. If T_1 and T_2 are switched ON at a phase lag of 90° from the positive going zero crossing and negative going zero crossing respectively of the voltage across the capacitor (bus voltage in this case), then the inductor current will be continuous. This is clear from Fig. 2.21(a).

On the other hand, if the switching is done at a phase lag of 180°, no current will flow through the inductor at all. For switching at a phase lag angle α ($90° < \alpha < 180°$), a discontinuous current flows through the inductor, as shown in Fig. 2.21(b). This angle α is also termed as the delay angle or the firing angle of the thyristors. It can be shown that the amplitude of the fundamental component of this discontinuous current is a function of α. Hence, the effective reactance offered by the thyristor controlled reactor (TCR) denoted as x_{tcr}, is also a function of α. Neglecting the harmonics, the total impedance of the FC-TCR combination, x_{svc} is given as [24]:

$$x_{svc} = \beta_1 (x_C + \beta_2) - \beta_4 \beta_5 - x_c \tag{2.83}$$

where,

$$\beta_1 = \frac{2(\pi - \alpha) + \sin 2(\pi - \alpha)}{\pi}, \quad \beta_2 = \frac{x_c x_l}{x_c - x_l}, \quad \beta_3 = \sqrt{\frac{x_c}{x_l}},$$

$$\beta_4 = \beta_3 \tan[\beta_3 (\pi - \alpha)] - \tan(\pi - \alpha), \quad \beta_5 = \frac{4\beta_2^2 \cos^2(\pi - \alpha)}{\pi x_l}$$

Thus, the FC-TCR can be equivalently represented by a variable reactance, x_{svc} as shown in Fig. 2.20(b). This reactance can be inductive or capacitive depending on the value of α. The variation of x_{svc} with α is shown in Fig. 2.22.

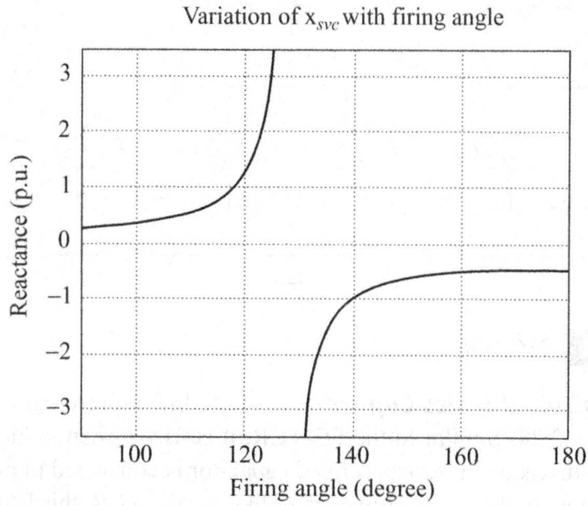

Fig. 2.22 *Variation of x_{svc} with α*

The values of x_c and x_l are 0.5 p.u. and 0.1667 p.u. respectively. A positive value of x_{svc} means the reactance is inductive, whereas a negative value of x_{svc} means it is capacitive. The minimum value of the capacitive reactance offered is equal to x_c, when α is 180°. On the other hand, the minimum inductive reactance offered is equal to the parallel combination of x_c and x_l, when α is 90°. The reactance becomes infinite for a value of $\alpha = 130°$. Actually at this point, the effective reactance of the TCR, x_{tcr}, becomes equal to the fixed value of x_c. Usually operation in the very high reactance region (e.g. $120° < \alpha < 140°$ for this set of values of x_c and x_l) is avoided due to stability issues.

Thyristor Switched Capacitor–Thyristor Controlled Reactor Here, anti-parallel thyristors are also provided in series with the capacitor (Fig. 2.23), so that the capacitor can be switched ON or OFF the circuit. However, the switching of the thyristors to bring the capacitor in the circuit can be done only at specific instants when the instantaneous value of the supply voltage matches the voltage on the capacitor due to its pre-charge. Any mismatch of voltage may lead to huge transients. Therefore, a capacitor will be/ will not be present in the circuit for the complete cycle and hence, the delay angle control of the reactance is not possible for the capacitor. Several capacitors are placed in parallel, with anti-parallel thyristors in series with each capacitor. Depending on the required value of the capacitive reactance (denoted by x_{tsc}), the required number of capacitors are kept in the circuit. Value of the inductive reactance provided by the single TCR (x_{tcr}) is controlled using the delay angle (α) control as in FC-TCR. The net reactance of TSC-TCR can be computed using Eq. (2.83) with x_c replaced by x_{tsc}.

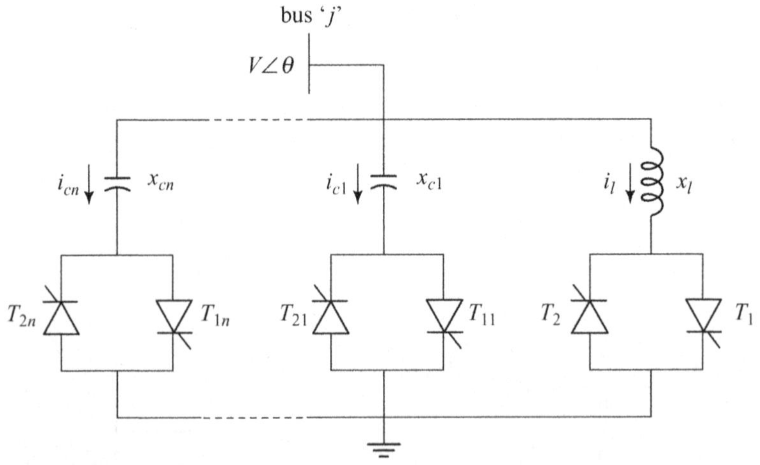

Fig. 2.23 *TSC–TCR*

Thyristor Controlled Series Capacitor Single-line representation of the TCSC is shown in Fig. 2.24. Similar to the FC-TCR, it consists of an inductor with a pair of anti-parallel thyristors in series. A fixed capacitor is connected in parallel with the inductor-thyristor combination. However, unlike the FC-TCR, this FACTS controller is connected in series with a transmission line, say line *i-j* between the buses *i* and *j*.

As a consequence, the voltage across the TCSC (and the fixed capacitor) is not the phase value of the supply voltage. Rather it is a function of the current flowing through the capacitor. The zero of the delay angle (α) of the thyristors is counted from the positive going zero crossing and the negative going zero crossing of the voltage across the capacitor. The net reactance offered by the TCSC, x_{tcsc}, is a function of α and can be computed using Eq. (2.83). The equivalent representation of the TCSC, shown in Fig. 2.24, has this variable reactance, x_{tcsc}, in series with the impedance of the line *i*-*j* (z_{line}).

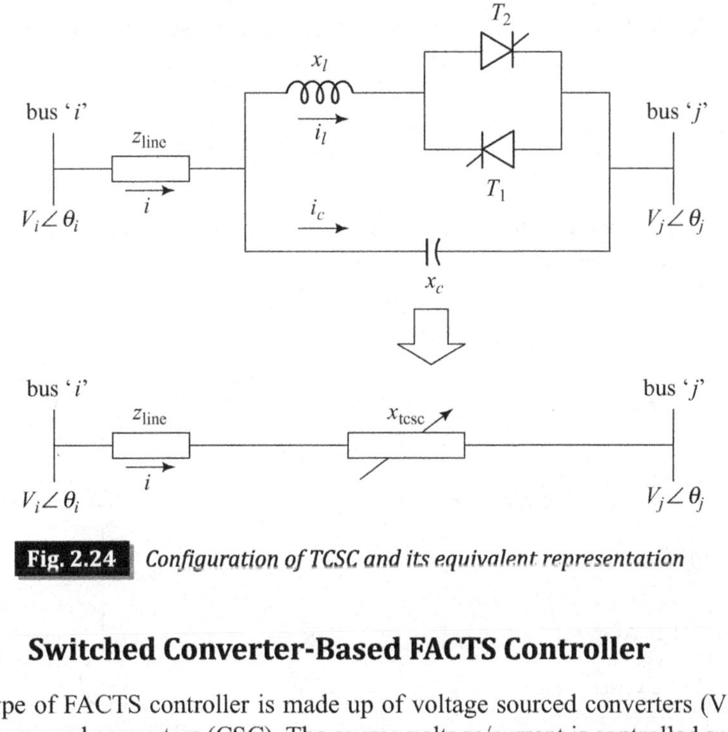

Fig. 2.24 *Configuration of TCSC and its equivalent representation*

2.7.2 Switched Converter-Based FACTS Controller

This type of FACTS controller is made up of voltage sourced converters (VSC) or current sourced converters (CSC). The source voltage/current is controlled such that the converter exchanges reactive power with the power system. This reactive power support does not depend on the system bus voltage or the line current. Moreover, the use of devices with turn-off capacity increases the speed of response and hence improves the dynamic performance of these controllers.

Depending on the connection, switched converter based FACTS controllers can be of (i) shunt type, (ii) series type, (iii) shunt-series type, and (iv) series-series type. The shunt connected controllers are called the static synchronous compensators (STATCOM) and the series connected controllers are called the static synchronous series compensators (SSSC). The shunt-series FACTS controller consists of two converters—one connected in shunt with one of the buses of the power system, and the other connected in series with a transmission line connected to that bus. The converters themselves are connected back to back through a common DC capacitor. The configuration is called unified power flow controller (UPFC). The series-series controller also has two back to back connected converters sharing a DC bus, but the

AC sides of both the converters are connected in series with two different transmission lines. The configuration is called interline power flow controller (IPFC). In addition to the reactive power support provided by the converters, both the UPFC and the IPFC can allow exchange of active power with the power system because of the presence of the back to back connection through the DC bus.

Static Synchronous Compensator A single-line representation of the STATCOM is shown in Fig. 2.25(a). It consists of a DC-AC converter whose AC side is connected to a power system bus 'm' through a coupling transformer of leakage reactance jx_t. Usually, a capacitor forms the DC side of the converter. However, if the STATCOM has to provide active power support for a short duration, then some energy source (like battery) is connected at the DC side. The current injected by the STATCOM at bus 'm', $|I_{sh}|\angle\psi$, can be obtained as:

$$|I_{sh}|\angle\psi = \frac{(|V_{sh}|\angle\theta_{sh} - |V_m|\angle\theta_m)}{jx_t} \tag{2.84}$$

where $|V_{sh}|\angle\theta_{sh}$ is the output voltage of the DC to AC converter, $|V_m|\angle\theta_m$ is the voltage of the bus 'm', and x_t is the leakage reactance of the coupling transformer.

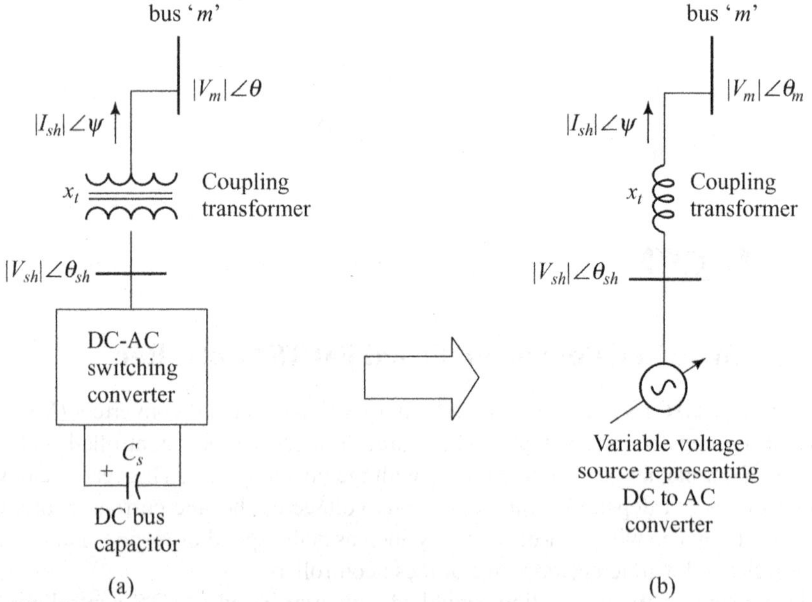

Fig. 2.25 *(a) Single line diagram of STATCOM connection, (b) Its representation*

Hence, the power injection at bus 'm' is given as

$$P_{sh} - jQ_{sh} = V_m^* I_{sh} = (|V_m|\angle -\theta_m)(|I_{sh}|\angle\psi)$$

$$= (|V_m|\angle -\theta_m)\left(\frac{|V_{sh}|\angle\theta_{sh} - |V_m|\angle\theta_m}{jx_t}\right) \tag{2.85}$$

On simplification,

$$P_{sh} = \frac{|V_m||V_{sh}|}{x_t}\sin(\theta_{sh} - \theta_m) \tag{2.86}$$

$$Q_{sh} = \frac{|V_m||V_{sh}|}{x_t}\cos(\theta_{sh} - \theta_m) - \frac{|V_m|^2}{x_t} \tag{2.87}$$

From this, it is clear that the active power exchange will be zero under the condition

$$\theta_{sh} = \theta_m \tag{2.88}$$

It is assumed that the active power loss in the converter and the coupling transformer is zero. Applying Eq. (2.88) to Eq. (2.87), the reactive power exchange between the STATCOM and the bus '*m*' becomes:

$$Q_{sh} = \frac{|V_m|}{x_t}(|V_{sh}| - |V_m|) \tag{2.89}$$

The value of $|V_{sh}|$ determines the direction (and hence nature) of the reactive power flow. If it is higher than the magnitude of the bus '*m*' voltage ($|V_m|$), then the reactive power is injected into the power system, whereas if the line mid-point voltage magnitude is higher, then reactive power is drawn from the power system. $|V_{sh}|$ and θ_{sh} can be varied by controlling the switching of the converter so that the STATCOM exchanges power as per the requirement of the system. Thus, the converter can be represented by a variable voltage source as shown in Fig. 2.25(b).

Static Synchronous Series Compensator Similar to STATCOM, SSSC also consists of a DC-AC converter whose AC side is connected to the power system using a coupling transformer. However, unlike STATCOM, the connection is in series with a transmission line *m-n* (between buses '*m*' and '*n*') as shown in the single-line representation, Fig. 2.26(a). The voltages of the two buses are $|V_m|\angle\theta$ and $|V_n|\angle 0°$. The transmission line reactance is x_{line}, the leakage reactance of the transformer is jx_t and the output (ac) voltage of the converter is $|V_{se}|\angle\theta_{se}$.

The current flowing through the line *m-n*, $|I|\angle\phi$, can be obtained as

$$|I|\angle\phi = \frac{(|V_m|\angle\theta + |V_{se}|\angle\theta_{se} - |V_n|\angle 0°)}{j(x_t + x_{\text{line}})} \tag{2.90}$$

The power exchanged by the converter with the power system is given as

$$P_{se} + jQ_{se} = V_{se}I^* = (|V_{se}|\angle\theta_{se})(|I|\angle -\phi)$$

$$\Rightarrow P_{se} - jQ_{se} = V_{se}^* I = (|V_{se}|\angle -\theta_{se})\left(\frac{|V_m|\angle\theta + |V_{se}|\angle\theta_{se} - |V_n|\angle 0°}{j(x_t + x_{\text{line}})}\right) \tag{2.91}$$

From which the expressions of the active and the reactive power exchanged are obtained as

$$P_{se} = \frac{|V_{se}|}{(x_t + x_{\text{line}})}[|V_m|\sin(\theta - \theta_{se}) - |V_n|\sin(-\theta_{se})] \tag{2.92}$$

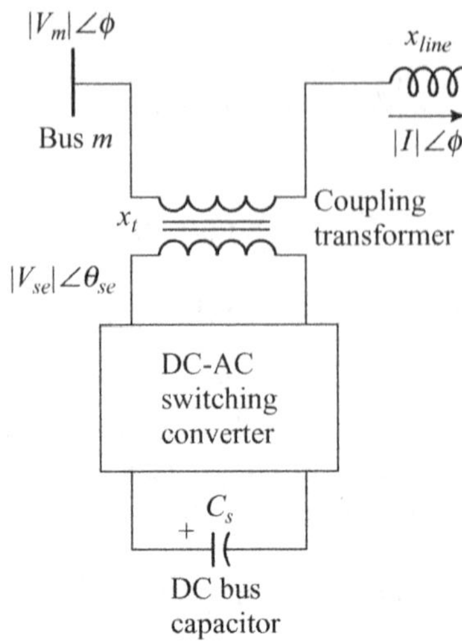

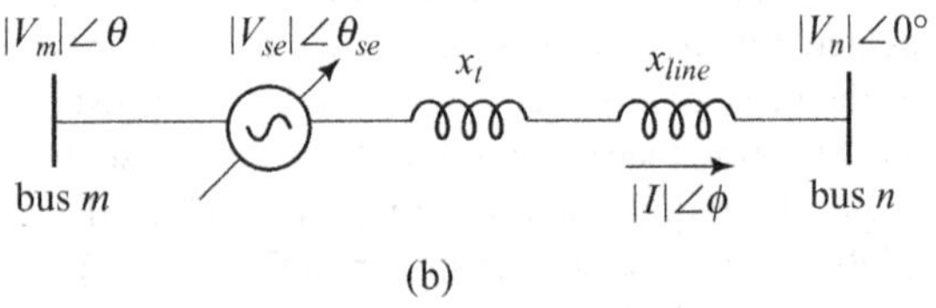

Fig. 2.26 *(a) Single line diagram of SSSC connection, (b) Power system representation of SSSC*

$$Q_{se} = \frac{|V_{se}|}{(x_t + x_{\text{line}})}[|V_m|\cos(\theta - \theta_{se}) - |V_n|\cos(-\theta_{se}) + |V_{se}|] \tag{2.93}$$

Also, the power flowing through the line into bus 'n',

$$P_{mn} + jQ_{mn} = V_n I^* = (|V_n|\angle 0°)(|I|\angle -\phi)$$

$$\Rightarrow P_{mn} - jQ_{mn} = V_n^* I = (|V_n|\angle 0°)\left(\frac{|V_m|\angle\theta + |V_{se}|\angle\theta_{se} - |V_n|\angle 0°}{j(x_t + x_{\text{line}})}\right) \tag{2.94}$$

Hence,

$$P_{mn} = \frac{|V_m||V_n|}{(x_t + x_{\text{line}})}\sin\theta + \frac{|V_{se}||V_n|}{(x_t + x_{\text{line}})}\sin\theta_{se} \tag{2.95}$$

$$Q_{mn} = \frac{|V_m||V_n|}{(x_t + x_{\text{line}})}\cos\theta - \frac{|V_n|^2}{(x_t + x_{\text{line}})} + \frac{|V_{se}||V_n|}{(x_t + x_{\text{line}})}\cos\theta_{se} \tag{2.96}$$

It is clear from Eqs. (2.95) and (2.96), that the power flow through the line can be controlled by varying the values of $|V_{se}|$ and θ_{se}, the VSC output voltage magnitude, and its angle. In the process, the power exchanged by the VSC also changes, as is evident from Eqs. (2.92) and (2.93). The VSC is represented by a variable voltage source in the equivalent representation shown in Fig. 2.26(b).

Usually, the SSSC is not provided with any source of energy and hence does not supply any active power. It is clear from Eq. (2.91) that the SSSC will exchange reactive power only (and no real power) if the voltage injected by it is in quadrature with the current flowing through the line, i.e.,

$$\theta_{se} = \phi \pm \frac{\pi}{2} \tag{2.97}$$

Using Eq. (2.97) in Eq. (2.90) and subsequent rearrangement,

$$(|I|(x_t + x_{\text{line}}) \mp |V_{se}|)\angle\phi = [|V_m|\sin\theta + j(|V_n| - |V_m|\cos\theta)] \tag{2.98}$$

from which we can get,

$$|I| = \frac{\sqrt{(|V_m|\sin\theta)^2 + (|V_n| - |V_m|\cos\theta)^2} \pm |V_{se}|}{(x_t + x_{\text{line}})} \tag{2.99}$$

$$\phi = \tan^{-1}\left(\frac{|V_n| - |V_m|\cos\theta}{|V_m|\sin\theta}\right) \tag{2.100}$$

Hence,
$$\theta_{se} = \tan^{-1}\left(\frac{|V_n| - |V_m|\cos\theta}{|V_m|\sin\theta}\right) \pm \frac{\pi}{2} \tag{2.101}$$

The power flow through the line becomes

$$P_{mn} = \frac{|V_m||V_n|}{(x_t + x_{\text{line}})}\sin\theta + \frac{|V_{se}||V_n|}{(x_t + x_{\text{line}})}\sin\left(\phi \pm \frac{\pi}{2}\right) \tag{2.102}$$

$$Q_{mn} = \frac{|V_m||V_n|}{(x_t + x_{\text{line}})}\cos\theta - \frac{|V_n|^2}{(x_t + x_{\text{line}})} + \frac{|V_{se}||V_n|}{(x_t + x_{\text{line}})}\cos\left(\phi \pm \frac{\pi}{2}\right) \tag{2.103}$$

Now if the reactive power flow is to be changed without changing the active power flow, the required condition is

$$\theta_{se} = \phi \pm \frac{\pi}{2} = 0 \quad \Rightarrow \phi = \mp\frac{\pi}{2} \tag{2.104}$$

Using this in Eq. (2.100) gives:

$$|V_m|\sin\theta = 0, \quad \text{i.e., either } |V_m| = 0 \text{ or } \theta = 0 \tag{2.105}$$

$\theta = 0$ means $P_{mn} = 0$. Hence none of these is a valid operating condition. Therefore, an independent control of the active and the reactive power flow through the line is not possible with the SSSC.

Unified Power Flow Controller A single-line representation of the UPFC and its connection to a power system is shown in Fig. 2.27(a). It consists of one converter connected in shunt and another converter connected in series with a transmission line. Individually, the converters act as variable voltage sources just like the STATCOM and the SSSC. But the converters are connected by a common DC bus capacitor, thus providing a path for active power flow. This allows each converter to exchange both the active and the reactive power with the power system. The reactive power is generated/absorbed by the converters. On the other hand, the active power supplied (to the system) by one converter comes from the other converter through the DC bus as the other converter draws an equal amount of active power (from the system). Thus the algebraic sum of the active power exchanged by the two converters is zero (neglecting active power loss of the converters and the coupling transformers). Hence, the angles of the output voltages of the converters are not restricted by the relations in Eqs. (2.88) and (2.97) and can be anything between 0° and 360°. This enables the UPFC to control the active and the reactive power flow through the line independently.

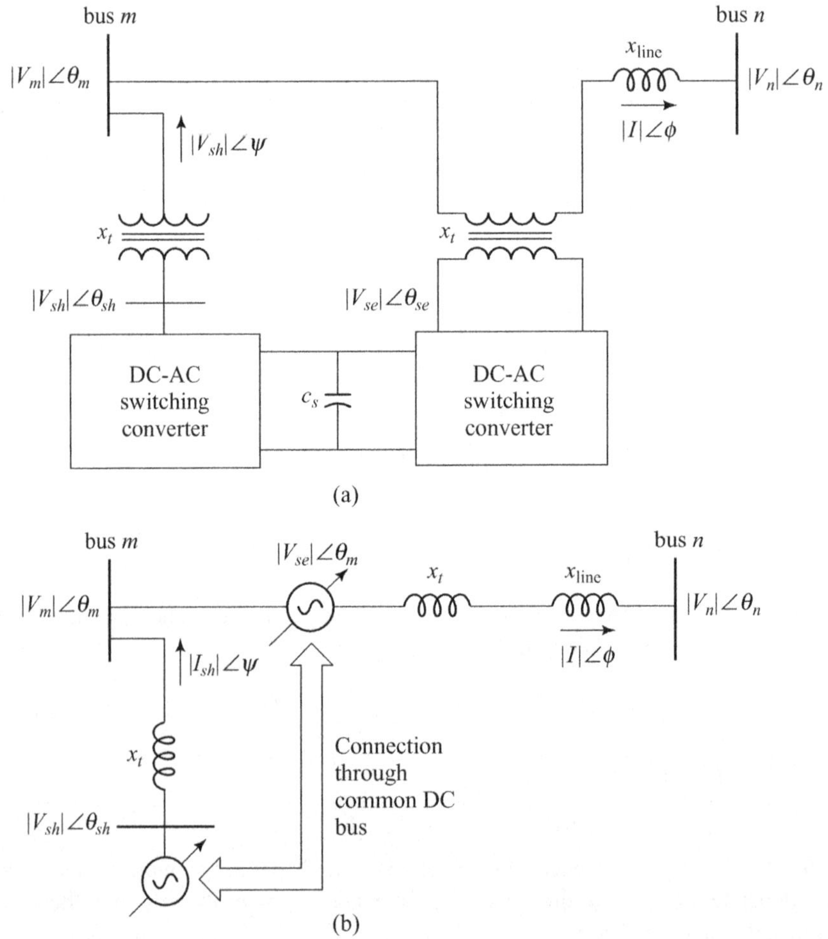

Fig. 2.27 *(a) Single line diagram of UPFC connection, (b) Power system representation of UPFC*

The equivalent representation of the UPFC, as shown in Fig. 2.27(b), has two variable voltage sources connected in shunt and series with the transmission line, and the sources have coupling among themselves. The voltages of these shunt and series sources are $|V_{sh}|\angle\theta_{sh}$ and $|V_{se}|\angle\theta_{se}$ respectively. The current, the active power, the and reactive power injected by the shunt converter of the UPFC can be obtained using Eqs. (2.84), (2.86) and (2.87) respectively. Similarly, the current flowing through the transmission line, the active and the reactive power exchanged by the series converter with the power system, and the active and the reactive power flow through the transmission line can be obtained using Eqs. (2.90), (2.92), (2.93), (2.94) and (2.95) respectively.

2.8 REPRESENTATION OF HVDC TRANSMISSION SYSTEM [20-22]

A typical HVDC system structure is shown in Fig. 2.28. It consists of converters at both ends, connected to the AC system through the converter transformers. The DC sides of the converters are connected by the HVDC transmission line. However, if the establishment of an asynchronous link is the only objective, then the line may not be there, both the converters may be placed in the same sub-station with their DC sides connected by a link. The HVDC line/link has two conductors—one having positive polarity and the other having negative polarity. This is called bipolar structure.

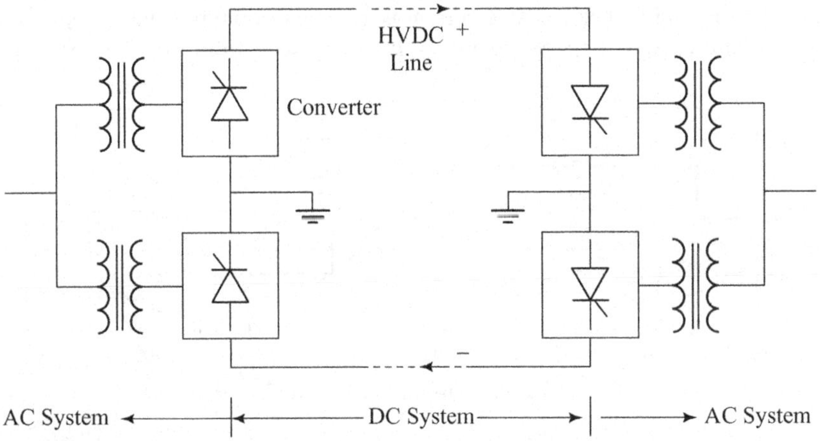

Fig. 2.28 *HVDC system structure*

The basic converter structure is shown in Fig. 2.29. It consists of six thyristors, $T_1 - T_6$. The AC side is represented by a voltage source with per phase source reactance, L_S. Two thyristors are connected to each phase, one for the positive half cycle and the other for the negative half cycle.

The thyristors are switched ON by supplying a firing pulse to its gate terminal, with positive voltage across the thyristors (V_{sw}). The phase angular delay between the instant when the V_{sw} becomes positive and the instant of supplying the firing pulses, is termed as the firing angle, α. The thyristors gets switched OFF automatically when the current through them becomes approximately zero. If the source reactance is neglected (L_s = 0), then the switching ON of the thyristors and hence the change

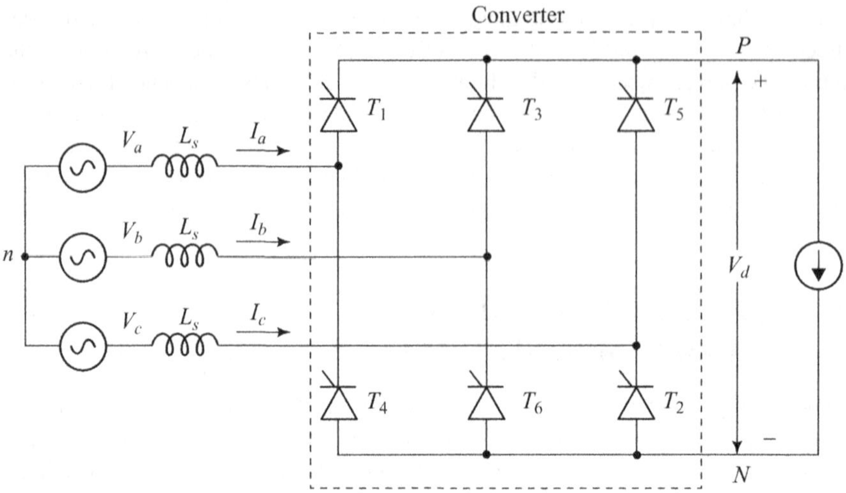

Fig. 2.29 *Six-pulse converter structure*

in the current through the thyristors takes place instantaneously. This is shown in Fig. 2.30(a), where the phase voltage and the phase/line current of the source is denoted by V_s and I_s. The DC side current is assumed ripple free and is represented by a constant current source of value I_d. It can be seen from Fig. 2.30(a) that the

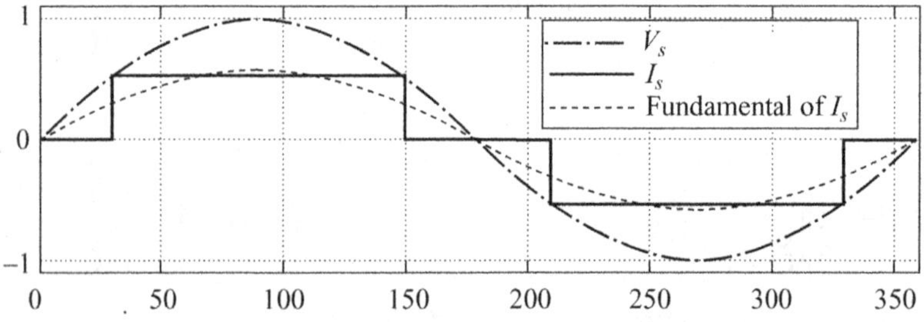

(a) V_s, I_s, fundamental of I_s for $L_s = 0$

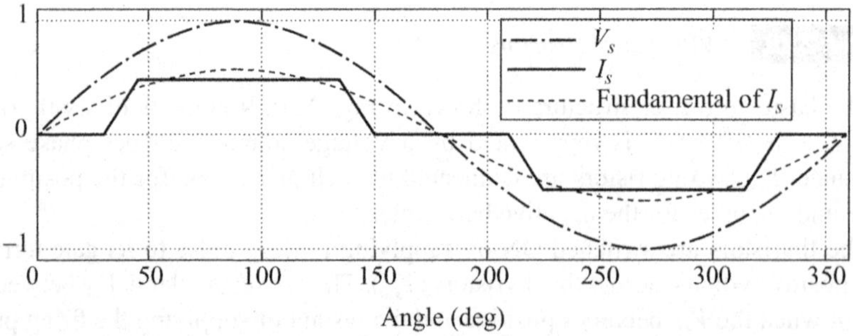

(b) V_s, I_s, fundamental of I_s for L_s non-zero

Fig. 2.30 *The AC-side phase voltage, current and the fundamental of the phase current*

current supplied by the AC source takes the shape of a stepped square waveform, with the peak value being equal to I_d. The fundamental of this current is also shown in the figure. Now if $L_s \neq 0$, then the switching cannot take place instantaneously. So the change of current I_s takes place over a small time and the current waveform takes the shape of a trapezoid, as shown in Fig. 2.30(b). This period over which the current changes, is called the overlap period and the corresponding angle is called the overlap angle (u). The angle at which the current finally becomes zero is called the extinction angle (δ). Therefore it is clear that $\delta = \alpha + u$. The extinction angle can be found by the following relation:

$$\cos\delta = \cos\alpha - \frac{2\omega L_s}{\sqrt{2}V_{L-L}} I_d \tag{2.106}$$

The amplitude of the fundamental of the current is not same for the two cases, but the difference is very small and can be neglected. The r.m.s. value of this fundamental component (I_{s1}) can be found using Fourier analysis as:

$$\sqrt{2}I_{s1} = \frac{2\sqrt{3}}{\pi} I_d \Rightarrow I_{s1} = \frac{\sqrt{6}}{\pi} I_d = 0.78 I_d \tag{2.107}$$

The average DC voltage available at the converter terminal V_d, is given by

$$V_d = \frac{3\sqrt{2}}{\pi} V_{L-L}\cos\alpha - \frac{3}{\pi}\omega L_s I_d = V_{d0}\cos\alpha - R_s I_d \tag{2.108}$$

where $V_{d0} = \frac{3\sqrt{2}}{\pi} V_{L-L}$ and $R_s = \frac{3}{\pi}\omega L_s$ and $V_{L\text{-}L}$ is the AC line-to-line voltage.

The AC-side power (P_{ac}) is obtained from the voltage and the fundamental of the current. This must be equal to the DC-side power (P_{dc}).

$$P_{ac} = \sqrt{3}V_{L-L}I_{s1}\cos\phi = P_{dc} = V_d I_d = I_d\left(V_{d0}\cos\alpha - \frac{3}{\pi}\omega L_s I_d\right) \tag{2.109}$$

For $L_s = 0$, and using Eqs. (2.107) and (2.108), we get:

$$\sqrt{3}V_{L-L}I_{s1}\cos\phi = \frac{\pi}{\sqrt{6}} I_{s1} \cdot \frac{3\sqrt{2}}{\pi} V_{L-L}\cos\alpha \Rightarrow \cos\phi = \cos\alpha \tag{2.110}$$

The reactive power drawn by the converter is:

$$Q_{ac} = \sqrt{3}V_{L-L}I_{s1}\sin\phi = \sqrt{3}V_{L-L}I_{s1}\sin\alpha \tag{2.111}$$

However, for $L_s \neq 0$, the relation gets modified to

$$\cos\phi = \frac{1}{2}(\cos\alpha + \cos\delta) \tag{2.112}$$

2.8.1 Inverter Mode of Operation

With $L_s = 0$, if $90° < \alpha \leq 180°$, V_d becomes negative. The current direction remaining the same, the power flow becomes negative. Thus, the power flows from the DC side to the AC side and thus the converter works as an inverter. For $L_s \neq 0$, this transition from a rectifier to an inverter mode takes place at a firing angle little less than 90°. In the case of the inverter, it is the usual practice to use an angle β instead of α where β is defined as

$$\beta = \pi - \alpha \tag{2.113}$$

Using this in Eq. (2.107), we get the expression for average DC voltage of the inverter as

$$V_d = V_{d0}\cos(\pi - \beta) - R_s I_d = -(V_{d0}\cos\beta + R_s I_d) \tag{2.114}$$

The negative sign indicates that the current enters the positive terminal as opposed to the case of the rectifier operation, where the current comes out of the positive terminal. The overall HVDC system consisting of one rectifier, one inverter joined by a line of resistance R_l, can be represented by the equivalent circuit shown in Fig. 2.31. From this, the expression of the current flowing through the HVDC line can be found as:

$$I_d = \frac{V_{d0r}\cos\alpha - V_{d0i}\cos\beta}{R_{cr} + R_l + R_{ci}} \tag{2.115}$$

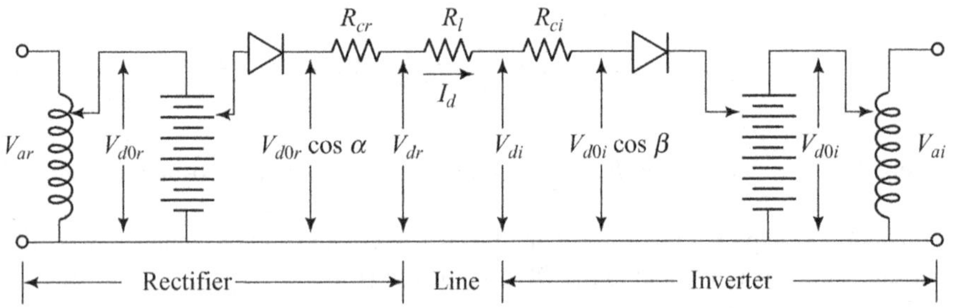

Fig. 2.31 *The equivalent circuit of the HVDC system*

2.9 CONCLUSION

This chapter has briefly considered the representation of various components of a power system, such as the transmission lines, the transformers, the loads, and the generators. All these representations are adequate for the basic system studies of short circuit, load flow, and transient stability. Power system models of different FACTS devices, and HVDC transmission system are also presented. We next proceed to interconnect these components into a dynamical interconnected power system and then analyze it.

Example 2.1 A thermal generating station has two 250 MVA, 22 kV, 50 Hz units. The two units are represented by a single equivalent generator modeled by the Classical model having parameters as: $x'_d = 0.4$ pu, $H = 2$ s on the equivalent machine base. The station is connected to the rest of the power system (represented by an infinite bus) through a transformer and a double circuit line as shown in Fig. 2.32.

The transformer is rated at 600 MVA, 22 kV/220 kV with leakage reactance of 10%. The lumped reactances of the 220 kV parallel transmission lines are 82.3 Ω (line 1) and 118.6 Ω (line 2). The infinite bus (bus 3) voltage is 220 kV and the generator terminal bus (bus 1) voltage (V_t) is 23.1 kV. The power supplied by the generator at the infinite bus is—P = 400 MW, Q = 55.1 MVAR. With 100 MVA as the common system base, and the infinite bus as the reference,

(a) Find the phase angle (θ_t) of the generator terminal bus voltage and the magnitude (in pu) and the angle of the voltage behind the transient reactance of the generator.

(b) Write the swing equation of the generator under this condition.

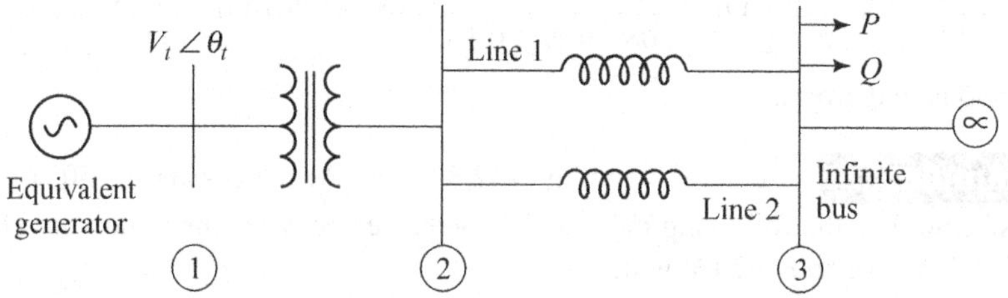

Fig. 2.32 *Power system of Example 2.1*

Solution: First, all the system quantities are to be converted to per unit on the common 100 MVA base.

Using Eq. (2.3), the base impedance for the transmission lines = $z_{\text{base}} = \dfrac{220^2}{100} = 484\ \Omega$

Line 1 reactance in p.u. = $\dfrac{82.3}{484} = 0.17$; Line 2 reactance in p.u. = $\dfrac{118.6}{484} = 0.245$;

Equivalent reactance, $x_{eq} = \dfrac{0.17 \times 0.245}{0.17 + 0.245} = 0.10$

Using Eq. (2.6), leakage reactance of the transformer = $x_t = 0.12 \times \dfrac{100}{600} = 0.02$ p.u.

The infinite bus voltage magnitude = 220 kV = 1.0 p.u.
The generator terminal bus voltage magnitude = 23.1 kV = 23.1/22 pu = 1.05 p.u.
The active power transferred = 400 MW = 4.0 p.u.
Angle of the generator terminal bus voltage

$$= \theta_t = \sin^{-1}\left(\frac{4.0 \times (0.02 + 0.10)}{1.05 \times 1.0}\right) = 27.203°$$

The current flowing out of bus 1 = $I = \dfrac{1.05\angle 27.203° - 1.0\angle 0°}{j(0.02 + 0.1)}$ (4.00 + j 0.55) p.u.

Transient reactance of the equivalent machine = $x'_d = 0.4 \times \dfrac{100}{2 \times 250} = 0.08$ p.u.

Voltage behind the transient reactance,

$$E'_{q_0}\angle\delta = 1.05\angle 27.203° + j0.08\times(4.0 + j0.55) = 1.1965\angle 41.96° \text{ p.u.}$$

For writing the swing equation, the inertia constant of the equivalent machine in the system base,

$$H_{\text{system}} = H_{\text{machine}} \times \frac{\text{Machine MVA base}}{\text{System MVA base}} = 2\times\frac{500}{100} = 10$$

So the swing equation of the machine becomes:

$$\frac{d^2\delta}{dt^2} = \frac{180\times 50}{10}\left[4.0 - \frac{1.1965\times 1.0}{0.08+0.02+0.10}\sin\delta\right] = 900(4.0 - 5.9825\sin\delta)$$

where δ is in degrees.

Example 2.2 A three-winding 132/33/6.6 kV, three-phase, 50 cycles/s transformer has the following measured impedances between the windings referred to 30 MVA base and 132 kV winding.

$$Z_{ps} = 0 + j\,0.15 \text{ p.u. } Z_{pt} = 0 + j\,0.09 \text{ pu; } Z_{st} = 0 + j\,0.08 \text{ p.u.}$$

The 6.6 kV secondary winding supplies a balanced load, taking a current of 2000 A at 0.8 lagging power factor and the 33 kV tertiary supplies a star-connected inductive reactor of (0 + j 50) Ω per phase. Determine the voltage required at the 132 kV primary terminals to maintain 6.6 kV at the secondary terminals.

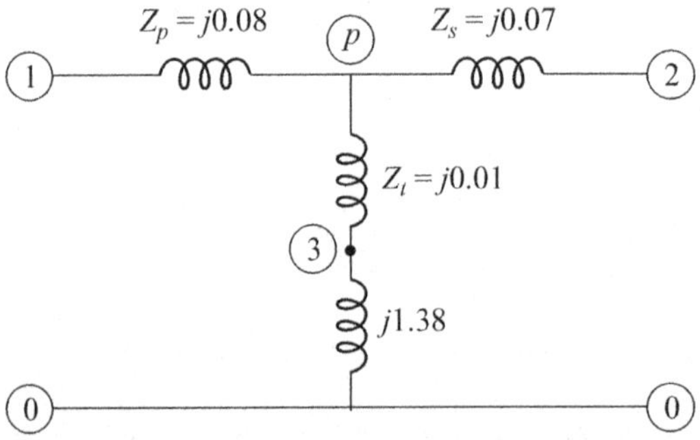

Fig. 2.33 *Three-winding transformer for Example 2.2*

Solution: Using Eqs. (2.74) – (2.76) in the equivalent circuit of Fig. 2.33, we get

$$Z_s = \frac{1}{2}(Z_{ps} + Z_{st} - Z_{pt}) = \frac{1}{2}(j0.15 + j0.08 - j0.09) = j0.07 \text{ p.u.}$$

$$Z_t = \frac{1}{2}(Z_{pt} + Z_{st} - Z_{ps}) = \frac{1}{2}(j0.09 + j0.08 - j0.15) = j.01 \text{ p.u.}$$

$$Z_p = \frac{1}{2}(Z_{ps} + Z_{pt} - Z_{st}) = \frac{1}{2}(j0.15 + j0.09 - j0.08) = j0.08 \text{ p.u.}$$

$$\text{Base current} = \frac{\text{Base MVA}}{\text{Base voltage}} = \frac{30 \times 10^3}{3} \bigg/ \left(\frac{6 \cdot 6}{\sqrt{3}}\right) = 2624 \text{ A}$$

$$\text{Load current in p.u.} = \frac{2000}{2624} = 0.76$$

The P.F. is 0.8 lagging, i.e., $\cos(\theta) = 0.8$. Hence, $\theta = 36.78°$. Taking the secondary voltage as reference $= 1\angle 0°$, the current is $0.76 \angle -36.78° = 0.76\,(0.8 - j\,0.6) = 0.608 - j\,0.456$.

The voltage at node p is $V_p = 1\angle 0° + j\,0.07\,(0.608 - j\,0.456) = 1.032 + j\,0.043$

33 kV tertiary winding

$$\text{Base current} = \frac{30 \times 10^3}{\sqrt{3}\,(33)} = 525 \text{ A}$$

$$\text{Base impedance} = \frac{\text{Base voltage}}{\text{Base current}} = \frac{33 \times 10^3}{\sqrt{3}\,(525)} = 36.3\,\Omega$$

$$\text{Hence, 50 } \Omega \text{ impedance is equal to } \frac{50}{36.3} = 1.38 \text{ p.u.}$$

$$\text{Current in the tertiary winding} = \frac{1.032 + j\,0.043}{j(1.38 + 0.01)}$$

$$= 0.031 - j0.742 \text{ p.u.}$$

Current in the primary winding = Current in the secondary winding + Current in the tertiary winding

$$= (0.608 - j\,0.456) + (.031 - j\,0.742)$$

$$= 0.639 - j\,1.198 \text{ p.u.}$$

Voltage V_1 across the primary winding is given by

$V_1 = (1.032 + j\,0.043 + j\,0.08\,(0.639 - j\,1.198)) = 1.13 + j\,0.09 = 1.133\,\angle 4.55°$ p.u.

Problems

2.1 Consider a transposed three-phase line whose parameters are $z = 0.9 \times 10^{-4} + j4 \times 10^{-4}$ Ω/m/phase and $y = j0.25 \times 10^{-8}$ Ω/m/phase. The line is 300 km long. Compute (1) the characteristic impedance Z_c, (2) the propagation constant γ, (3) the parameters of the equivalent π and T representations using the first two terms of expansion of the hyperbolic terms (Eq. (2.27)).

2.2 In Problem 2.1, for a receiving end voltage of 220 kV between phases and a receiving end power of 200 MVA at a power factor of 0.8 lagging, compute, using the π representation, the sending end voltage, and the current.

2.3 For the line in Problem 2.1, compute the parameters in the nominal π and T circuits. Calculate the sending end voltage and the current for both cases for the same terminal conditions as in Problem 2.2. Compare with the results of Problem 2.2.

2.4 Compute the A, B, C, and D parameters for the nominal π network of Problem 2.3, and using these, compute the Y parameters of the line. Verify the Y parameters by writing the node equations at the sending and the receiving ends in the π network. From the A, B, C, and D parameters, compute the Z parameters and verify that $Z = Y^{-1}$.

2.5 A three-winding transformer with primary and secondary rated voltages as 15 kV and 66 kV respectively, has a tertiary winding rated at 5 kV. The measured impedances referred to a 15 MVA base are as follows:
$Z_{ps} = j7.0$ p.u., $Z_{pt} = j15.0$ p.u., and $Z_{st} = j10.0$ p.u
Construct the equivalent circuit.

2.6 In Problem 2.5, another transformer is connected in parallel with the primary and secondary terminals having a series impedance of $j6.0$ p.u. on a 15 MVA base. Assume a symmetric three-phase to ground short circuit on the 5 kV bus. Compute the fault current. Assume a voltage source of $1 + j0$ p.u. connected to the 15 kV bus.

2.7 (a) Consider an off-nominal turns ratio transformer with tap ratio $a = 0.8$ and an impedance of j 0.05 referred to the unity tap side. Construct the equivalent circuit and identify the shunt elements as capacitive/reactive.
(b) Repeat (a) if $a = 1.2$.

3 TOPOLOGICAL ANALYSIS OF POWER NETWORKS

3.1 INTRODUCTION

The purpose of this chapter is to derive, via a circuit-theoretic viewpoint, the commonly used transmission network representations in power systems. These derivations also provide a logical basis to understand easily the solution of large-scale power networks both for steady state and dynamic analysis. In the new restructured scenario, such a system viewpoint is essential.

For most power system studies such as short circuit, load flow, and transient stability, the transmission network (inclusive of transformers) can be treated as a passive network. The dynamic phenomena in the transmission network are much faster than in the rest of the components so that a steady state representation of the network is reasonable and this fact can be rigorously established via the theory of singular perturbations. The dynamics of the transmission network are relevant in the 0.005 – 0.05 s range and are known as *electro-magnetic transients*. For deriving the representations of the transmission network, linear graph theory is found to be a very useful tool. The concept of cut-sets, to be discussed, is useful in system stability studies also. There are several texts on linear graph theory.

3.2 ORIENTED GRAPHS

In the electric transmission network, we are concerned with the interconnection of transmission lines, transformers, and shunt reactors/capacitors that can be modeled in terms of two terminal passive components called *elements* as discussed in Chapter 2. The points of interconnection are called *buses*. The graph of a network represents the manner in which the passive elements and the buses are interconnected. Each two terminal element is represented by a line segment called the *edge* or the *branch*. The edges represent the interconnections or *topology* of the network. In the resulting graph, the buses are termed as *nodes*.

Figure 3.1(a) shows a network consisting of nine elements. Its graph is shown in Fig. 3.1(b) where the five nodes are numbered in parentheses. A direction may be associated with each edge of the graph in which case it is called an oriented or a directed graph (Fig. 3.1(c)). The directions assigned are consistent with the concept of associated reference directions for a two terminal passive element in circuit theory.

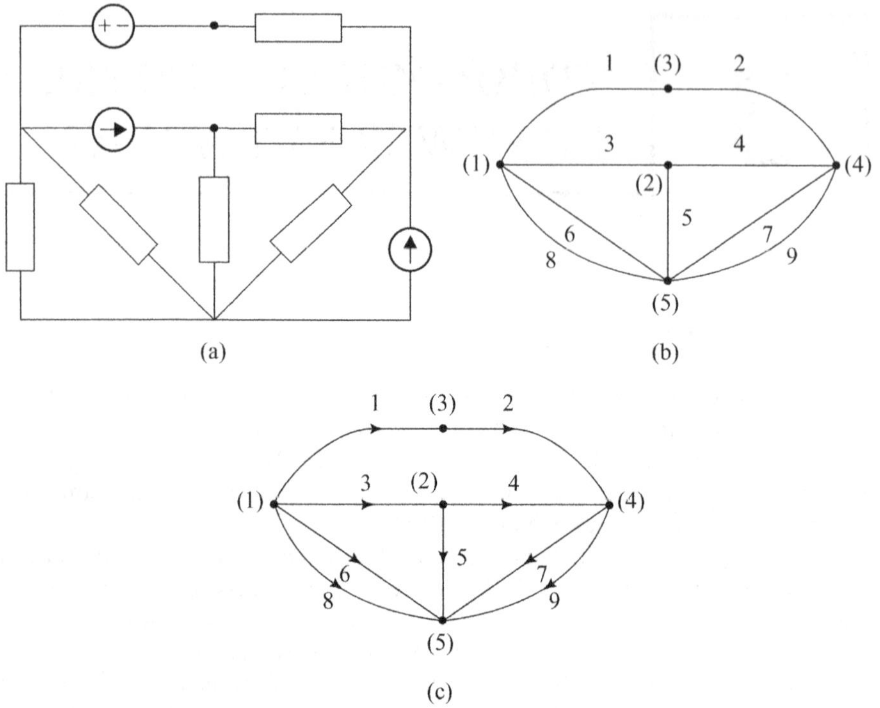

Fig. 3.1 *A network and its oriented graph*

3.2.1 Associated Reference Direction

Consider the element in Fig. 3.2(a) which may be a passive element, current, or a voltage source. The associated reference directions are such that a positive current enters the '+' terminal of the voltage reference and leaves the '–' terminal of the voltage reference. The oriented graph is shown in Fig. 3.2(b). If the element is purely passive as in Fig. 3.3(a), and v and i are the voltage and the current phasors, then $v = zi$ or $i = yv$, where z and y are the complex impedance and the admittance of the element respectively. The reference direction for the branch in the oriented graph is chosen to agree with the current direction (Fig. 3.3 (b)). If the element is a current source (Fig. 3.4(a)), then the positive orientation of the current source is chosen to agree with the reference direction in the graph (Fig. 3.4(b)). Note that terminal (1) has the + sign and terminal (2), the – sign for the voltage across the current source. If the element is a voltage source (Fig. 3.5(b)), the orientation in the graph is chosen such that the arrow in the graph goes from the positive to the negative terminal of the voltage reference. The current, unlike in conventional circuit analysis, goes from the '+' to the '–' terminal. Thus, while for the passive element the orientation of the graph is consistent with the associated reference directions of circuit theory, for the current and the voltage source, it is not.

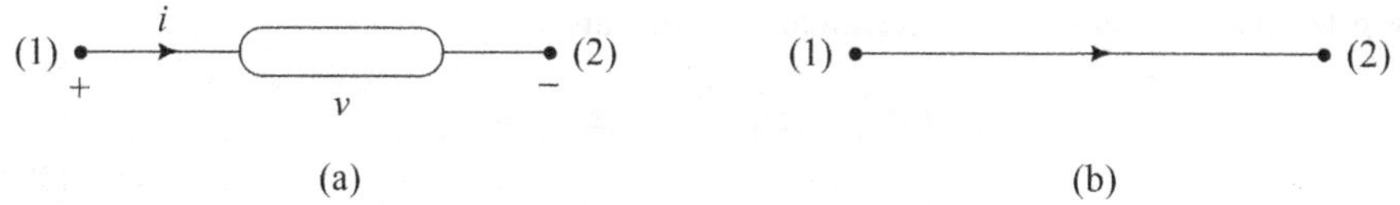

Fig. 3.2 *Generalized circuit element and its oriented graph*

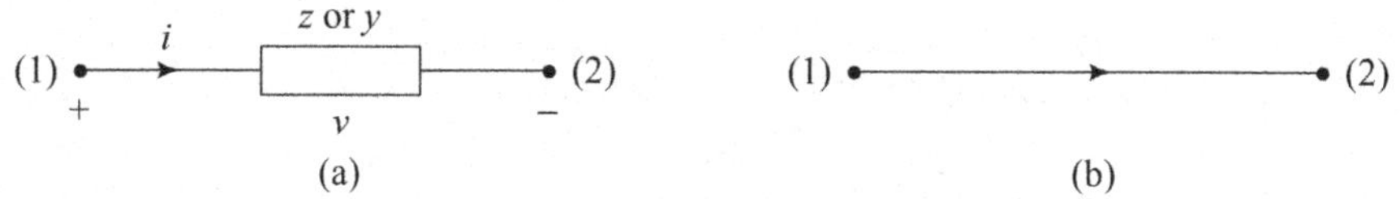

Fig. 3.3 *Passive element and its oriented graph*

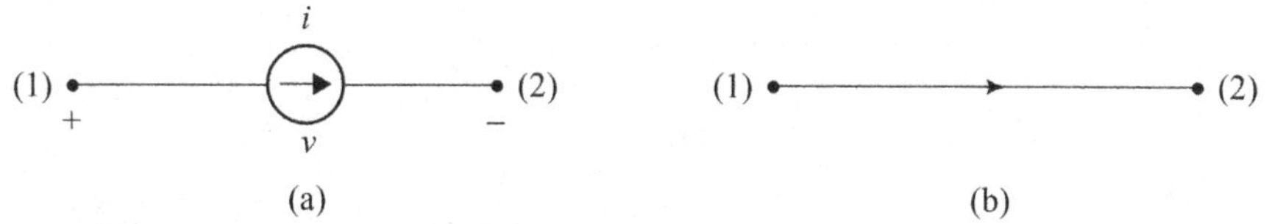

Fig. 3.4 *Current source and its oriented graph*

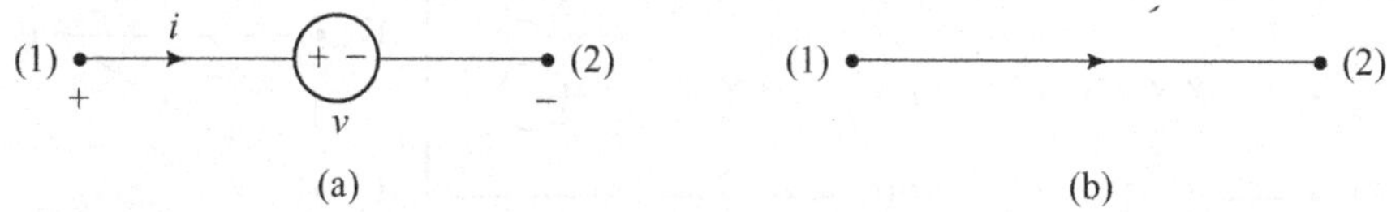

Fig. 3.5 *Voltage source and its oriented graph*

3.3 PRIMITIVE IMPEDANCE AND ADMITTANCE MATRICES

Consider a network of interconnected components. The passive components may be mutually coupled. The primitive impedance and admittance representation is: $\boldsymbol{v} = \boldsymbol{z}\,\boldsymbol{i}$ where $\boldsymbol{v}$ and $\boldsymbol{i}$ are vectors, $\boldsymbol{z}$ is the impedance matrix with $\boldsymbol{y}$ as the inverse of $\boldsymbol{z}$. The diagonal elements of $\boldsymbol{z}$ are self-impedances and the off-diagonal elements are mutual impedances. If the ith and the jth elements are mutually coupled, then the corresponding (i–j and j–i) elements are non-zero.

Example 3.1 Consider a four terminal network (e.g., three phases of a generator which are mutually coupled) shown in Fig. 3.6, with all unequal mutual impedances.

For the passive network, the terminal relations are:

$$\begin{bmatrix} v_1 \\ v_2 \\ v_3 \end{bmatrix} = \begin{bmatrix} z_{11} & z_{12} & z_{13} \\ z_{21} & z_{22} & z_{23} \\ z_{31} & z_{32} & z_{33} \end{bmatrix} \begin{bmatrix} i_1 \\ i_2 \\ i_3 \end{bmatrix} \tag{3.1}$$

$$\boldsymbol{v} = \boldsymbol{z}\,\boldsymbol{i} \tag{3.2}$$

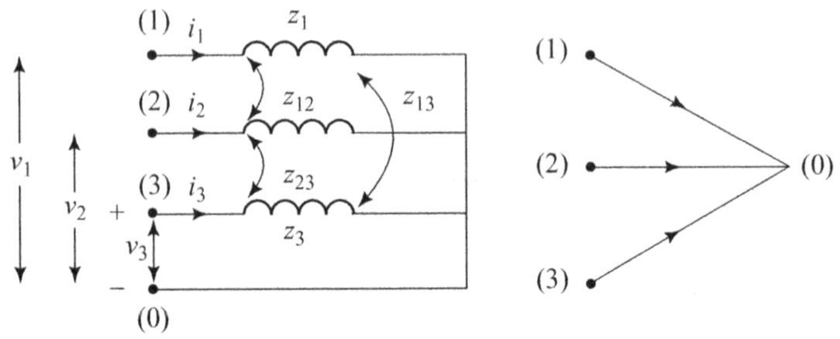

(a)

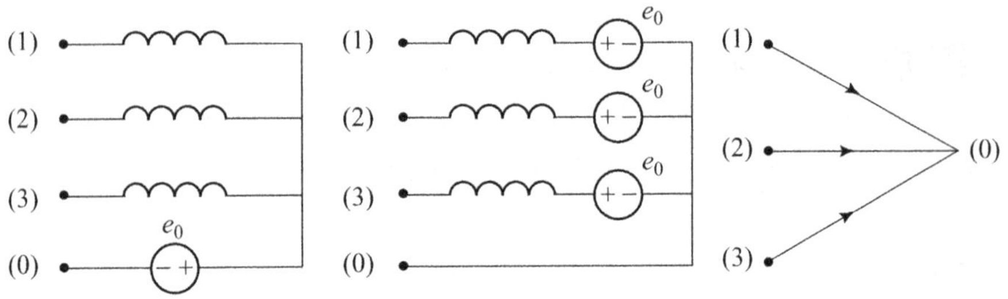

(b)

Fig. 3.6 **(a)** *A three-phase network and its oriented graph,* **(b)** *Modified network*

Suppose an identical voltage source e_0 is introduced between the node (0) and the common terminal with the polarity shown in Fig. 3.6(a). Then it is equivalent to moving the voltage source in series with all the three coils (Fig. 3.6(b)). The graph will remain the same with each edge of the graph representing the Thevenin source. The terminal relations are now

$$\boldsymbol{v} = \boldsymbol{z}\,\boldsymbol{i} + \begin{bmatrix} 1 \\ 1 \\ 1 \end{bmatrix} e_0 \tag{3.3}$$

The admittance formulations for Eqs. (3.2) and (3.3) are respectively,

$$\boldsymbol{i} = \boldsymbol{y}\boldsymbol{v} \tag{3.4}$$

$$\boldsymbol{i} = \boldsymbol{y}\,\boldsymbol{v} - \boldsymbol{y}\begin{bmatrix} 1 \\ 1 \\ 1 \end{bmatrix} e_0 \tag{3.5}$$

3.4 SYSTEM GRAPH FOR TRANSMISSION NETWORK

A power system is generally analyzed on a per-phase basis with balanced three-phase loads. Hence, only the positive sequence network is considered. The impedances in the per-phase equivalent are known as positive sequence impedances. The calculation of these positive sequence impedances for a transmission line (both series impedance and shunt admittance) can be found in the standard texts as a first course in power system analysis. Topologically the positive sequence network is the same as the original single-line diagram of the network. Consider the graph of a certain passive network shown in Fig. 3.7.

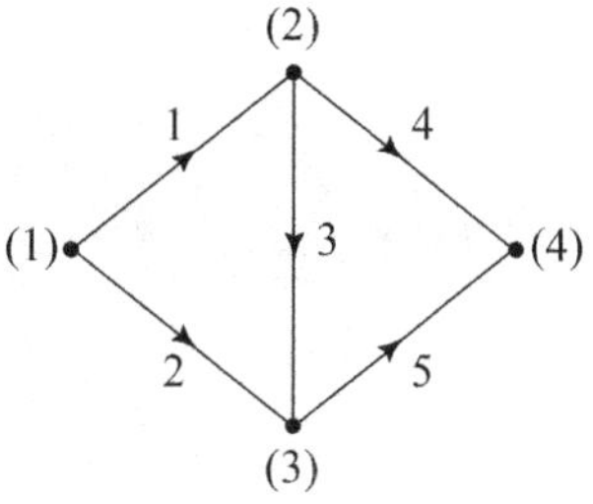

Fig. 3.7 *Graph of a network*

The primitive impedance $v-i$ relationship is given by

$$\begin{bmatrix} v_1 \\ v_2 \\ v_3 \\ v_4 \\ v_5 \end{bmatrix} = \begin{bmatrix} z_{11} & 0 & 0 & 0 & 0 \\ 0 & z_{22} & 0 & 0 & 0 \\ 0 & 0 & z_{33} & 0 & 0 \\ 0 & 0 & 0 & z_{44} & 0 \\ 0 & 0 & 0 & 0 & z_{55} \end{bmatrix} \begin{bmatrix} i_1 \\ i_2 \\ i_3 \\ i_4 \\ i_5 \end{bmatrix} \quad (3.6)$$

The primitive admittance $i-v$ relationship is: $\boldsymbol{i} = \boldsymbol{yv}$ where $\boldsymbol{y} = \boldsymbol{z}^{-1}$

$$\boldsymbol{y} = \begin{bmatrix} z_{11}^{-1} & 0 & 0 & 0 & 0 \\ 0 & z_{22}^{-1} & 0 & 0 & 0 \\ 0 & 0 & z_{33}^{-1} & 0 & 0 \\ 0 & 0 & 0 & z_{44}^{-1} & 0 \\ 0 & 0 & 0 & 0 & z_{55}^{-1} \end{bmatrix} \quad (3.7)$$

3.5 RELEVANT CONCEPTS IN GRAPH THEORY

Graph theory is a vast mathematical discipline with applications in various engineering fields. We need only a few basic concepts for our work in power systems.

A graph consisting of finite edges and nodes is called a *finite* graph. It is said to be *connected* if there exists a path between any two nodes of the graph. A subset of edges of the graph is called a *subgraph*. Certain degenerate subsets such as an edge by itself or a node by itself are also considered subgraphs. A few subgraphs for the graph in Fig. 3.7 are shown in Fig. 3.8.

The number of edges incident at a node gives the *degree* of the node. In Fig. 3.7, the degree of node (2) is 3. A subgraph with two endpoints (which are the nodes) and all other nodes of degree two in the subgraph is called a *path*. A path can traverse an edge at the most once. For example, the subgraphs shown in Fig. 3.9 form the paths between nodes (1) and (4). The direction of the path that is arbitrarily drawn for each path is independent of the orientation of its edges.

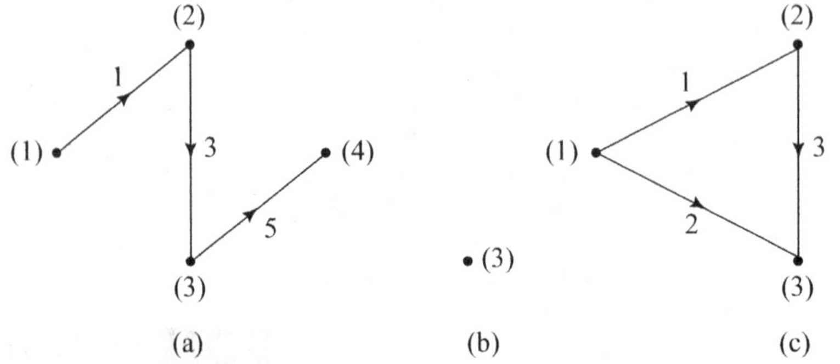

Fig. 3.8 *Some subgraphs of the graph in Fig. 3.7*

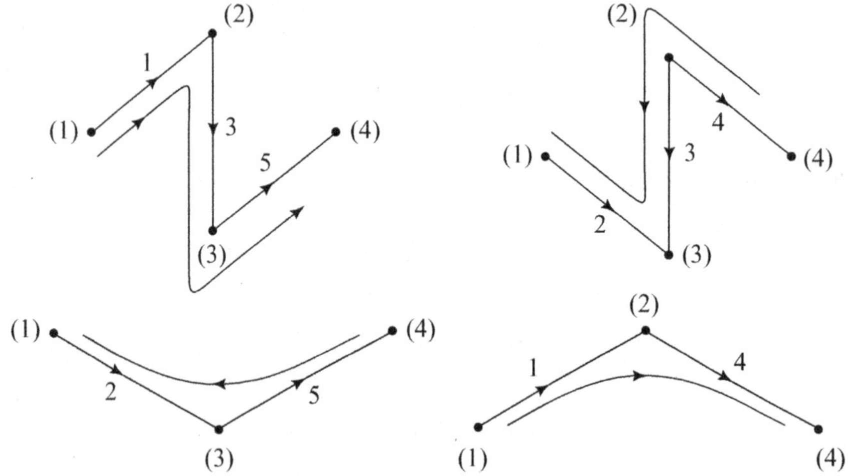

Fig. 3.9 *Some paths of the graph in Fig. 3.7*

3.5.1 Loop

A *loop* (circuit) is a connected subgraph with the degree of each of the nodes in the subgraph equal to two. The number of nodes and edges in a loop are equal. A loop is also referred to as a closed path. For Fig. 3.1(c), some of the loops are (1, 2, 4, 3), (3, 5, 6), (6, 8), (1, 2, 7.5, 3) (shown in Fig. 3.10). A loop may also have an orientation that points away from one node and finally goes back towards the same node along the elements of the loop. For the graph in Fig. 3.7, two of the loops are shown in Fig. 3.11 (a) and (b) along with their orientations. Figure 3.11(c) is not a loop since the degree of node (2) in that subgraph is three.

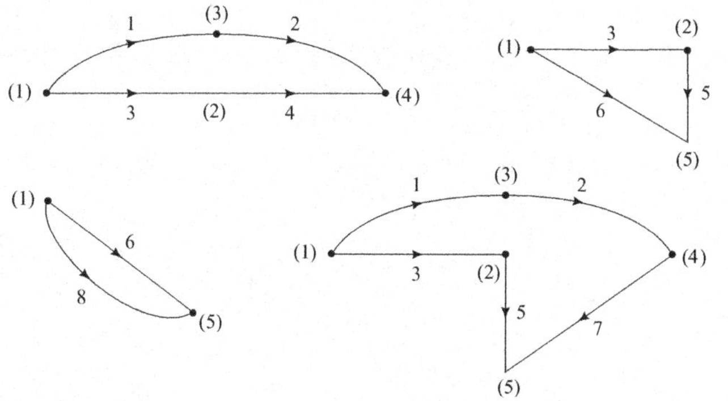

Fig. 3.10 *Loops for the graph in Fig. 3.1(c)*

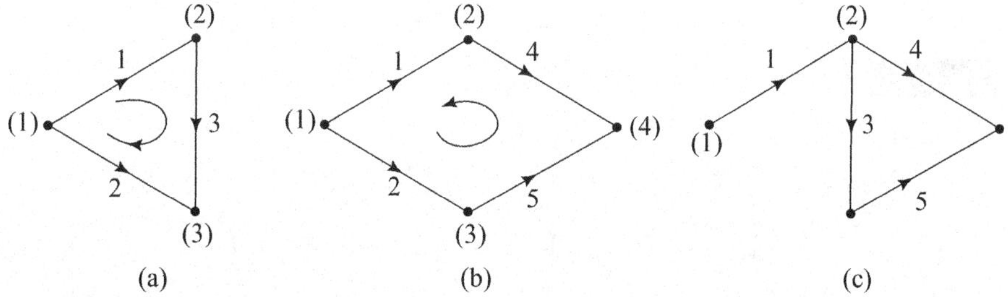

Fig. 3.11 *Illustration of loops and subgraphs that are not loops*

3.5.2 Tree and Co-tree

One of the important concepts in a linear graph is that of a tree. A *tree* is a subgraph that is connected, contains all nodes, and has no loops. For example, in Fig. 3.1, a tree can be formed by the elements (2, 5, 6, 7) or (2, 3, 4, 9). A few trees for Fig. 3.7 are shown in Fig. 3.12. In a tree, there is exactly one path between any two nodes. If the number of nodes in a graph is n, there are exactly $(n-1)$ edges in a tree. The proof of this observation is obvious. The elements of the tree are called *tree-branches*.

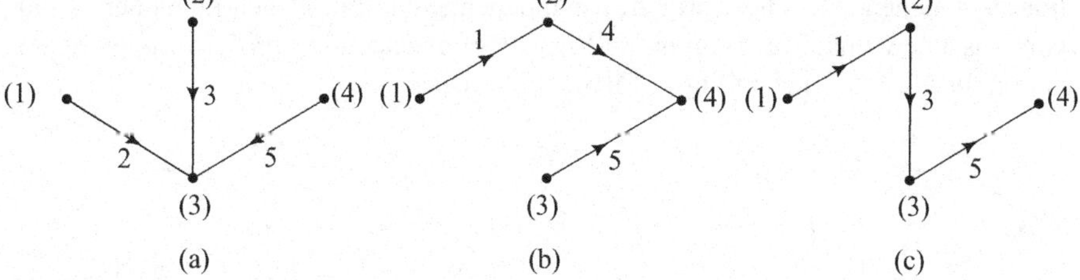

Fig. 3.12 *Trees for the graph in Fig. 3.7*

Those edges of the graph that are not in a tree form a *co-tree*, and the edges of the co-tree are called *links* or *chords*. We use the term links. For each chosen tree, there is a *co-tree*. For the three trees chosen in Fig. 3.12, the corresponding co-trees are shown in Fig. 3.13. In general, a co-tree does not contain all the nodes of the graph as illustrated in Fig. (3.13(a) and (b)). A co-tree may be connected or it may consist of several subgraphs (Fig. 3.13(c)). Figure 3.14 shows another example of a graph, a tree, and the corresponding co-tree. If the total number of edges in a graph is e, then the number of links $\ell = e - (n - 1) = e - n + 1$.

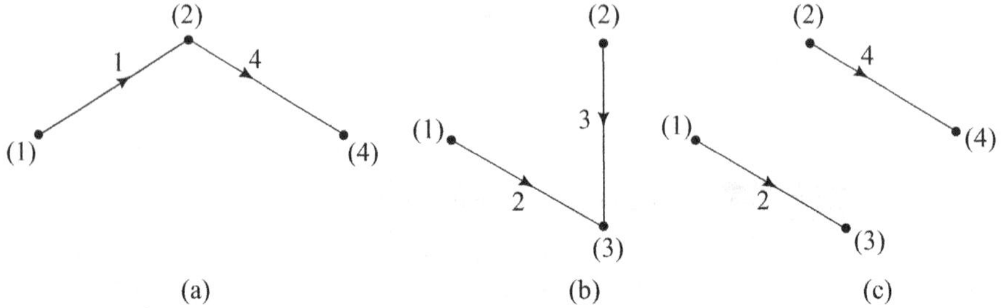

Fig. 3.13 *Co-trees of the trees in Fig. 3.12*

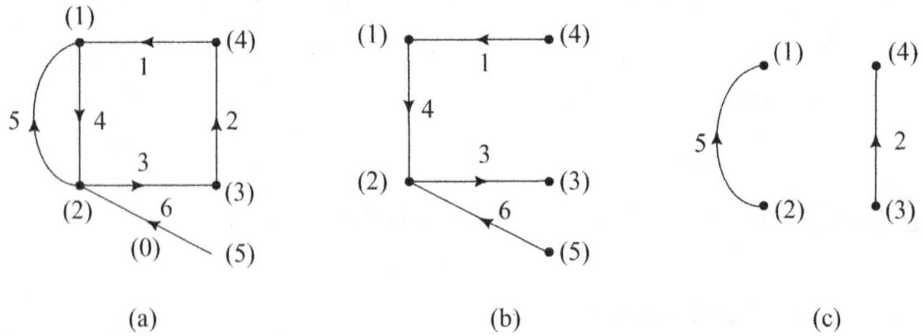

Fig. 3.14 **(a)** *A graph,* **(b)** *a tree, and* **(c)** *the corresponding co-tree*

3.5.3 Fundamental Loop

A *fundamental loop* for a graph is formed from the tree of the graph by inserting an appropriate link. For each link inserted, we create a new fundamental loop in the tree. There will be in all $(e - n + 1)$ fundamental loops for a chosen tree, all of these being linearly independent. These are linearly independent because each fundamental loop contains a new link. For the graph of Fig. 3.1(c), repeated in Fig. 3.15(a), let the tree chosen be (1, 4, 8, 9) as in Fig. 3.15(b) (solid lines).

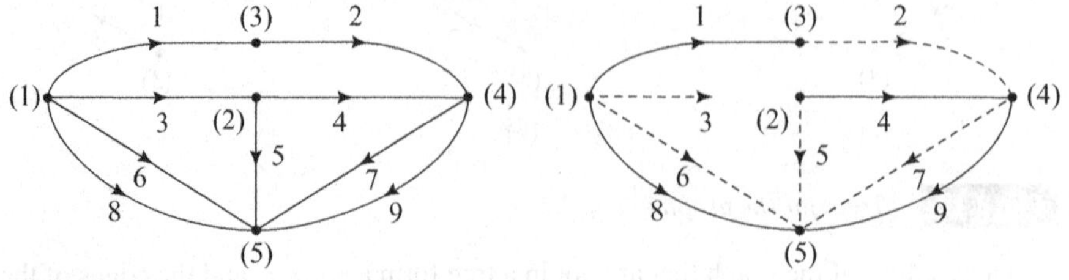

Fig. 3.15 *Graph, tree (solid), and the links (dotted)*

By inserting the links 2, 3, 5, 6, 7 *one at a time*, the following $e - n + 1 = 9 - 5 + 1 = 5$ fundamental loops are generated. (The links are underlined.)

(1, 2, 4, 9, 8), (3, 4, 9, 8), (5, 9, 4), (6, 8), (9, 7)

3.5.4 Kirchhoff's Voltage Law and the Fundamental Loop Matrix

We now state an important topological property of a graph, namely the *Fundamental Loop Matrix*, through the application of Kirchhoff's voltage law (KVL). KVL states that for any closed path or loop, the algebraic sum of voltages around the loop is zero. We write KVL systematically for the fundament loops as follows:

1. Select a tree.
2. For each fundamental loop, assign a positive reference direction to agree with the orientation associated with the link for that loop.
3. Going around the loop along the reference direction, assign a + sign to the voltage of the edge if the orientation of the edge agrees with the reference direction, a – sign if it is the opposite, and a zero if the edge is not contained in that loop.
4. Repeat step 3 for all the fundamental loops.
5. Arrange the voltage vector such that the tree-branch voltages appear first and the link voltages afterwards.
6. The resulting matrix of +1, –1, and 0 entries is called the *Fundamental Loop Matrix*.

Example 3.2 Consider the graph of Fig. 3.16(a) and the tree (1, 3, 4) in Fig. 3.16(b). Draw the corresponding fundamental loops and hence write the KVL in matrix form.

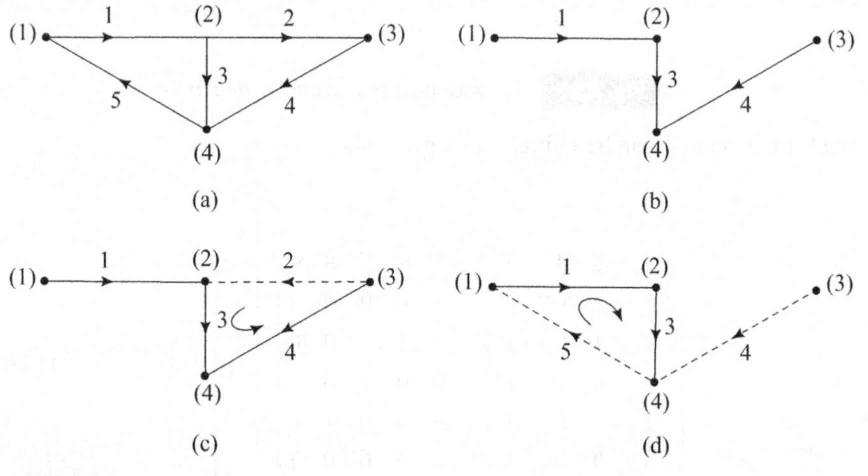

Fig. 3.16 **(a)** *Graph,* **(b)** *tree,* **(c)** *fundamental loop 1,* and **(d)** *fundamental loop 2*

Solution: The fundamental loops obtained by inserting links 2 and 5 are shown in Fig. 3.16 (c) and (d). The positive reference direction for each fundamental loop is shown dotted, to coincide with that of the defining link for that loop. Thus, the KVL for the two loops are written as

$$v_2 + v_3 - v_4 = 0$$

$$v_1 + v_3 - v_5 = 0 \tag{3.8}$$

The voltage vector is defined in the order of the tree branches and links as $[v_1\ v_3\ v_4 \vdots v_2\ v_5\ ]^{\mathrm{T}}$.

The KVL equations can now be put in a matrix form as

$$\begin{array}{c} \text{Tree-branches Links} \\ \begin{array}{c c} & \begin{array}{ccccc} 1 & 3 & 4 & 2 & 5 \end{array} \\ \begin{array}{c} 2 \\ 5 \end{array} & \left[\begin{array}{ccc|cc} 0 & 1 & -1 & 1 & 0 \\ 1 & 1 & 0 & 0 & 1 \end{array}\right] \end{array} \end{array} \begin{bmatrix} v_1 \\ v_3 \\ v_4 \\ \hline v_2 \\ v_5 \end{bmatrix} \tag{3.9}$$

Example 3.3 For Fig. 3.1(c), choose the tree (1, 2, 4, 5) and write the KVL in matrix form.

Solution: The tree is shown in Fig. 3.17 in solid lines and the links in dotted lines.

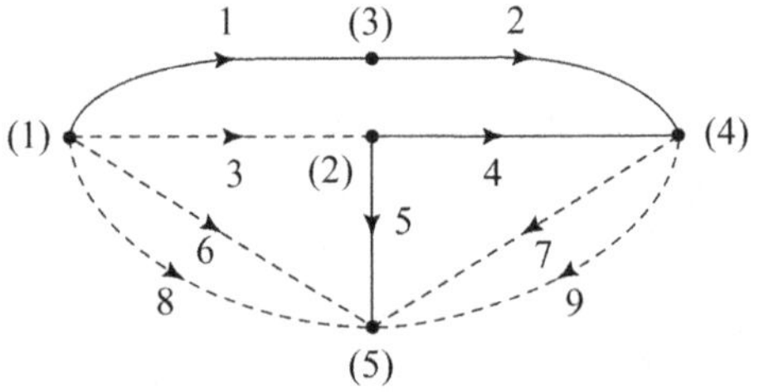

Fig. 3.17 *Network with tree-branches and links*

The KVL equations can be written by inspection as

$$\begin{array}{c} \begin{array}{ccccccccc} 1 & 2 & 4 & 5 & 3 & 6 & 7 & 8 & 9 \end{array} \\ \left(\begin{array}{cccc|ccccc} -1 & -1 & 1 & 0 & 1 & 0 & 0 & 0 & 0 \\ -1 & -1 & 1 & -1 & 0 & 1 & 0 & 0 & 0 \\ 0 & 0 & 1 & -1 & 0 & 0 & 1 & 0 & 0 \\ -1 & -1 & 1 & -1 & 0 & 0 & 0 & 1 & 0 \\ 0 & 0 & 1 & -1 & 0 & 0 & 0 & 0 & 1 \end{array}\right) \end{array} \begin{bmatrix} v_1 \\ v_2 \\ v_4 \\ v_5 \\ \hline v_3 \\ v_6 \\ v_7 \\ v_8 \\ v_9 \end{bmatrix} = 0 \tag{3.10}$$

Generalization

If the preceding procedure is followed for a general finite graph, then the KVL equations can be written in the form

$$e-n+1\begin{bmatrix} \overset{n-1}{C_b} : \overset{e-n+1}{U} \end{bmatrix} \begin{bmatrix} v_b \\ \hline v_\ell \end{bmatrix} = 0 \tag{3.11}$$

i.e., $C.v = 0$

where C_b is a $(e - n + 1) \times (n - 1)$ matrix.

U is a $(n - 1)$ square matrix.

v_b is a subvector of order $n - 1$ corresponding to the tree-branch variables.

$v\ell$ is a subvector of order $e - n + 1$ corresponding to the link variables.

C is called the *fundamental loop matrix*.

The existence of the unity submatrix in C is easily verified from the fact that

(i) Each fundamental loop contains one link only, and

(ii) The positive orientation of the loop coincides with the orientation of the link for that particular loop.

In general, the entries of C are such that

1. $C_{ij} = +1$ if the element corresponding to the jth column is in the fundamental loop defined by the link in the ith row and their orientations agree.
2. $C_{ij} = -1$ if the element corresponding to the jth column is in the fundamental loop defined by the link in the ith row but their orientations are opposite.
3. $C_{ij} = 0$ if the element corresponding to the jth column is not in the fundamental loop defined by the link in the ith row.

Example 3.4 For the transmission network shown in Fig. 3.18(a), assume that the shunt admittances at each bus are lumped into a single admittance. The oriented graph is shown in Fig. 3.18(b) with (0) representing the ground bus. Pick a tree and write the fundamental loop matrix C.

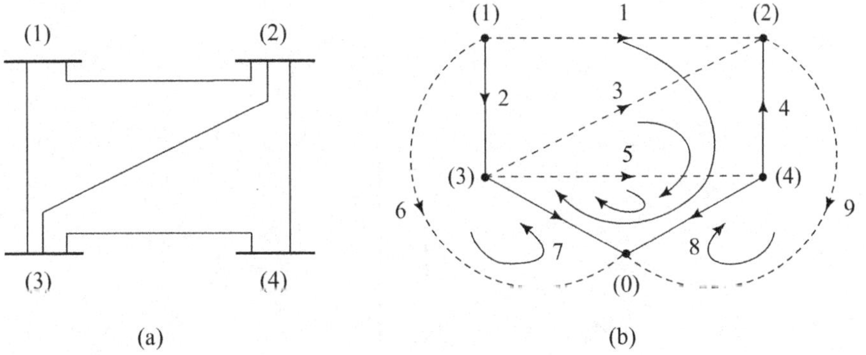

Fig. 3.18 **(a)** *Transmission network,* **(b)** *graph tree, and co-tree*

Solution: The following tree is chosen with the tree-branches (2, 4, 7, 8), shown by solid lines. The links are: (1, 3, 5, 6, 9) shown by dotted lines. The orientations for the fundamental loops are shown dotted.

The *C* matrix is written as

$$
\begin{array}{r} \\ \\ 1 \\ 3 \\ C = 5 \\ 6 \\ 9 \end{array}
\begin{array}{c}
\begin{array}{cccc} \multicolumn{4}{c}{\text{tree-branches}} \\ 2 & 4 & 7 & 8 \end{array} \quad \begin{array}{ccccc} \multicolumn{5}{c}{\text{links}} \\ 1 & 3 & 5 & 6 & 9 \end{array} \\
\left(\begin{array}{cccc|ccccc}
-1 & -1 & -1 & 1 & 1 & 0 & 0 & 0 & 0 \\
0 & -1 & -1 & 1 & 0 & 1 & 0 & 0 & 0 \\
0 & 0 & -1 & 1 & 0 & 0 & 1 & 0 & 0 \\
-1 & 0 & -1 & 0 & 0 & 0 & 0 & 1 & 0 \\
0 & 1 & 0 & -1 & 0 & 0 & 0 & 0 & 1
\end{array}\right)
\end{array}
\qquad (3.12)
$$

3.5.5 Fundamental Cutset

Another basic concept in graph theory is that of a cutset. A cutset of a connected graph is defined as the minimal set of elements whose removal leaves the graph in *exactly* two parts. Consider the graph in Fig. 3.19(a). Removal of the elements (3, 4, 5, 6, 7) (Fig. 3.19(b)) leaves the graph in three parts as shown in Fig. 3.19(c). Note that node (5) by itself constitutes a subgraph. Hence, (3, 4, 5, 6, 7) does not form a cutset. On the other hand, removal of (4, 6, 7) (Fig. 3.19(d)) leaves it in

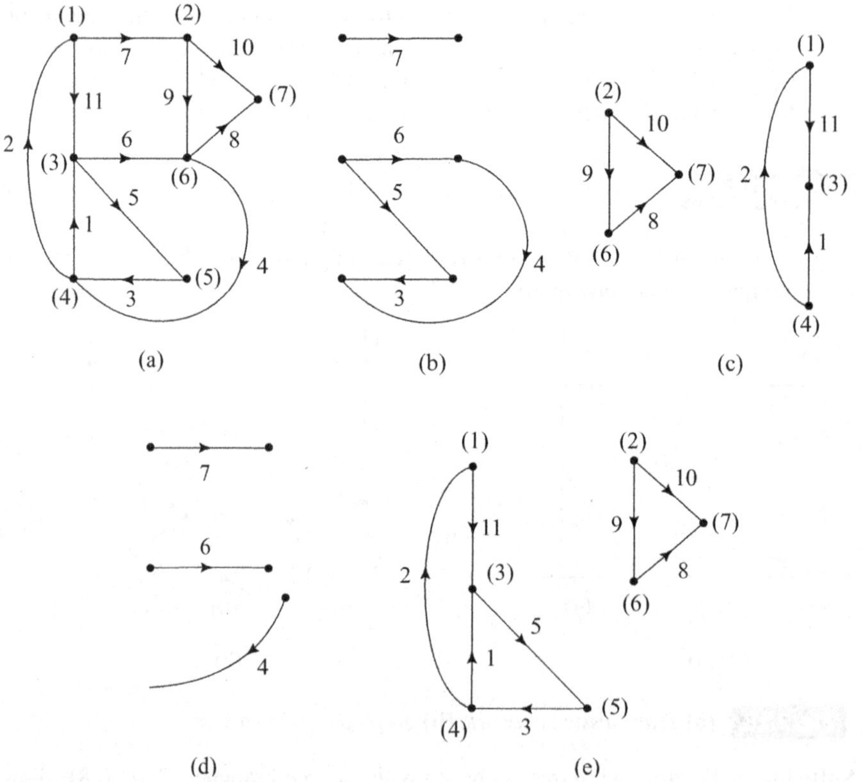

Fig. 3.19 *Graph and cutset concept*

two parts as shown in Fig. 3.19(e). Hence (4, 6, 7) is a cutset. The elements of the cutset can also be selected by "cutting" the graph with a curved (dotted) line not passing through any node and dividing the graph in two connected subgraphs. It can also be noted that the nodes of the graph also get divided in two groups by the cutset (4, 6, 7)—one group consisting of nodes (1), (3), (4), and (5), and the other group consisting of nodes (2), (6), and (7). The edges of the cutset connect the nodes between the two groups as shown in Fig. 3.20. The reader may verify the other cutsets in Fig. 3.19(a) as (2, 11, 7), (1, 2, 3, 4), (4, 6, 9, 10), (2, 4, 6, 11), etc. Just as the concept of fundamental loops is associated with a link, so is the concept of fundamental cutsets associated with a tree-branch that we discuss next.

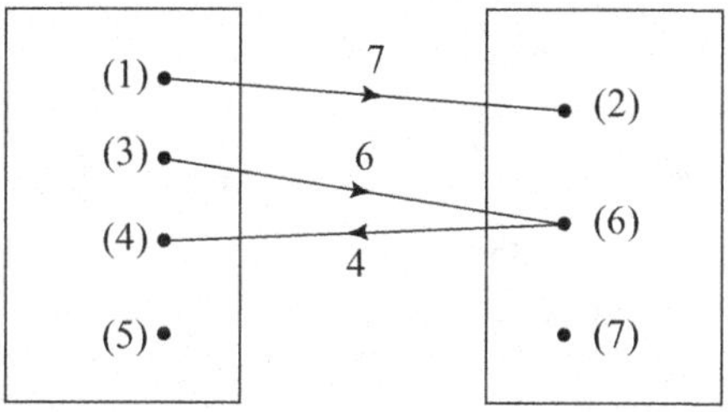

Fig. 3.20 *Further illustration of cutset of Fig. 3.19(d)*

The tree is a connected subgraph of a given graph. Removal of any tree-branch leaves the tree in two parts, each part having a certain number of nodes. We thus have two groups of nodes. The edges of the graph connecting these two groups of nodes are called *fundamental cutsets* and they correspond to that particular tree branch. The edges of the cutset are the particular tree-branch and other links that connect the two groups of nodes. Thus, for each tree-branch we have an associated fundamental cutset. In all, we have $(n - 1)$ fundamental cutsets since a tree in an n node graph has $(n - 1)$ edges.

Consider the graph in Fig. 3.21(a). Let the tree-branches be (2, 4, 5, 7) which constitutes a connected graph (Fig. 3.21(b)). Removal of tree-branch 2 in the tree divides the nodes into two groups as shown in Fig. 3.21(c). We then insert all the possible links of the graph between the two nodes. This constitutes a fundamental cutset associated with branch 2. For convenience, tree-branch 2 is shown in a solid line and the other links are shown dotted. The fundamental cutsets corresponding to the other tree-branches, 4, 5, and 7 are similarly shown in Figs. 3.21(d), (e), and (f) respectively. To avoid this laborious procedure, we can follow the simple rule of cutting the graph by a curve not crossing any node such that it cuts only one tree-branch at a time. This is shown in Fig. 3.21(g). Thus the fundamental cutsets for the graph in Fig. 3.21(a) and the chosen tree in Fig. 3.21(b) are ($\underline{2}$, 1, 6), ($\underline{4}$, 1, 3), ($\underline{5}$, 1, 3), and (7, 6). (The tree-branches are underlined.)

It is of interest to remark here that a set of linearly independent cutsets can also exist which cannot be determined by a tree. As an example, consider the graph in Fig. 3.22. The elements incident on each node is a cutset, and the edges of each cutset are the ones cut by a curved line. But we shall see later that only $(n - 1)$ cutsets in an n node graph constitute a linearly independent set of cutsets.

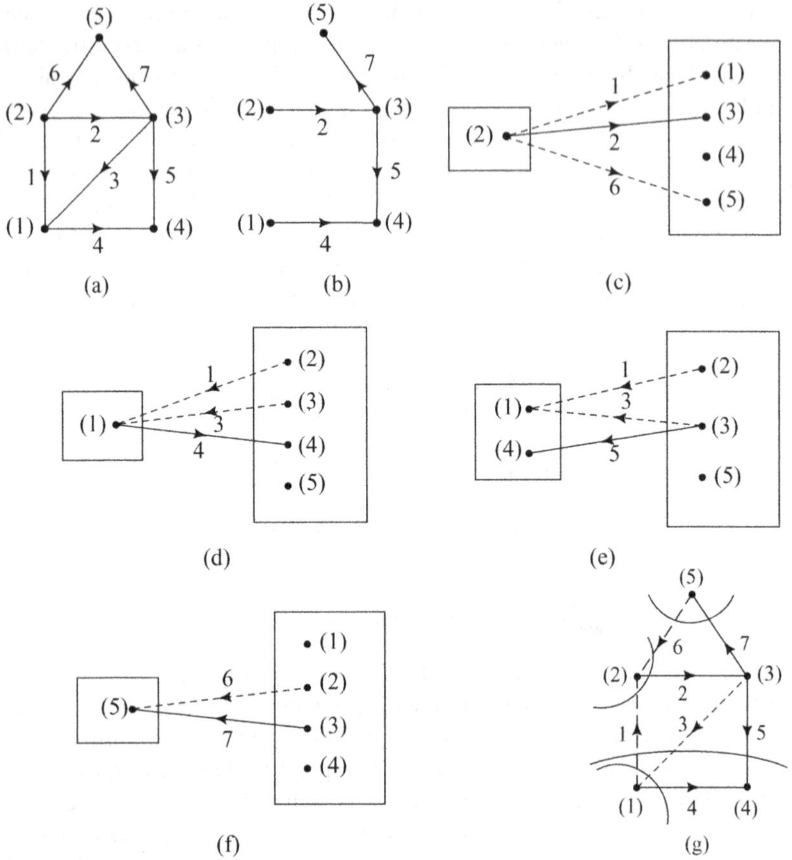

Fig. 3.21 *Network and fundamental cutsets*

3.5.6 Kirchhoff's Current Law (KCL) and the Fundamental Cutset Matrix

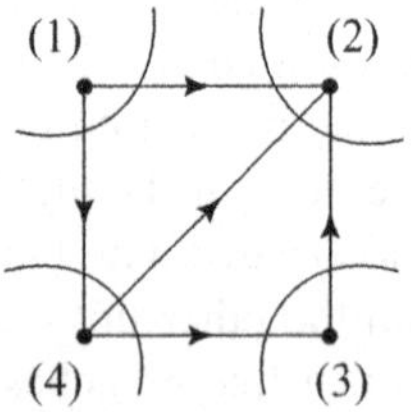

Fig. 3.22 *Cutsets not generated by a tree*

Just as in the case of fundamental loops, we shall use KCL to derive another important topological relationship. We shall use the same graph (Fig. 3.16) as for KVL, and it is reproduced in Fig. 3.23. Choose (1, 3, 4) as the tree.

The three fundamental cutsets associated with tree-branches are shown by the dotted curved lines. These are (1, 5), (2, 3, 5), and (2, 4). The underlined element corresponds to the tree-branch. If corresponding to each fundamental cutset, the curved dotted line were extended to form a closed surface, then KCL states that the algebraic sum of the currents leaving a closed surface is zero. To apply KCL systematically, we define the orientation of each cutset (shown by farm arrow) to coincide with the orientation of the associated tree-branch. While writing KCL, we give a + sign to an edge of the cutset if its orientation agrees with the orientation of the cutset and a – sign if it is the opposite. The application of KCL to each of the three cutsets in Fig. 3.23 gives

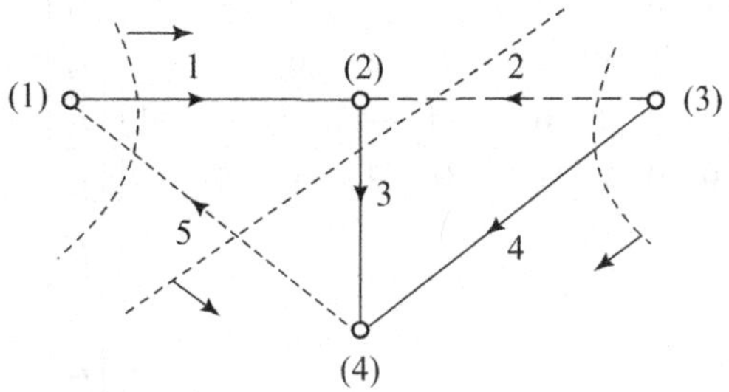

Fig. 3.23 *Fundamental cutsets and KCL*

$$i_1 - i_5 = 0; \quad -i_2 + i_3 - i_5 = 0; \quad i_2 + i_4 = 0 \tag{3.13}$$

Arranging Eq. (3.20) in matrix form, we get

$$\begin{array}{c} \\ \\ \text{Tree-branches} \end{array} \begin{array}{c} \\ 1 \\ 3 \\ 4 \end{array} \overset{\begin{array}{ccccc} & \text{Tree-branches} & & \text{Links} & \\ 1 & 3 & 4 & 2 & 5 \end{array}}{\begin{bmatrix} 1 & 0 & 0 & | & 0 & -1 \\ 0 & 1 & 0 & | & -1 & -1 \\ 0 & 0 & 1 & | & 1 & 0 \end{bmatrix}} \begin{bmatrix} i_1 \\ i_3 \\ i_4 \\ \hline i_2 \\ i_5 \end{bmatrix} = 0 \tag{3.14}$$

Example 3.5 For the graph in Fig. 3.24 and tree (1, 2, 4, 5), write the KCL.

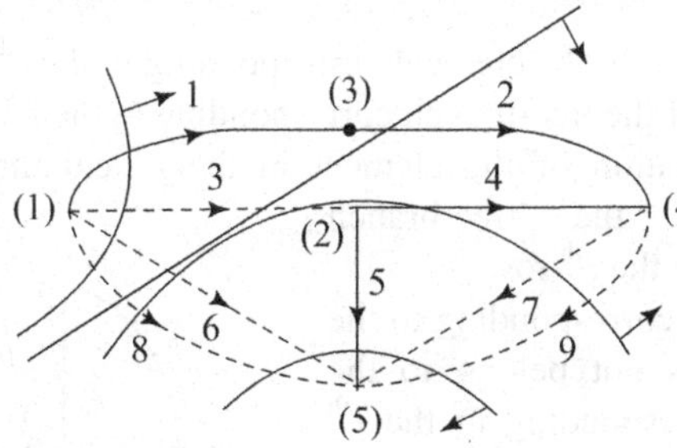

Fig. 3.24 *Fundamental cutsets*

Solution: The fundamental cutsets are shown in Fig. 3.24 along with their positive orientations shown by an arrow in the direction coinciding with that of the tree branch.

The KCL is written as

$$\begin{array}{cccc|ccccc} \multicolumn{4}{c}{\text{Tree-branches}} & \multicolumn{5}{c}{\text{Links}} \\ 1 & 2 & 4 & 5 & 3 & 6 & 7 & 8 & 9 \end{array}$$

$$\begin{bmatrix} 1 & 0 & 0 & 0 & 1 & 1 & 0 & 1 & 0 \\ 0 & 1 & 0 & 0 & 1 & 1 & 0 & 1 & 0 \\ 0 & 0 & 1 & 0 & -1 & -1 & -1 & -1 & -1 \\ 0 & 0 & 0 & 1 & 0 & 1 & 1 & 1 & 1 \end{bmatrix} \begin{bmatrix} i_1 \\ i_2 \\ i_4 \\ i_5 \\ \hline i_3 \\ i_6 \\ i_7 \\ i_8 \\ i_9 \end{bmatrix} = 0 \tag{3.15}$$

Generalization

For a general graph, we can write the KCL as

$$n-1\begin{bmatrix} \overset{n-1}{U} \,|\, \overset{e-n+1}{B_\ell} \end{bmatrix} \begin{bmatrix} \boldsymbol{i}_b \\ \boldsymbol{i}_\ell \end{bmatrix} = 0 \tag{3.16}$$

Since each fundamental cutset contains only one tree-branch, the nature of the unity matrix $\boldsymbol{U}$ is self-evident. In a more compact form,

$$\boldsymbol{Bi} = \boldsymbol{0} \tag{3.17}$$

where
$$\boldsymbol{B} = [\,\boldsymbol{U} \,|\, \boldsymbol{B}_\ell] \tag{3.18}$$

and $\boldsymbol{i}$ is the vector of currents arranged in the order of the tree branch and the link currents. $\boldsymbol{B}$ is called the *fundamental cutset matrix*. It has a unity submatrix of order $(n-1)$ in the leading position, and the matrix $\boldsymbol{B}_\ell$ of order $(n-1) \times (e-n+1)$ in the trailing position. Each row is identified with a tree-branch. The entries of the matrix $\boldsymbol{B}$ are such that

b_{ij} = 1 if the orientation of the element corresponding to the jth column agrees with the orientation of the tree-branch corresponding to the ith row.

b_{ij} = –1 if the orientation of the element in the jth column is opposite to the orientation of the tree-branch corresponding to the ith row.

b_{ij} = 0 if the element corresponding to the jth column does not belong to the tree-branch corresponding to the ith row.

Example 3.6 For the graph in Fig. 3.18(b) and the chosen tree (2, 4, 7, 8), write the $\boldsymbol{B}$ matrix.

Solution: The graph is redrawn in Fig. 3.25 with the curved lines defining the fundamental cutsets.

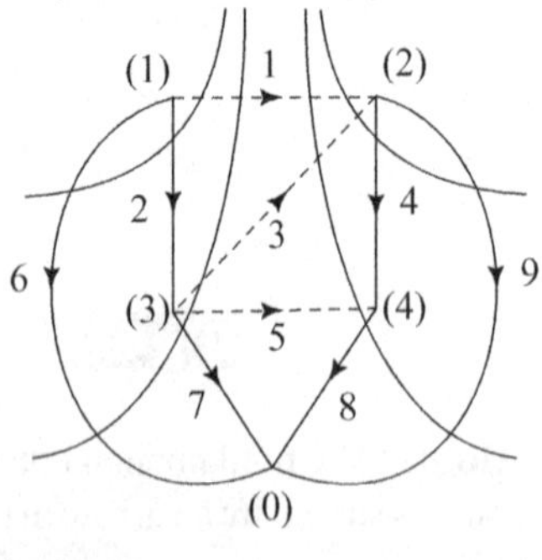

Fig. 3.25 *Graph and fundamental cutsets for the transmission network of Fig. 3.18*

The ***B*** matrix is written by inspection as

$$\mathbf{B} = \begin{array}{c} \\ \\ 2 \\ 4 \\ 7 \\ 8 \end{array} \begin{array}{c} \text{tree branches} \qquad\qquad \text{links} \\ \begin{array}{cccc|ccccc} 2 & 4 & 7 & 8 & 1 & 3 & 5 & 6 & 9 \end{array} \\ \left(\begin{array}{cccc|ccccc} 1 & 0 & 0 & 0 & 1 & 0 & 0 & 1 & 0 \\ 0 & 1 & 0 & 0 & -1 & -1 & 0 & 0 & 1 \\ 0 & 0 & 1 & 0 & 1 & 1 & 1 & 1 & 0 \\ 0 & 0 & 0 & 1 & -1 & -1 & -1 & 0 & 1 \end{array}\right) \end{array} \qquad (3.19)$$

3.5.7 Incidence or Vertex Matrix

One of the characterizations of a graph is the *incidence* matrix. The edges incident to a node in a graph is called the *incidence set*. Thus a connected graph has as many incidence sets as there are nodes. We can write KCL at each of these nodes giving a + sign to the currents leaving a node and a – sign to the currents entering the node. Alternatively, we can interpret each incidence set as a cutset with a line enclosing the node and the positive orientation of the cutset outwards from the dotted closed line (see Fig. 3.26).

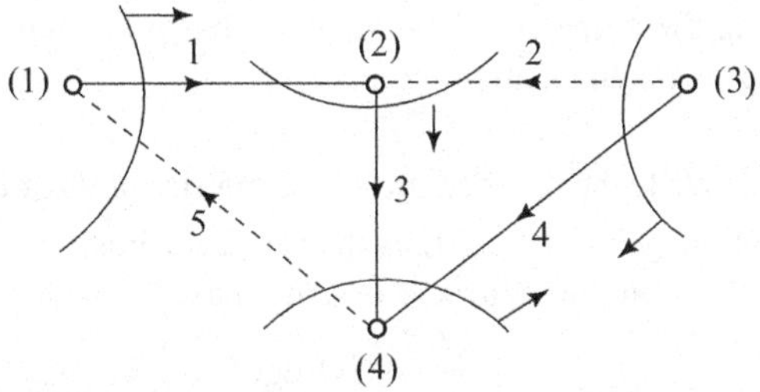

Fig. 3.26 *Incidence sets in a graph*

The KCL equations for nodes (1) – (4) can be written as

$$\begin{aligned} i_1 - i_5 &= 0 \\ -i_1 - i_2 + i_3 &= 0 \\ i_2 + i_4 &= 0 \\ -i_3 - i_4 + i_5 &= 0 \end{aligned} \qquad (3.20)$$

In matrix form, these can be written as

$$\text{Nodes} \begin{array}{c} \\ (1) \\ (2) \\ (3) \\ (4) \end{array} \begin{array}{c} \text{Edges} \\ \begin{array}{ccccc} 1 & 2 & 3 & 4 & 5 \end{array} \\ \begin{pmatrix} 1 & 0 & 0 & 0 & -1 \\ -1 & -1 & 1 & 0 & 0 \\ 0 & 1 & 0 & 1 & 0 \\ 0 & 0 & -1 & -1 & 1 \end{pmatrix} \end{array} \begin{bmatrix} i_1 \\ i_2 \\ i_3 \\ i_4 \\ i_5 \end{bmatrix} = 0 \qquad (3.21)$$

In a more compact form, $\boldsymbol{A_a i} = 0$

In general, the order of $\boldsymbol{A_a}$ is n × e where n = number of nodes and e = number of edges in the graph. $\boldsymbol{A_a}$ is called the *node to branch incidence matrix* or the *augmented incidence matrix*. The entries of $\boldsymbol{A_a}$ are such that

$(a_{ij})_a$ = +1 if the edge corresponding to the jth column is incident to the node corresponding to the ith row and is directed away from it.

$(a_{ij})_a$ = –1 if the edge corresponding to the jth column is incident to the node corresponding to the ith row and is directed towards the node.

$(a_{ij})_a$ = 0 if the edge corresponding to the jth column is not incident to the node corresponding to the ith row.

It may be observed that since each element is incident on two nodes, the columns of the $\boldsymbol{A_a}$ matrix have each a +1 and a –1 entry. If we add up all the rows of $\boldsymbol{A_a}$ matrix, we get a zero row. This indicates that the rows are linearly dependent. The number of linearly independent rows are $n - 1$, or we say that the rank of the matrix $\boldsymbol{A_a}$ is $(n - 1)$. We can delete any one row and the resulting matrix $\boldsymbol{A}$ is called the *reduced-incidence* or simply the *incidence* matrix. The order $\boldsymbol{A}$ is $(n - 1) \times e$. In power networks, if a ground bus is present, it is generally taken as the reference bus and the node corresponding to it is generally deleted while writing the $\boldsymbol{A}$ matrix. If the network has no connection to the ground, any one of the nodes is taken as the reference.

Example 3.7 Write the reduced incidence matrix for the transmission network in Fig. 3.18. Choose the ground bus (0) as the reference bus.

Solution: By inspection, the matrix $\boldsymbol{A}$ is written as

$$A = \text{Nodes}\ \begin{matrix} & \text{Edges} \\ & \begin{matrix} 1 & 2 & 3 & 4 & 5 & 6 & 7 & 8 & 9 \end{matrix} \\ \begin{matrix} (1) \\ (2) \\ (3) \\ (4) \end{matrix} & \begin{pmatrix} 1 & 1 & 0 & 0 & 0 & 1 & 0 & 0 & 0 \\ -1 & 0 & -1 & -1 & 0 & 0 & 0 & 0 & 1 \\ 0 & -1 & 1 & 0 & 1 & 0 & 1 & 0 & 0 \\ 0 & 0 & 0 & -1 & -1 & 0 & 0 & 1 & 0 \end{pmatrix} \end{matrix}$$

3.5.8 Interrelationships between the Matrices *A*, *B*, *C*, and the Network Graph

In the matrix $\boldsymbol{A}$, the columns corresponding to the edges were arranged sequentially. They can be written in any particular order. In fact, one of the ways is to arrange the columns in the order of tree branches and links for a given tree in the graph. Thus, we can write $\boldsymbol{A}$ as

$$\begin{matrix} & \text{Tree branches} & \text{Links} \\ A = [& A_b & \mid \; A_\ell \;] \end{matrix} \qquad (3.22)$$

The following properties are now stated without a rigorous proof and illustrated with some examples. The proofs can be found in texts on graph theory.

Property 1

For a given tree of a graph, each row of the fundamental loop matrix C is orthogonal to each row of the fundamental cutset matrix B. Mathematically, this relationship implies

$$BC^T = C^T B = 0 \tag{3.23}$$

Since $B = [U \mid B_\ell]$ and $C = [C_b \mid U]$, it follows

$$[U \mid B_\ell]\left[\frac{C_b^T}{U}\right] = 0 \tag{3.24}$$

Therefore, $C_b^T = -B_\ell$ which is the same as

$$C_b = -B_\ell^T \tag{3.25}$$

This is a very important result. It tells us that for a given tree of a graph, if the fundamental loop matrix C is known, the fundamental cutset matrix is also known and vice-versa. This relationship can be verified from Eq. (3.25).

Property 2

Let the incidence matrix A be arranged in the order of tree-branches and links for a given tree, i.e.,

$$A = \begin{matrix} & n-1 & e-n+1 \\ n-1 & [\ A_b & \mid \ A_\ell\ ] \end{matrix} \tag{3.26}$$

It can be shown that A_b is nonsingular. Furthermore, the fundamental cutset matrix for the given tree is given by

$$B = A_b^{-1} A = A_b^{-1}[A_b \mid A_l] = [U \mid A_b^{-1} A_l] \tag{3.27}$$

since $B = [U \mid B_l]$, $\therefore B_l = A_b^{-1} A_l$

This important result tells us that by choosing a tree and writing the incidence matrix by inspection (or computer generated), we can obtain the fundamental cutset matrix B and from property 1, also the fundamental loop matrix C.

Example 3.8 Consider the graph shown in Fig. 3.27. Choose the tree whose branches are (1, 3, 5). Find the fundamental cutset and loop matrices B and C using the incidence matrix A.

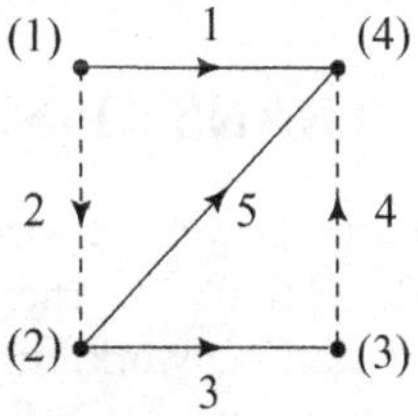

Fig. 3.27 *Oriented graph*

Solution: Choosing (2) as the reference node, we write the reduced incidence matrix A as

$$A = \begin{matrix} & \text{Tree branches} & \text{Links} \\ \text{Nodes} & 1 \quad 3 \quad 5 & 2 \quad 4 \\ (1) \\ (3) \\ (4) \end{matrix} \begin{pmatrix} 1 & 0 & 0 & | & 1 & 0 \\ 0 & -1 & 0 & | & 0 & 1 \\ -1 & 0 & -1 & | & 0 & -1 \end{pmatrix} = [A_b \mid A_\ell] \tag{3.28}$$

$$A_b^{-1} = \begin{bmatrix} 1 & 0 & 0 \\ 0 & -1 & 0 \\ -1 & 0 & -1 \end{bmatrix}^{-1} = \begin{bmatrix} 1 & 0 & 0 \\ 0 & -1 & 0 \\ -1 & 0 & -1 \end{bmatrix} \tag{3.29}$$

Therefore,

$$A_b^{-1} A_\ell = \begin{bmatrix} 1 & 0 & 0 \\ 0 & -1 & 0 \\ -1 & 0 & -1 \end{bmatrix} \begin{bmatrix} 1 & 0 \\ 0 & 1 \\ 0 & -1 \end{bmatrix} = \begin{bmatrix} 1 & 0 \\ 0 & -1 \\ -1 & 1 \end{bmatrix} \tag{3.30}$$

Hence,

$$B = \begin{matrix} & \text{Tree branches} & \text{Links} \\ & 1 \quad 3 \quad 5 & 2 \quad 4 \\ 1 \\ 3 \\ 5 \end{matrix} \begin{pmatrix} 1 & 0 & 0 & | & 1 & 0 \\ 0 & 1 & 0 & | & 0 & -1 \\ 0 & 0 & 1 & | & -1 & 1 \end{pmatrix} = [U \mid B_\ell] \tag{3.31}$$

Since $C_b = -B_\ell^T$, we have

$$C = \begin{matrix} & \text{Tree branches} & \text{links} \\ & 1 \quad 3 \quad 5 & 2 \quad 4 \\ 2 \\ 4 \end{matrix} \begin{pmatrix} -1 & 0 & 1 & | & 1 & 0 \\ 0 & 1 & -1 & | & 0 & 1 \end{pmatrix} \tag{3.32}$$

The nature of matrices B and C can be independently verified from the graph.

3.6 TRANSMISSION NETWORK REPRESENTATIONS

In electric power networks, the most useful representations are the multiport characterizations of the transmission network. The network consists of the transmission network, transformers (off-nominal or phase shifting), shunt VAr devices (capacitors or inductors), and the loads. The terminals are the n buses plus the additional ground bus, i.e., $(n + 1)$ buses in all. In many cases such as load flow, short circuit, or transient stability study, one is interested in the terminal quantities only, namely the voltages of the buses with respect to ground, and the current entering the buses (injected currents). Thus the ground bus is implicitly assumed as a common reference bus. We therefore have n terminal pairs or n ports as shown in Fig. 3.28.

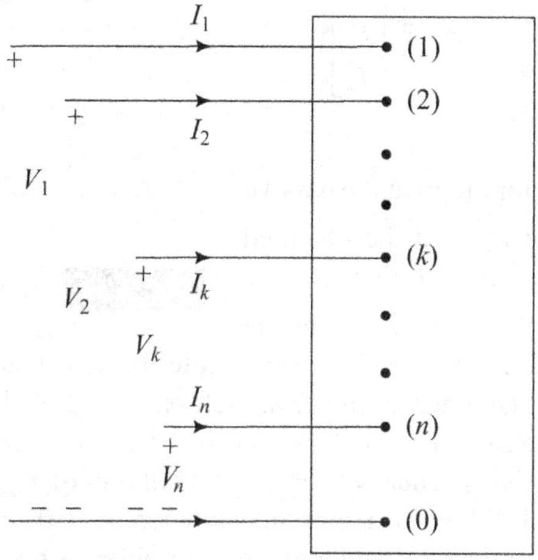

Fig. 3.28 *An n-port network*

Let the currents and voltages be phasors so that the voltage and current vectors denoted by $\boldsymbol{V}_{\text{bus}}$ and $\boldsymbol{I}_{\text{bus}}$ respectively, are

$$\boldsymbol{V}_{\text{bus}} = \begin{bmatrix} V_1 \\ \vdots \\ V_n \end{bmatrix}, \quad \boldsymbol{I}_{\text{bus}} = \begin{bmatrix} I_1 \\ \vdots \\ I_n \end{bmatrix} \tag{3.33}$$

The relationship between V_{bus} and I_{bus} can be characterized either in an impedance or an admittance form as

$$\boldsymbol{I}_{\text{bus}} = \boldsymbol{Y}_{\text{bus}} \boldsymbol{V}_{\text{bus}} \tag{3.34}$$

$$\boldsymbol{V}_{\text{bus}} = \boldsymbol{Z}_{\text{bus}} \boldsymbol{I}_{\text{bus}} \tag{3.35}$$

where

$$\boldsymbol{Z}_{\text{bus}} = \boldsymbol{Y}_{\text{bus}}^{-1} \tag{3.36}$$

$\boldsymbol{Z}_{\text{bus}}$ is called the open-circuit impedance matrix in circuit theory with the elements of the matrix known as the driving point and transfer impedances. The reason for this designation will be explained shortly. In power literature, it is known as the *bus impedance matrix*. $\boldsymbol{Y}_{\text{bus}}$ is called the *short-circuit admittance matrix*, or *nodal admittance matrix*, with the elements of the matrix known as the driving point and transfer admittances. In power system literature, it is known as the *bus admittance matrix*.

3.6.1 Impedance Parameters

Consider a three-terminal network with an additional ground bus (Fig. 3.29). We assume that the graph of the four nodes is connected.

The three-port description can be explicitly written as

$$\begin{bmatrix} V_1 \\ V_2 \\ V_3 \end{bmatrix} = \begin{bmatrix} Z_{11} & Z_{12} & Z_{13} \\ Z_{21} & Z_{22} & Z_{23} \\ Z_{31} & Z_{32} & Z_{33} \end{bmatrix} \begin{bmatrix} I_1 \\ I_2 \\ I_3 \end{bmatrix}$$

i.e., $\boldsymbol{V}_{bus} = \boldsymbol{Z}_{bus}\, \boldsymbol{I}_{bus}$

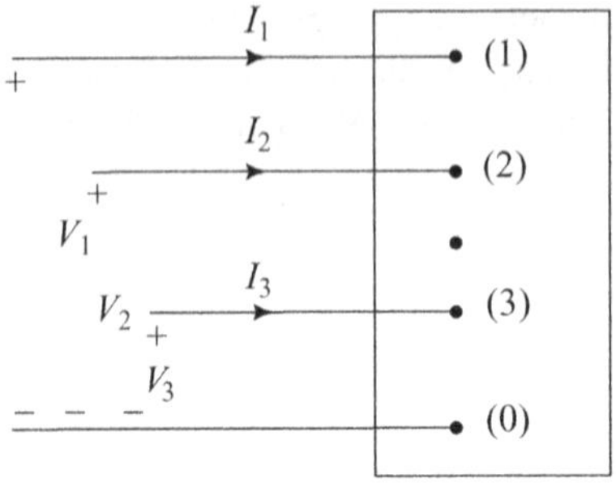

Fig. 3.29 *Three-port network*

Without knowing the topology of the passive network, the element $\boldsymbol{Z}_{bus}$ can be obtained by external measurements as follows.

1. Excite terminal (1) with a current source $I_1 = 1.0 \angle 0°$ with the current entering terminal (1) and leaving the reference bus. The other terminals are left open-circuited.
2. Compute or measure V_1, V_2, V_3 which will be numerically equal to Z_{11}, Z_{12}, and Z_{31}, respectively. Thus we get the first column of $\boldsymbol{Z}_{bus}$.
3. Excite terminal (2) with a current source $I_2 = 1.0 \angle 0°$ with thc current entering into terminal (2) and leaving the reference bus. The other terminals are open. Compute or measure V_1, V_2, and V_3 to get the second column of $\boldsymbol{Z}_{bus}$, namely Z_{12}, Z_{22}, and Z_{32}.
4. Repeat step 3 with terminal (3) to get the third column of $\boldsymbol{Z}_{bus}$. Thus, we get all the impedance parameters.

Example 3.9 Determine the $\boldsymbol{Z}_{bus}$ of the π-section of the transmission line shown in Fig. 3.30.

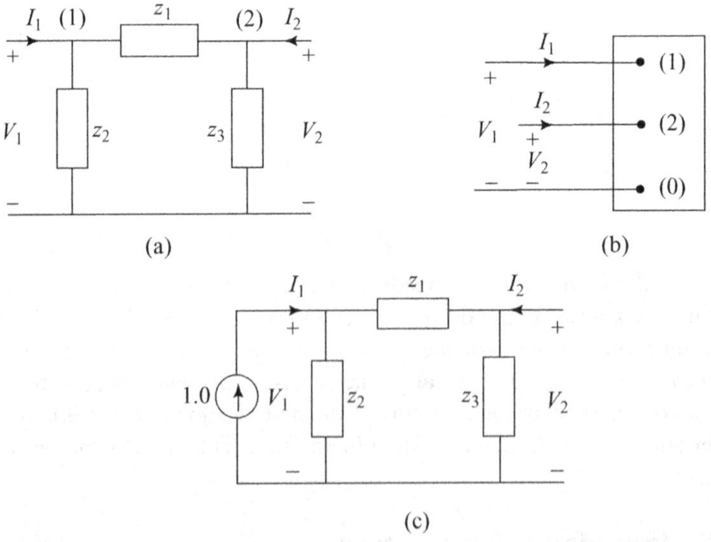

Fig. 3.30 *The bus impedance matrix $\boldsymbol{Z}_{bus}$ of a transmission line*

Solution: Excite terminal (1) with a unit current as shown in Fig. 3.30 leaving terminal (2) open. It can be easily verified that

$$V_1 = Z_{11} = \frac{z_2(z_1 + z_3)}{z_1 + z_2 + z_3} \tag{3.37}$$

$$V_2 = Z_{21} = \frac{z_2 z_3}{z_1 + z_2 + z_3} \tag{3.38}$$

Similarly, by exciting terminal (2) and leaving terminal (1) open-circuited, we can get

$$Z_{12} = \frac{z_2 z_3}{z_1 + z_2 + z_3}; \quad Z_{22} = \frac{z_3(z_1 + z_2)}{z_1 + z_2 + z_3};$$

Therefore,

$$\boldsymbol{Z}_{\text{bus}} = \frac{1}{z_1 + z_2 + z_3}\begin{bmatrix} z_2(z_1 + z_3) & z_2 z_3 \\ z_2 z_3 & z_3(z_1 + z_2) \end{bmatrix}; \tag{3.39}$$

3.6.2 Admittance Parameters

For the four-terminal network in Fig. 3.29, the admittance representation will be

$$I_1 = Y_{11}V_1 + Y_{12}V_2 + Y_{13}V_3$$

$$I_2 = Y_{21}V_1 + Y_{22}V_2 + Y_{23}V_3$$

$$I_3 = Y_{31}V_1 + Y_{32}V_2 + Y_{33}V_3 \tag{3.40}$$

Y_{ij}'s can be computed as follows from the external measurements.

1. Excite terminal (1) with voltage source = $1.0 \angle 0°$ with the positive terminal of the source at bus (1). The terminals (2) and (3) are short-circuited to the ground (0).
2. Compute or measure I_1, I_2, and I_3 which will be numerically equal to Y_{11}, Y_{21}, and Y_{31}, respectively. Thus, we get the first column of $\boldsymbol{Y}_{\text{bus}}$.
3. Repeat steps 1 and 2 at terminals (2) and (3) to get the second and third columns, respectively.

Example 3.10 Suppose the π-section parameters of Fig. 3.30 are given as admittances as in Fig. 3.31. Express the elements of the $\boldsymbol{Y}_{\text{bus}}$ matrix in terms of these admittances.

Solution: Excite terminal (1) with a voltage source equal to unity and short-circuit terminal (2) as shown in Fig. 3.31(c). It is easily seen that

$$I_1 = Y_{11} = y_1 + y_2, \qquad I_2 = Y_{21} = -y_1 \tag{3.41}$$

Similarly, exciting terminal (2) with a unit voltage source and short-circuiting terminal (1), we get

$$Y_{12} = -y_1; \qquad Y_{22} = y_1 + y_3;.$$

Thus
$$Y_{bus} = \begin{bmatrix} y_1 + y_2 & -y_1 \\ -y_1 & y_1 + y_3 \end{bmatrix} \tag{3.42}$$

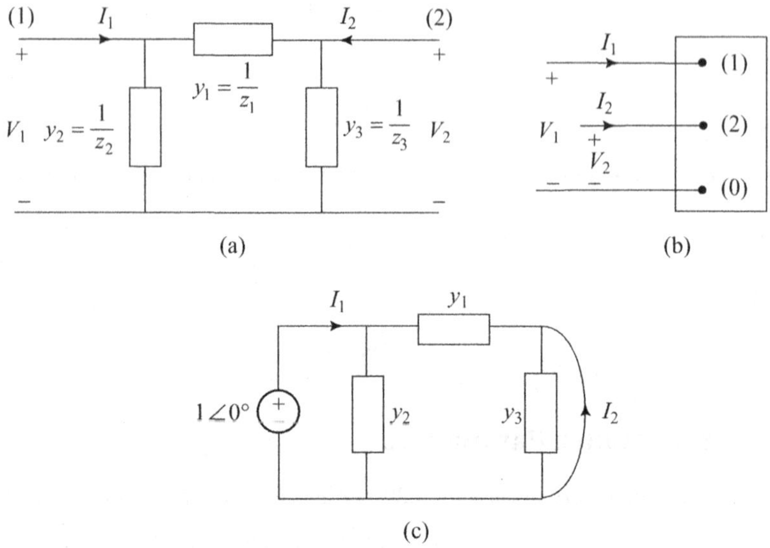

Fig. 3.31 *Admittance parameters of a transmission line*

3.7 NETWORK MATRICES

3.7.1 Link, Branch and Nodal Transformations

From KCL, the fundamental cutset equations are

$$\boldsymbol{B}\,\boldsymbol{i} = 0 \tag{3.43}$$

i.e.,
$$[U \mid \boldsymbol{B}_\ell]\begin{bmatrix} \boldsymbol{i}_b \\ \boldsymbol{i}_\ell \end{bmatrix} = 0 \tag{3.44}$$

Thus
$$\boldsymbol{i}_b = -\mathbf{B}_\ell\,\boldsymbol{i}_\ell \tag{3.45}$$

Since $\boldsymbol{B}_\ell = -\boldsymbol{C}_b^T$ (from Eq. (3.25)), so $\boldsymbol{i}_b = \boldsymbol{C}_b^T\,\boldsymbol{i}_\ell$

Adding the trivial equation $\boldsymbol{i}_\ell = \boldsymbol{i}_\ell$ to Eq. (3.45), we get

$$\begin{bmatrix} \boldsymbol{i}_b \\ \boldsymbol{i}_\ell \end{bmatrix} = \begin{bmatrix} \boldsymbol{C}_b^T \\ \hline \boldsymbol{U} \end{bmatrix}\boldsymbol{i}_\ell = \boldsymbol{C}^T\boldsymbol{i}_\ell \tag{3.46}$$

$\boldsymbol{C}^T$ is known as the *connection matrix* since it expresses all the currents in terms of the link currents.

Hence the *link transformation* is

$$\boldsymbol{i} = \boldsymbol{C}^T \boldsymbol{i}_\ell \tag{3.47}$$

where $\boldsymbol{i}$ represents currents in all the elements of the network. Note the ordering of elements in the vector $\boldsymbol{i}$.

The KVL corresponding to the fundamental loops can be written as

$$[\boldsymbol{C}_b \mid \boldsymbol{U}]\begin{bmatrix} \boldsymbol{v}_b \\ \boldsymbol{v}_\ell \end{bmatrix} = 0 \tag{3.48}$$

$$-\boldsymbol{C}_b \boldsymbol{v}_b = \boldsymbol{v}_\ell \tag{3.49}$$

Since $\boldsymbol{C}_b = -\boldsymbol{B}_\ell^T$, we have from Eq. (3.49),

$$\boldsymbol{v}_\ell = \boldsymbol{B}_\ell^T \boldsymbol{v}_b \tag{3.50}$$

Attaching the trivial equation $\boldsymbol{v}_b = \boldsymbol{v}_b$, we have

$$\begin{bmatrix} \boldsymbol{v}_b \\ \hline \boldsymbol{v}_\ell \end{bmatrix} = \begin{bmatrix} \mathbf{U} \\ \hline \mathbf{B}_\ell^T \end{bmatrix} \boldsymbol{v}_b \tag{3.51}$$

i.e.,

$$\boldsymbol{v} = \boldsymbol{B}^T \boldsymbol{v}_b \tag{3.52}$$

Thus in Eq. (3.52), the voltages in all the elements of the network are expressed in terms of the tree-branch voltages. This is known as the *tree-branch transformation.* Again, note the ordering of the elements in V. Hence, the fundamental cutset and loop matrices contain important information regarding the dependence of link voltages on tree-branch voltages, and tree-branch currents on link currents, respectively. To be specific,

$$\boldsymbol{v}_\ell = \boldsymbol{B}_\ell^T \boldsymbol{v}_b \tag{3.53}$$

$$\boldsymbol{i}_b = \boldsymbol{C}_b^T \boldsymbol{i}_\ell \tag{3.54}$$

The nodal transformation to be discussed next exploits the incidence matrix $\boldsymbol{A}$ discussed earlier. Consider an n bus network with an additional ground bus (0) so that there are $(n + 1)$ buses in total. Let the bus voltages be measured with respect to the ground bus. The voltages of each branch of the network may be expressed in terms of the bus voltages as follows.

For the kth branch,

$v_k = V_i - V_j$ if in the graph, the edge k is directed from node (i) to (j).

$= V_j - V_i$ if in the graph, the edge k is directed from node (j) to (i).

$= \pm V_i$ if node (i) is connected to the ground bus, with a + sign if it is directed from node (i) towards the ground bus, and a – sign if it is directed from the ground bus to node (i).

In matrix form, we can write it as

$$\boldsymbol{v}_k = \begin{bmatrix} 0\cdots \overset{i}{1}\cdots \overset{j}{-1}\cdots 0 \end{bmatrix} \begin{bmatrix} V_1 \\ \vdots \\ V_i \\ \vdots \\ V_j \\ \vdots \\ V_n \end{bmatrix} \tag{3.55}$$

or

$$\boldsymbol{v}_k = \begin{bmatrix} 0\cdots \overset{i}{-1}\cdots \overset{j}{1}\cdots 0 \end{bmatrix} \begin{bmatrix} V_1 \\ \vdots \\ V_i \\ \vdots \\ V_j \\ \vdots \\ V_n \end{bmatrix} \tag{3.56}$$

or

$$\boldsymbol{v}_k = \begin{bmatrix} 0\ldots \overset{i}{\pm 1}\ldots 0 \end{bmatrix} \begin{bmatrix} V_1 \\ V_2 \\ \vdots \\ V_i \\ \vdots \\ V_n \end{bmatrix} \tag{3.57}$$

For the e elements in the network, we have

$$\begin{bmatrix} v_1 \\ \vdots \\ v_e \end{bmatrix} = \overset{e}{\text{edges}} \overset{n \text{ nodes}}{\begin{bmatrix} 1, \ -1, \ \text{or } 0 \text{ entry} \end{bmatrix}} \begin{bmatrix} V_1 \\ \vdots \\ V_n \end{bmatrix} \tag{3.58}$$

It is easy to see that the $e \times n$ matrix is precisely the transpose of the incidence matrix $\boldsymbol{A}$,

i.e.,
$$\boldsymbol{v} = \boldsymbol{A}^{\mathrm{T}} \boldsymbol{V}_{\text{bus}} \tag{3.59}$$

This is the nodal transformation.

Alternatively, suppose the network is excited by current sources as in Fig. 3.32(a), then at the nodes, we can write KCL as

$$\boldsymbol{A}\boldsymbol{i} = \boldsymbol{I}_{\text{bus}} \tag{3.60}$$

where $\boldsymbol{I}$ is the vector of currents in the network of dimension n.

Also, in the graph of Fig. 3.32(b), the voltage vector associated with the augmented current graph is

$$V'_{\text{bus}} = -V_{\text{bus}} \tag{3.61}$$

Now from the conservation of power law,

$$v^T i^* + (V'_{\text{bus}})^T I^*_{\text{bus}} = 0 \tag{3.62}$$

Using Eqs. (3.60) and (3.61), we get

$$v^T i^* = (V_{\text{bus}})^T I^*_{\text{bus}} = (V_{\text{bus}})^T A i^* \tag{3.63}$$

Since this is true for all the values of ***i***, we have

$$v = A^T V_{\text{bus}} \tag{3.64}$$

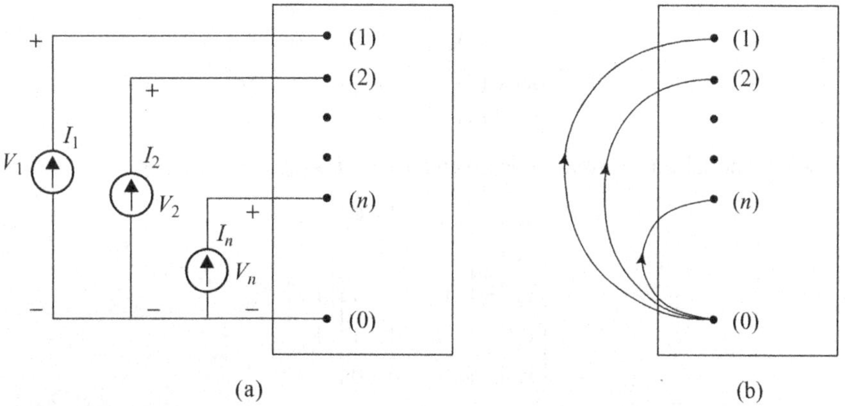

Fig. 3.32 **(a)** *Network augmented by current sources, and* **(b)** *the graph*

Example 3.11 Write the link, branch, and the nodal transformation for the graph in Fig. 3.16(a) for the tree (1, 3, 4). Choose (4) as the reference node.

Solution: We already have the ***C***, ***B***, and ***A*** matrices in Eqs. (3.9), (3.14), and (3.21) respectively. ***A*** is obtained from A_a by deleting the row corresponding to the reference bus, namely (4). We thus write by inspection from these equations.

Link Transformation

$$i = C^T i_\ell \tag{3.65}$$

$$\begin{bmatrix} i_1 \\ i_3 \\ \underline{i_4} \\ i_2 \\ i_5 \end{bmatrix} = \begin{bmatrix} 0 & 1 \\ 1 & 1 \\ \underline{-1} & \underline{0} \\ 1 & 0 \\ 0 & 1 \end{bmatrix} \begin{bmatrix} i_2 \\ i_5 \end{bmatrix} \tag{3.66}$$

Branch Transformation

$$V = B^T V_b$$

$$\begin{bmatrix} v_1 \\ v_3 \\ v_4 \\ \hline v_2 \\ v_5 \end{bmatrix} = \begin{bmatrix} 1 & 0 & 0 \\ 0 & 1 & 0 \\ 0 & 0 & 1 \\ \hline 0 & -1 & 1 \\ -1 & -1 & 0 \end{bmatrix} \begin{bmatrix} V_1 \\ V_3 \\ V_4 \end{bmatrix} \tag{3.67}$$

By arranging the rows and columns of $\boldsymbol{A}_a$ in Eq. (3.21) and omitting row (4), the incidence matrix $\boldsymbol{A}$ for this graph with (4) as the reference node, is given by

$$\begin{array}{c} \text{Edges} \\ \begin{array}{cc} & \begin{array}{ccccc} 1 & 3 & 4 & 2 & 5 \end{array} \\ \boldsymbol{A} = \text{Nodes} \begin{array}{c} (1) \\ (2) \\ (3) \end{array} & \begin{pmatrix} 1 & 0 & 0 & 0 & -1 \\ -1 & 1 & 0 & -1 & 0 \\ 0 & 0 & 1 & 1 & 0 \end{pmatrix} \end{array} \end{array} \tag{3.68}$$

Hence the nodal transformation is given by $\boldsymbol{V} = \boldsymbol{A}^T \boldsymbol{V}_{\text{bus}}$

$$\begin{bmatrix} v_1 \\ v_3 \\ v_4 \\ \hline v_2 \\ v_5 \end{bmatrix} = \begin{bmatrix} 1 & -1 & 0 \\ 0 & 1 & 0 \\ 0 & 0 & 1 \\ \hline 0 & -1 & 1 \\ -1 & 0 & 0 \end{bmatrix} \begin{bmatrix} V_1 \\ V_2 \\ V_3 \end{bmatrix} \tag{3.69}$$

Note that the ordering of the edges in Eq. (3.68) is different from Eq. (3.21). However, the ordering of the voltages in Eq. (3.69) is consistent with the ordering of currents in Eq. (3.66).

3.7.2 Derivation of Y_{bus}

Consider the n-port network shown in Fig. 3.32(a) excited by current sources. Omitting the graph of the network, the additional edges to the graph corresponding to the current sources are shown in Fig. 3.32(b). The currents are denoted by the vector $\boldsymbol{I}_{\text{bus}}$ and noting the reference directions, the voltages associated with these edges are denoted by the vector $-\boldsymbol{V}_{\text{bus}}$. With the addition of the current sources, KCL at all the nodes can be written in matrix form as explained in the previous section as

$$\boldsymbol{A}\boldsymbol{i} - \boldsymbol{I}_{\text{bus}} = 0 \tag{3.70}$$

i.e.,

$$\boldsymbol{I}_{\text{bus}} = \mathbf{A}\boldsymbol{i} \tag{3.71}$$

Since $\boldsymbol{v} = \boldsymbol{A}^T \boldsymbol{V}_{\text{bus}}$ and $\boldsymbol{i} = \boldsymbol{y}\,\boldsymbol{v}$, where $\boldsymbol{y}$ is the primitive admittance matrix, we have

$$\boldsymbol{I}_{\text{bus}} = \boldsymbol{A}\,\boldsymbol{y}\boldsymbol{A}^T \boldsymbol{V}_{\text{bus}}$$

$$= \boldsymbol{Y}_{\text{bus}} \boldsymbol{V}_{\text{bus}} \tag{3.72}$$

Thus the topological formula for Y_{bus} is

$$Y_{bus} = A\,yA^T \tag{3.73}$$

Example 3.12 Consider the network in Fig. 3.33(a) where two branches have mutual coupling as shown. Find the primitive impedance matrices **z**, **y**, and the Y_{bus} matrix. Choose (0) as reference bus.

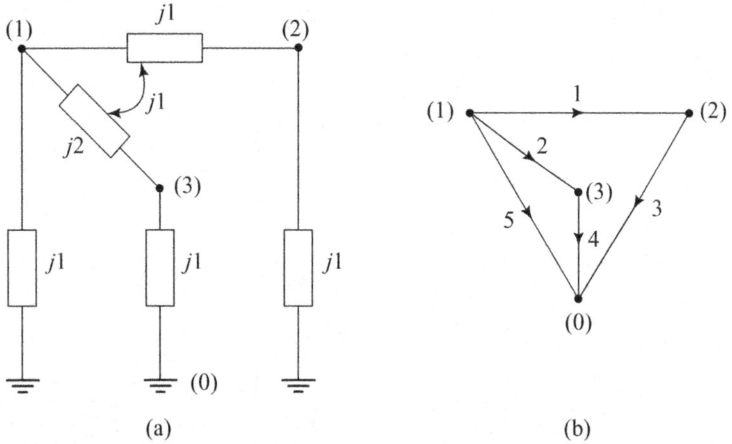

Fig. 3.33 *Network to illustrate Y_{bus}*

Solution: The graph is shown in Fig. 3.33(b). The primitive impedance matrix **z** is

$$z = j\begin{pmatrix} 1 & 1 & 0 & 0 & 0 \\ 1 & 2 & 0 & 0 & 0 \\ 0 & 0 & 1 & 0 & 0 \\ 0 & 0 & 0 & 1 & 0 \\ 0 & 0 & 0 & 0 & 1 \end{pmatrix} \tag{3.74}$$

(columns and rows numbered 1 2 3 4 5)

The primitive admittance matrix y is calculated as z^{-1}.

$$y = j\begin{pmatrix} 2 & 1 & 0 & 0 & 0 \\ 1 & -1 & 0 & 0 & 0 \\ 0 & 0 & -1 & 0 & 0 \\ 0 & 0 & 0 & -1 & 0 \\ 0 & 0 & 0 & 0 & -1 \end{pmatrix} \tag{3.75}$$

(columns and rows numbered 1 2 3 4 5)

The incidence matrix obtained by taking (0) as the reference node is

$$A = \begin{matrix} \\ (1) \\ (2) \\ (3) \end{matrix} \begin{matrix} 1 & 2 & 3 & 4 & 5 \\ \end{matrix}$$

$$A=\begin{matrix}(1)\\(2)\\(3)\end{matrix}\begin{pmatrix} 1 & 1 & 0 & 0 & 1 \\ -1 & 0 & 1 & 0 & 0 \\ 0 & -1 & 0 & 1 & 0 \end{pmatrix} \tag{3.76}$$

A^T is given by

$$A^T = \begin{matrix}1\\2\\3\\4\\5\end{matrix}\begin{pmatrix} 1 & -1 & 0 \\ 1 & 0 & -1 \\ 0 & 1 & 0 \\ 0 & 0 & 1 \\ 1 & 0 & 0 \end{pmatrix} \quad \text{columns } (1)\ (2)\ (3) \tag{3.77}$$

Hence,

$$\boldsymbol{Y}_{\text{bus}} = \boldsymbol{A}\,\boldsymbol{y}\,\boldsymbol{A}^T$$

$$= j\begin{bmatrix} -2 & 1 & 0 \\ 1 & -3 & 1 \\ 0 & 1 & -2 \end{bmatrix} \tag{3.78}$$

3.7.3 Bus Admittance Matrix when there is no Mutual Coupling

Consider the same network as in Fig. 3.33(a) but with no mutual coupling. Denote the network in terms of the admittances $y_{10}, y_{21}, y_{13}, y_{20}, y_{30}$ where y is the admittance to ground. The primitive admittance matrix **y** is

$$\boldsymbol{y} = \begin{pmatrix} y_{11} & 0 & 0 & 0 & 0 \\ 0 & y_{12} & 0 & 0 & 0 \\ 0 & 0 & y_{30} & 0 & 0 \\ 0 & 0 & 0 & y_{40} & 0 \\ 0 & 0 & 0 & 0 & y_{50} \end{pmatrix} \quad \text{columns } 1\ 2\ 3\ 4\ 5 \tag{3.79}$$

The matrix $\boldsymbol{A}$ remains the same. Performing the triple matrix product, we get the bus admittance matrix

$$\boldsymbol{Y}_{\text{bus}} = \boldsymbol{A}\,\boldsymbol{y}\boldsymbol{A}^{\text{T}}$$

$$= \begin{matrix}(1)\\(2)\\(3)\end{matrix}\begin{pmatrix} y_{12}+y_{13}+y_{10} & -y_{12} & -y_{13} \\ -y_{12} & y_{12}+y_{20} & 0 \\ -y_{13} & 0 & y_{13}+y_{30} \end{pmatrix} = \begin{bmatrix} Y_{11} & Y_{12} & Y_{13} \\ Y_{21} & Y_{22} & Y_{23} \\ Y_{31} & Y_{32} & Y_{33} \end{bmatrix} \quad \text{columns } (1)\ (2)\ (3) \tag{3.80}$$

From this pattern, we can generalize and say that when there is no mutual coupling,

1. The $(i - j)$ element Y_{ij} of the $\boldsymbol{Y}_{bus}$ matrix is the negative of the sum of admittances connected between nodes (i) and (j), and is called the *transfer admittance.*
2. The diagonal $(i - i)$ element is the sum of all admittances connected to node (i). It is called the *self-admittance*. Thus, $\boldsymbol{Y}_{bus}$ can be written by inspection in this case and also easily programmed on the computer.

In network theory literature, $\boldsymbol{Y}_{bus}$ is called the *Nodal Admittance matrix* and is obtained by writing the node equations at all the buses except the ground bus.

Example 3.13 For the transmission network in Fig. 3.34, the lines 2-3 and 2-4 have half line charging admittances of 0.005 mho and 0.01 mho, respectively. Write the $\boldsymbol{Y}_{bus}$ matrix. The shunt capacitor at bus (4) has an admittance of 0.02 mho.

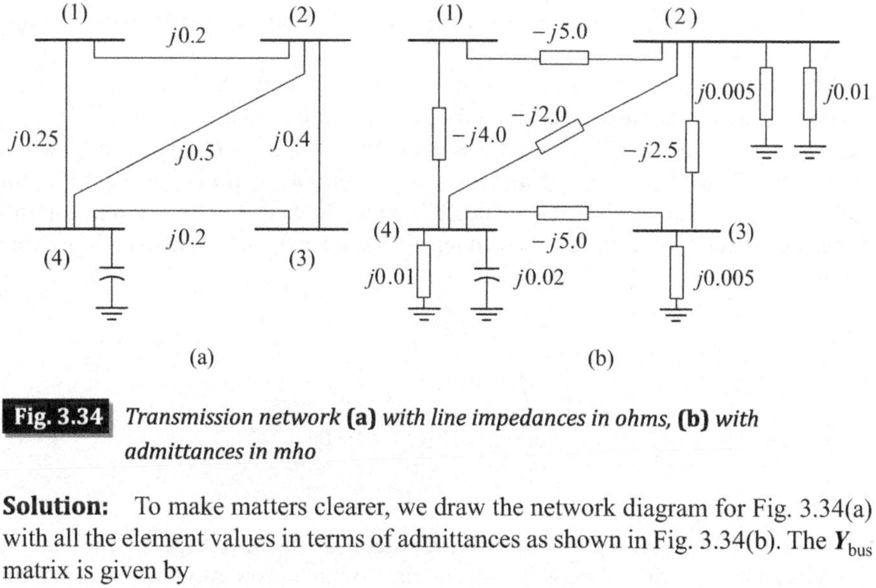

Fig. 3.34 ***Transmission network* (a) *with line impedances in ohms,* (b) *with admittances in mho***

Solution: To make matters clearer, we draw the network diagram for Fig. 3.34(a) with all the element values in terms of admittances as shown in Fig. 3.34(b). The $\boldsymbol{Y}_{bus}$ matrix is given by

$$\boldsymbol{Y}_{bus} = \begin{matrix} & (1) & (2) & (3) & (4) \\ (1) & -j9 & j5 & 0 & j4 \\ (2) & j5 & -j9.485 & j2.5 & j2 \\ (3) & 0 & j2.5 & -j7.495 & j5 \\ (4) & j4 & j2 & j5 & -j10.97 \end{matrix}$$

The reader can verify the entries Y_{22}, Y_{33} and Y_{44} by running all the admittances at nodes (2), (3), and (4) respectively. The off-diagonal entries are the negative of admittances connected between the two nodes.

3.7.4 Derivation of the Bus Impedance Matrix Z

Earlier in Section 3.6.1, it was mentioned that the elements Z_{ij} of $\boldsymbol{Z}$ can be interpreted as the open-circuit impedance parameters, and computed by exciting the buses with current sources one at a time. To derive an analytical expression for Z_{ij}, we shall generalize this procedure by exciting all the buses with current sources simultaneously, and solving for the bus voltages in terms of these current sources.

Let the graph of the transmission network consisting of $(n + 1)$ buses, including the ground bus, be augmented by elements corresponding to the current sources connected between the buses and the ground bus, with positive bus current orientation into the buses. Hence, $\boldsymbol{I}_{\text{bus}} = [I_1, I_2, ..., I_n]^T$. In accordance with our convention of the associated reference directions for the elements, the voltages of these augmented edges will be negative of the bus voltages. If $\boldsymbol{V}'_{\text{bus}}$ is the vector of the voltages of these elements, then $\boldsymbol{V}_{\text{bus}} = -\boldsymbol{V}'_{\text{bus}}$. We now proceed to solve this network by topological techniques. The overall network now has $n + 1$ nodes and $e + n$ elements, of which e = number of elements of the network, and n = number of elements corresponding to the current sources.

First we choose a tree for the overall augmented network. It will have n tree-branches. The tree is chosen such that the current sources constituting the vector $\boldsymbol{I}_{\text{bus}}$ belong to the co-tree. This choice of assigning the current sources to the co-tree is a logical one. From the link transformations, all currents are expressible in terms of the link currents, and hence this choice decreases the number of unknown link currents in the augmented network. The fundamental loop equations for the overall network are written as

$$\boldsymbol{CU} = 0$$

$$e\overset{\quad n \qquad e}{\begin{bmatrix} \boldsymbol{C}_b & \vdots & \boldsymbol{U} \end{bmatrix}} \begin{bmatrix} \boldsymbol{v}_b \\ \cdots \\ \boldsymbol{v}_\ell \\ \boldsymbol{V}'_{\text{bus}} \end{bmatrix} = 0 \tag{3.81}$$

In Eq. (3.87), $\boldsymbol{v}_b$, $\boldsymbol{v}_\ell$, and $\boldsymbol{V}'_{\text{bus}}$ are of the order n, $e - n$ and n respectively, with the elements corresponding to $\boldsymbol{v}_b$ and $\boldsymbol{v}_\ell$ belonging to the transmission network, and the elements of $\boldsymbol{V}_{\text{bus}}$ corresponding to the voltages across the augmented current sources. Expanding Eq. (3.81),

$$\begin{matrix} & n & e-n & n \end{matrix}$$

$$e\begin{bmatrix} & \vdots & \boldsymbol{U} & 0 \\ \boldsymbol{C}_b & \vdots & \cdots \quad \cdots & \cdots \\ & \vdots & 0 & \boldsymbol{U} \end{bmatrix} \begin{bmatrix} \boldsymbol{v}_b \\ \cdots \\ \boldsymbol{v}_\ell \\ \boldsymbol{V}'_{\text{bus}} \end{bmatrix} = 0 \tag{3.82}$$

$\boldsymbol{C}_b$ is further partitioned as

$$\boldsymbol{C}_b = \overset{n}{\begin{bmatrix} \boldsymbol{C}_{b11} \\ \boldsymbol{C}_{b22} \end{bmatrix}} \begin{matrix} e-n \\ n \end{matrix} \tag{3.83}$$

After re-arrangement of terms, Eq. (3.82) becomes

$$\begin{matrix} & n \\ e-n & \\ n & \end{matrix}\begin{bmatrix} \mathbf{0} \\ V'_{\text{bus}} \end{bmatrix} = -\overset{\begin{matrix} n & e-n \end{matrix}}{\begin{bmatrix} C_{b11} & U \\ C_{b22} & \mathbf{0} \end{bmatrix}}\begin{bmatrix} v_b \\ v_\ell \end{bmatrix} \tag{3.84}$$

The primitive network equations are arranged in the usual order of tree-branches and links of the transmission network as

$$v = \begin{bmatrix} v_b \\ \cdots \\ v_\ell \end{bmatrix} = [z]\begin{bmatrix} i_b \\ \cdots \\ i_1 \end{bmatrix} \tag{3.85}$$

Substituting Eq. (3.85) in Eq. (3.84) and replacing $V'_{\text{bus}} = -V_{\text{bus}}$, we get

$$\begin{bmatrix} \mathbf{0} \\ \cdots \\ V_{\text{bus}} \end{bmatrix} = \begin{bmatrix} C_{b11} & \vdots & U \\ \cdots & \cdots & \cdots \\ C_{b22} & \vdots & \mathbf{0} \end{bmatrix}[z]\begin{bmatrix} i_b \\ \cdots \\ i_\ell \end{bmatrix} \tag{3.86}$$

From the link transformation Eq. (3.47), we have

$$\begin{bmatrix} i_b \\ i_\ell \\ I_{\text{bus}} \end{bmatrix} = \left[C^T\right]\begin{bmatrix} i_\ell \\ I_{\text{bus}} \end{bmatrix} \tag{3.87}$$

Using the same partitioning of C as in Eq. (3.84), we get

$$\begin{bmatrix} i_b \\ i_\ell \\ I_{\text{bus}} \end{bmatrix} = \begin{bmatrix} C_{b11}^T & C_{b22}^T \\ U & \mathbf{0} \\ \mathbf{0} & U \end{bmatrix}\begin{bmatrix} i_\ell \\ I_{\text{bus}} \end{bmatrix} \tag{3.88}$$

Hence,

$$\begin{bmatrix} i_b \\ i_\ell \end{bmatrix} = \begin{bmatrix} C_{b11}^T & C_{b22}^T \\ U & 0 \end{bmatrix}\begin{bmatrix} i_\ell \\ I_{\text{bus}} \end{bmatrix} \tag{3.89}$$

Substituting Eq. (3.89) in Eq. (3.86), we get

$$\begin{bmatrix} \mathbf{0} \\ V_{\text{bus}} \end{bmatrix} = \begin{bmatrix} C_{\text{b}11} & U \\ C_{\text{b}22} & \mathbf{0} \end{bmatrix}[z]\begin{bmatrix} C_{b11}^T & C_{b22}^T \\ U & \mathbf{0} \end{bmatrix}\begin{bmatrix} i_\ell \\ I_{\text{bus}} \end{bmatrix} \tag{3.90}$$

The triple matrix product in Eq. (3.90) is then performed to get

$$\begin{bmatrix} \mathbf{0} \\ V_{\text{bus}} \end{bmatrix} = \begin{bmatrix} Z_{11} & Z_{12} \\ Z_{21} & Z_{22} \end{bmatrix}\begin{bmatrix} i_\ell \\ I_{\text{bus}} \end{bmatrix} \tag{3.91}$$

Solving for $i\ell$ from the first set of equations in Eq. (3.91), we get

$$i_\ell = -Z_{11}^{-1} Z_{12} I_{\text{bus}} \tag{3.92}$$

Eq. (3.92) is then substituted in the second set of equations in Eq. (3.91).

$$V_{bus} = -Z_{21} Z_{11}^{-1} Z_{12} I_{bus} + Z_{22} I_{bus} \tag{3.93}$$

Hence,

$$V_{bus} = \left(Z_{22} - Z_{21} Z_{11}^{-1} Z_{12}\right) I_{bus} = Z_{bus} I_{bus} \tag{3.94}$$

Z_{bus} is a square matrix and has the same order as i_ℓ, i.e.,, $e - n$ which is the number of links in the transmission network alone. This is also equal to the number of fundamental loops in the transmission network, i.e., network without the augmented current sources. Hence, to compute Z_{bus}, we need an inverse of the same order as the number of links in the network, i.e., $e - n$. In the power network, this number could be very large since there are a number of shunt paths to the ground due to the transmission lines. Therefore, for a large network, the computation of Z_{bus} needs an inverse of a very high order. The Z_{bus} building algorithm avoids this problem by progressively building the Z_{bus} by taking one element at a time. This is a specialized topic and is omitted. Z_{bus} can also be computed from Y_{bus} by some efficient numerical inversion schemes such as the LU decomposition technique. Hence, with these fast numerical techniques, building Z_{bus} or inverting Y_{bus} has comparable computing requirements. Since Y_{bus} is generally sparse and we may not require all the elements of Z_{bus} such as in short-circuit studies, the LU decomposition technique has some advantages. But in small systems, it is nice to have a program for building Z_{bus} as each element is added or deleted.

Example 3.14 Consider the network of Fig. 3.33. Its graph augmented with the current source edges (6, 7, 8) is shown in Fig. 3.35. Find the Z_{bus} considering the tree with tree-branches (2, 3, 4).

Solution: The links are (1, 5, 6, 7, 8). The fundamental loop matrix for the augmented graph is

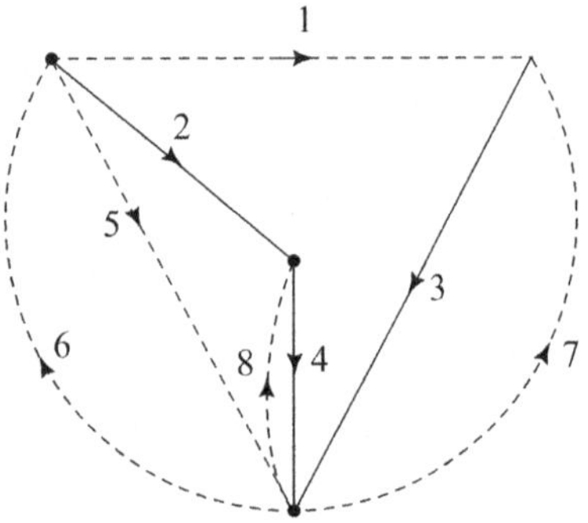

Fig. 3.35 *Network to compute* Z_{bus}

$$C = \begin{array}{c} \\ 1 \\ 5 \\ 6 \\ 7 \\ 8 \end{array} \begin{array}{c} \begin{array}{ccccccccc} 2 & 3 & 4 & 1 & 5 & 6 & 7 & 8 \end{array} \\ \left(\begin{array}{ccc|cc|ccc} -1 & 1 & -1 & 1 & 0 & 0 & 0 & 0 \\ -1 & 0 & -1 & 0 & 1 & 0 & 0 & 0 \\ 1 & 0 & 1 & 0 & 0 & 1 & 0 & 0 \\ 0 & 1 & 0 & 0 & 0 & 0 & 1 & 0 \\ 0 & 0 & 1 & 0 & 0 & 0 & 0 & 1 \end{array}\right) \end{array} \tag{3.95}$$

Hence,

$$C_{b11} = \begin{bmatrix} -1 & 1 & -1 \\ -1 & 0 & -1 \end{bmatrix}, \quad C_{b22} = \begin{bmatrix} 1 & 0 & 1 \\ 0 & 1 & 0 \\ 0 & 0 & 1 \end{bmatrix} \tag{3.96}$$

The primitive impedance matrix is given by (note that the ordering of elements has to be in the same order as the tree-branches and links of the network graph)

$$z = j\begin{matrix} & 2 & 3 & 4 & 1 & 5 \\ 2 & 2 & 0 & 0 & 1 & 0 \\ 3 & 0 & 1 & 0 & 0 & 0 \\ 4 & 0 & 0 & 1 & 0 & 0 \\ 1 & 1 & 0 & 0 & 1 & 0 \\ 5 & 0 & 0 & 0 & 0 & 1 \end{matrix} \tag{3.97}$$

The triple matrix product corresponding to Eq. (3.90) is given by

$$\left[\frac{0}{V_{\text{bus}}}\right] = \begin{matrix} & 2 & 3 & 4 & & 1 & 5 \\ 1 & -1 & 1 & -1 & | & 1 & 0 \\ 5 & -1 & 0 & -1 & | & 0 & 1 \\ 6 & 1 & 0 & 1 & | & 0 & 0 \\ 7 & 0 & 1 & 0 & | & 0 & 0 \\ 8 & 0 & 0 & 1 & | & 0 & 0 \end{matrix} [z] \begin{pmatrix} 1 & 5 & 6 & 7 & 8 \\ -1 & -1 & 1 & 0 & 0 \\ 1 & 0 & 0 & 1 & 0 \\ -1 & -1 & 1 & 0 & 1 \\ 1 & 0 & 0 & 0 & 0 \\ 0 & 1 & 0 & 0 & 0 \end{pmatrix} \left[\frac{i_\ell}{I_{\text{bus}}}\right] \tag{3.98}$$

Performing the triple matrix product, we get

$$\left[\frac{0}{V_{\text{bus}}}\right] = j\left[\begin{array}{cc|ccc} 3 & 2 & -2 & 1 & -1 \\ 2 & 4 & -3 & 0 & -1 \\ \hline -2 & -3 & 3 & 0 & 1 \\ 1 & 0 & 0 & 1 & 0 \\ -1 & -1 & 1 & 0 & 1 \end{array}\right] = \begin{bmatrix} Z_{11} & Z_{12} \\ Z_{21} & Z_{22} \end{bmatrix}\left[\frac{i_\ell}{I_{\text{bus}}}\right] \tag{3.99}$$

Hence, performing the elimination of $[i_\ell]$ in Eq. (3.99)

$$V_{\text{bus}} = \left(Z_{22} - Z_{21}Z_{11}^{-1}Z_{12}\right) I_{\text{bus}} = Z_{\text{bus}}\, I_{\text{bus}} \tag{3.100}$$

Hence,

$$Z_{\text{bus}} = j\begin{bmatrix} 3 & 0 & 1 \\ 0 & 1 & 0 \\ 1 & 0 & 1 \end{bmatrix} - j\begin{bmatrix} -2 & -3 \\ 1 & 0 \\ -1 & -1 \end{bmatrix}\begin{bmatrix} 3 & 2 \\ 2 & 4 \end{bmatrix}^{-1}\begin{bmatrix} -2 & 1 & -1 \\ -3 & 0 & -1 \end{bmatrix}$$

$$= j\begin{bmatrix} 3 & 0 & 1 \\ 0 & 1 & 0 \\ 1 & 0 & 1 \end{bmatrix} - j\begin{bmatrix} -2 & -3 \\ 1 & 0 \\ -1 & -1 \end{bmatrix}\frac{1}{8}\begin{bmatrix} 4 & -2 \\ -2 & 3 \end{bmatrix}\begin{bmatrix} -2 & 1 & -1 \\ -3 & 0 & -1 \end{bmatrix}$$

$$= j\begin{bmatrix} 3 & 0 & 1 \\ 0 & 1 & 0 \\ 1 & 0 & 1 \end{bmatrix} - \begin{bmatrix} 2.375 & -0.25 & 0.875 \\ -0.25 & 0.5 & -0.25 \\ 0.875 & -0.25 & 0.375 \end{bmatrix} = j\begin{bmatrix} 0.625 & 0.25 & 0.125 \\ 0.25 & 0.5 & 0.25 \\ 0.125 & 0.25 & 0.625 \end{bmatrix} \tag{3.101}$$

By using Y_{bus} from Eq. (3.78), it can be verified that

$$Y_{\text{bus}} Z_{\text{bus}} = U \tag{3.102}$$

Example 3.15 Obtain the Z_{bus} for the transmission network of Example 3.13 (Fig. 3.34).

Solution: In Fig. 3.34(b), we lump the shunt admittances at a bus into a single admittance, and then draw the graph of the network. The augmented current sources are shown dotted (9, 10, 11, 12) in Fig. 3.36. Choose the tree (1, 2, 4, 6). The tree-branches are hatched. The links are loop matrix for the augmented graph.

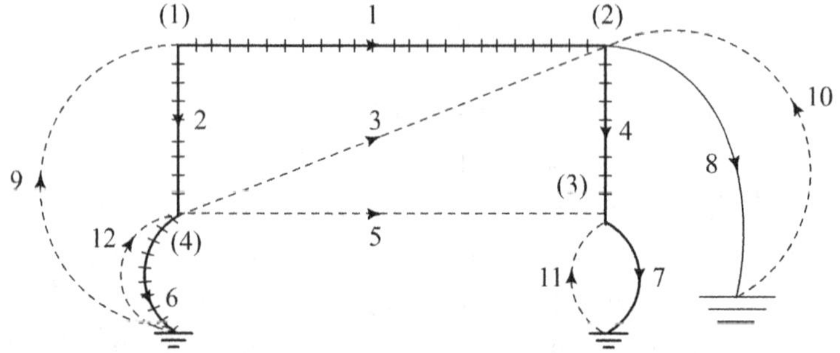

Fig. 3.36 *An example for Z_{bus}*

$$
C = \begin{array}{c} \\ 3 \\ 5 \\ 7 \\ 8 \\ 9 \\ 10 \\ 11 \\ 12 \end{array}
\begin{array}{c}
\begin{array}{cccc|cccc|cccc} 1 & 2 & 4 & 6 & 3 & 5 & 7 & 8 & 9 & 10 & 11 & 12 \end{array} \\
\left(\begin{array}{cccc|cccc|cccc}
-1 & 1 & 0 & 0 & 1 & 0 & 0 & 0 & 0 & 0 & 0 & 0 \\
-1 & 1 & -1 & 0 & 0 & 1 & 0 & 0 & 0 & 0 & 0 & 0 \\
1 & -1 & 0 & -1 & 0 & 0 & 1 & 0 & 0 & 0 & 0 & 0 \\
1 & -1 & 0 & -1 & 0 & 0 & 0 & 1 & 0 & 0 & 0 & 0 \\
0 & 1 & 0 & 1 & 0 & 0 & 0 & 0 & 1 & 0 & 0 & 0 \\
-1 & 1 & 0 & 1 & 0 & 0 & 0 & 0 & 0 & 1 & 0 & 0 \\
-1 & 1 & -1 & 1 & 0 & 0 & 0 & 0 & 0 & 0 & 1 & 0 \\
0 & 0 & 0 & 1 & 0 & 0 & 0 & 0 & 0 & 0 & 0 & 1
\end{array}\right)
\end{array}
$$

Hence,

$$
C_{b11} = \begin{bmatrix} -1 & 1 & 0 & 0 \\ -1 & 1 & -1 & 0 \\ 1 & -1 & 0 & -1 \\ 1 & -1 & 0 & -1 \end{bmatrix}
$$

$$
C_{b22} = \begin{bmatrix} 0 & 1 & 0 & 1 \\ -1 & 1 & 0 & 1 \\ -1 & 1 & -1 & 1 \\ 0 & 0 & 0 & 1 \end{bmatrix}
$$

The primitive impedance matrix written in an order consistent with the tree-branches and links of the network is

$$z = \begin{array}{c} \begin{matrix} 1 & 2 & 4 & 6 & 3 & 5 & 7 & 8 \end{matrix} \\ \begin{pmatrix} 0.2 & 0 & 0 & 0 & 0 & 0 & 0 & 0 \\ 0 & .25 & 0 & 0 & 0 & 0 & 0 & 0 \\ 0 & 0 & .04 & 0 & 0 & 0 & 0 & 0 \\ 0 & 0 & 0 & -33.33 & 0 & 0 & 0 & 0 \\ 0 & 0 & 0 & 0 & 0.5 & 0 & 0 & 0 \\ 0 & 0 & 0 & 0 & 0 & 0.2 & 0 & 0 \\ 0 & 0 & 0 & 0 & 0 & 0 & -200 & 0 \\ 0 & 0 & 0 & 0 & 0 & 0 & 0 & -66.67 \end{pmatrix} \end{array}$$

The rest of the details of deriving $\boldsymbol{Z}_{\text{bus}}$ is left to the reader. The reader may also verify that $\boldsymbol{Z}_{\text{bus}}\,\boldsymbol{Y}_{\text{bus}} = \boldsymbol{U}$.

3.8 NETWORK REDUCTION

In large-scale systems, it may be necessary to reduce the dimension of the network by eliminating certain buses. Let the description be of the form

$$\boldsymbol{I}_{\text{bus}} = \boldsymbol{Y}_{\text{bus}}\,\boldsymbol{V}_{\text{bus}} \tag{3.103}$$

We assume that the elements of $\boldsymbol{I}_{\text{bus}}$ and $\boldsymbol{V}_{\text{bus}}$ are ordered such that the buses to be retained appear first and are denoted by the subscript (r). Those to be eliminated are denoted by the subscript (s). Thus

$$\boldsymbol{I}_{\text{bus}} = \begin{bmatrix} \boldsymbol{I}_{\text{bus(r)}} \\ \boldsymbol{I}_{\text{bus(s)}} \end{bmatrix} \text{ and } \boldsymbol{V}_{\text{bus}} = \begin{bmatrix} \boldsymbol{V}_{\text{bus(r)}} \\ \boldsymbol{V}_{\text{bus(s)}} \end{bmatrix} \tag{3.104}$$

Partition $\boldsymbol{Y}_{\text{bus}}$ such that

$$\boldsymbol{I}_{\text{bus(r)}} = \boldsymbol{Y}_A\,\boldsymbol{V}_{\text{bus(r)}} + \boldsymbol{Y}_B\,\boldsymbol{V}_{\text{bus(s)}} \tag{3.105}$$

$$\boldsymbol{I}_{\text{bus(s)}} = \boldsymbol{Y}_C\,\boldsymbol{V}_{\text{bus(r)}} + \boldsymbol{Y}_D\,\boldsymbol{V}_{\text{bus(s)}} \tag{3.106}$$

If there are no sources at the buses to be eliminated, then $\boldsymbol{I}_{\text{bus}}\,(\text{s}) \equiv 0$. Hence, by elementary matrix manipulation from Eqs. (3.105) and (3.106), we get,

$$\boldsymbol{I}_{\text{bus(r)}} = \left[\boldsymbol{Y}_A - \boldsymbol{Y}_B\,\boldsymbol{Y}_D^{-1}\,\boldsymbol{Y}_C\right]\boldsymbol{V}_{\text{bus(r)}} = \boldsymbol{V}_{\text{bus(r)}}\,\boldsymbol{V}_{\text{bus(r)}} \tag{3.107}$$

Instead of taking the inverse of the matrix $\boldsymbol{Y}_D$, we can perform the elimination bus by bus. Suppose we first want to eliminate bus k in $\boldsymbol{Y}_{\text{bus}}$, then,

$$\boldsymbol{Y}_{\text{bus}} = \begin{array}{c} \begin{matrix} 1 & & k & & n \end{matrix} \\ \begin{pmatrix} Y_{11} & \cdots & Y_{1k} & \cdots & Y_{1n} \\ Y_{21} & \cdots & Y_{2k} & \cdots & Y_{2n} \\ \vdots & & \vdots & & \vdots \\ Y_{k1} & \cdots & Y_{kk} & \cdots & Y_{kn} \\ \vdots & & \vdots & & \vdots \\ Y_{n1} & \cdots & Y_{nk} & \cdots & Y_{nn} \end{pmatrix} \end{array} \tag{3.108}$$

The row and column corresponding to the bus k is exhibited explicitly. We can re-order the rows and columns so that the kth row and the kth column become the last row and the last column, respectively, resulting in Eq. (3.109). Hence,

$$\boldsymbol{Y}_{bus} = \begin{bmatrix} Y_{11} & \cdots & Y_{1n} & Y_{1k} \\ Y_{21} & \cdots & Y_{2n} & Y_{2k} \\ \vdots & & \vdots & \vdots \\ Y_{n1} & \cdots & Y_{nn} & Y_{nk} \\ & & & \\ Y_{k1} & \cdots & Y_{kn} & Y_{kk} \end{bmatrix} \tag{3.109}$$

Eliminating the last row and the last column in Eq. (3.110) using (3.107), we get,

$$Y_{ij(\text{new})} = Y_{ij\,(\text{old})} - \frac{Y_{ik(\text{old})}\,Y_{kj\,(\text{old})}}{Y_{kk}} \tag{3.110}$$

This can be repeated for each node to be eliminated. Kron reduction of this type is useful in transient stability studies.

Example 3.16 For the network in Fig. 3.33(a), eliminate node (2) and obtain the reduced 2×2 description at the nodes (1) and (3).

Solution:

$$\boldsymbol{Y}_{bus} = j\begin{pmatrix} -2 & 1 & 0 \\ 1 & -3 & 1 \\ 0 & 1 & -2 \end{pmatrix} \begin{matrix} 1 \\ 2 \\ 3 \end{matrix} \quad (\text{columns } 1\ 2\ 3)$$

Row 2 and column 2 are made as the last row and column first as follows

$$\boldsymbol{Y}_{bus} = j\,\begin{matrix} 1 \\ 2 \end{matrix}\begin{pmatrix} -2 & 1 & 0 \\ 1 & -3 & 1 \end{pmatrix} \Rightarrow j\begin{pmatrix} -2 & 0 & 1 \\ 0 & -2 & 1 \\ 1 & 1 & -3 \end{pmatrix}\begin{matrix} 1 \\ 3 \\ 2 \end{matrix}$$

(columns 1 2 3 on the left matrix; columns 1 3 2 on the right matrix)

Eliminating node 2 gives the elements of the reduced $\mathbf{Y}_{bus}$ matrix as

$$Y_{11\,\text{new}} = -2 - \frac{(1)(1)}{(-3)} = -2 + \frac{1}{3} = -\frac{5}{3}$$

$$Y_{12\,\text{new}} = 0 - \frac{(1)(1)}{(-3)} = \frac{1}{3}$$

$$Y_{21\,\text{new}} = 0 - \frac{(1)(1)}{(-3)} = \frac{1}{3}$$

$$Y_{22\,\text{new}} = -8 - \frac{(1)(1)}{(-3)} = -\frac{5}{3}$$

Hence, the reduced matrix is

$$
\begin{array}{c}
\quad\quad\quad\;\; 1 \quad\quad\;\; 3 \\
Y_{\text{bus}} = \begin{pmatrix} -\dfrac{5}{3} & \dfrac{1}{3} \\ \\ \dfrac{1}{3} & -\dfrac{5}{3} \end{pmatrix} \begin{matrix} 1 \\ \\ 3 \end{matrix}
\end{array}
$$

3.9 DISTRIBUTION FACTORS

In Eq. (3.106), if $\boldsymbol{I}_{\text{bus(s)}} \neq 0$, then we can still eliminate those buses as follows. From Eq. (3.106),

$$\boldsymbol{V}_{\text{bus(s)}} = \boldsymbol{Y}_D^{-1} \boldsymbol{I}_{\text{bus(s)}} - \boldsymbol{Y}_D^{-1} \boldsymbol{Y}_C \boldsymbol{V}_{\text{bus(r)}} \tag{3.111}$$

Substituting this in Eq. (3.105) yields

$$\boldsymbol{I}_{\text{bus(r)}} = \left[\boldsymbol{Y}_A - \boldsymbol{Y}_B \boldsymbol{Y}_D^{-1} \boldsymbol{Y}_C\right] \boldsymbol{V}_{\text{bus(r)}} + \boldsymbol{Y}_B \boldsymbol{Y}_D^{-1} \boldsymbol{I}_{\text{bus(s)}}$$

$$\Rightarrow \boldsymbol{I}_{\text{bus(r)}} - \boldsymbol{Y}_B \boldsymbol{Y}_D^{-1} \boldsymbol{I}_{\text{bus(s)}} = \left[\boldsymbol{Y}_A - \boldsymbol{Y}_B \boldsymbol{Y}_D^{-1} \boldsymbol{Y}_C\right] \boldsymbol{V}_{\text{bus(r)}} = \boldsymbol{Y}_{\text{bus(r)}} \boldsymbol{V}_{\text{bus(r)}}$$

The elements of the vector $-\boldsymbol{Y}_B \boldsymbol{Y}_D^{-1} \boldsymbol{I}_{\text{bus(s)}}$ reflect the effect of the current sources which are eliminated on the buses which are retained. The elements of the matrix $-\boldsymbol{Y}_B \boldsymbol{Y}_D^{-1}$ are called the distribution factors D_{ij}.

3.10 CONCLUSION

In this chapter we have discussed the systematic way in which the multiport representations of transmission networks are obtained using linear graph theory. We have also discussed Kron reduction and distribution factors.

Problems

3.1 For the transmission network in Fig. 3.18(a) and its corresponding oriented graph in Fig. 3.18(b), select the tree T (6, 7, 8, 9) and write the $\boldsymbol{B}$ and $\boldsymbol{C}$ matrices. Verify the orthogonality relations. Choosing ground as the reference bus, write the $\boldsymbol{A}$ matrix.

3.2 For the graph in Fig. P.3.1, select the tree T (2, 4, 5, 6).

(i) Write the fundamental loop matrix $\boldsymbol{C}$ and the fundamental cutset matrix $\boldsymbol{B}$. Verify the relation $\boldsymbol{B}\boldsymbol{C}^T = 0$ and $\boldsymbol{C}_b = -\boldsymbol{B}_\ell^T$.

(ii) Write the augmented incidence matrix $\boldsymbol{A}_a$ and the incidence matrix $\boldsymbol{A}$ by choosing 4 as the reference node. Arrange $\boldsymbol{A}$ as $\left[\boldsymbol{A}_b \vdots \boldsymbol{A}_\ell\right]$ corresponding to the tree T (2, 4, 5, 6) and verify the relation $\boldsymbol{B}_\ell = -\boldsymbol{A}_b^{-1}\boldsymbol{A}_\ell$.

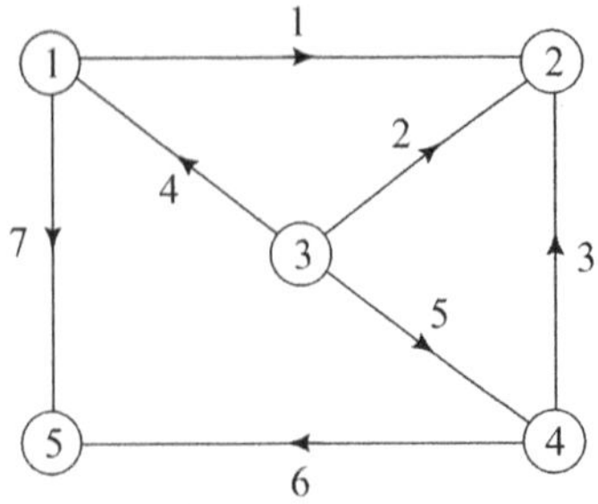

Fig. P.3.1

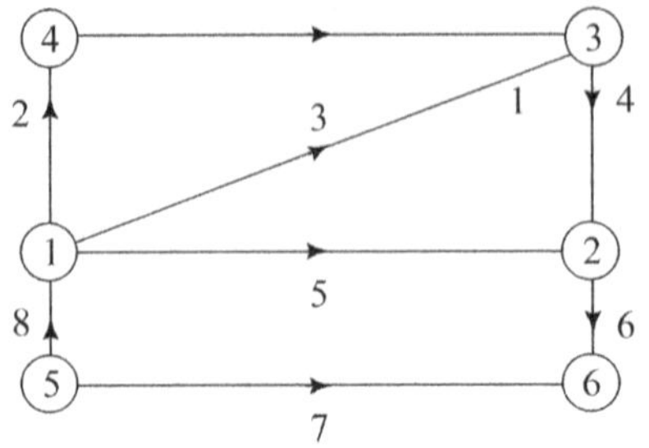

Fig. P.3.2

3.3 Repeat Problem 3.2 for the tree T (3, 5, 6, 7).

3.4 For the oriented graph in Fig. P.3.2, select the tree T (6, 7, 1, 4, 5). Write the $\boldsymbol{B}$ and $\boldsymbol{C}$ matrices and verify that $\boldsymbol{BC}^T = 0$. Write the incidence matrix $\boldsymbol{A}$ with 6 as the reference node and derive $\boldsymbol{B}$ from $\boldsymbol{A}$.

3.5 Consider the linear graph in Fig. P.3.3 which represents a 3-bus transmission system with all the shunt admittances at a bus lumped together. Each transmission line has a series impedance of $0.02 + j\ 0.08$ and a half line charging admittance of $j\ 0.02$ (all in p.u.). 0 is the ground bus.

(a) Compute $\boldsymbol{Y}_{\text{bus}}$ by inspection as well as by the analytical formula (3.73).

(b) Compute $\boldsymbol{Z}_{\text{bus}}$ analytically by the procedure of Section 3.7.4.

(c) Verify $\boldsymbol{Z}_{\text{bus}}\ \boldsymbol{Y}_{\text{bus}} = \boldsymbol{U}$.

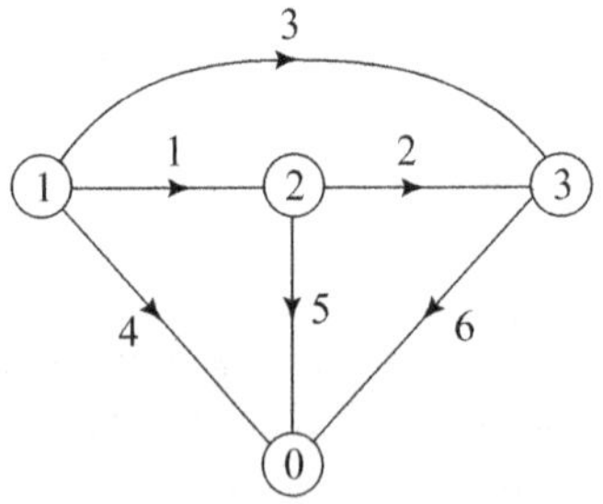

Fig. P.3.3

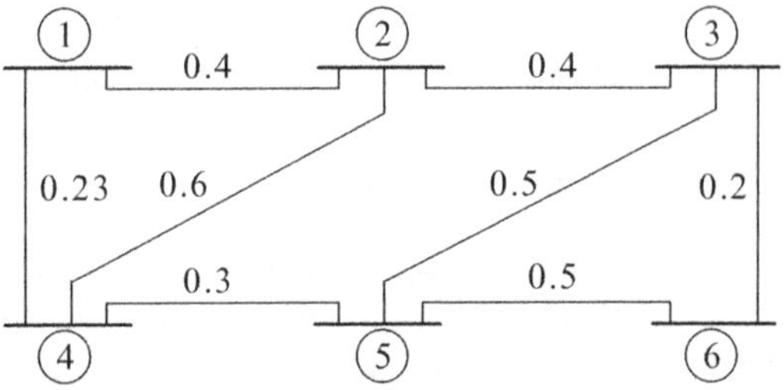

Fig. P.3.3

3.6 Consider the system shown in Fig. P.3.4. It shows a transmission network with series reactances of transmission lines all in p.u. as shown. The line charging and shunt impedances are neglected.

(a) By choosing the appropriate tree for the graph, write the $\boldsymbol{B}$ and $\boldsymbol{C}$ matrices. Write the incidence matrix $\boldsymbol{A}$ with 6 as the reference node.

(b) Verify that $\boldsymbol{BC}^{\mathrm{T}} = 0$ and deduce the fundamental cutset matrix from the matrix $\boldsymbol{A}$.

(c) Compute $\boldsymbol{Y}_{\text{bus}}$ matrix with ground as the reference.

(d) Choosing **6** as the reference bus, compute $\boldsymbol{Z}_{\text{bus}}$ using the procedure of Section 3.7.4.

(e) In the $\boldsymbol{Y}_{\text{bus}}$ derived in step 3, since the ground bus is isolated, delete the row and column corresponding to the ground bus to get a reduced $\boldsymbol{Y}_{\text{bus}}$. Verify that this reduced $\boldsymbol{Y}_{\text{bus}}$ is the inverse of $\boldsymbol{Z}_{\text{bus}}$ obtained in step 4.

3.7 In the $\boldsymbol{Y}_{\text{bus}}$ of Problem 3.6, eliminate the nodes **3** and **5** and obtain the reduced admittance matrix with ground bus as the reference.

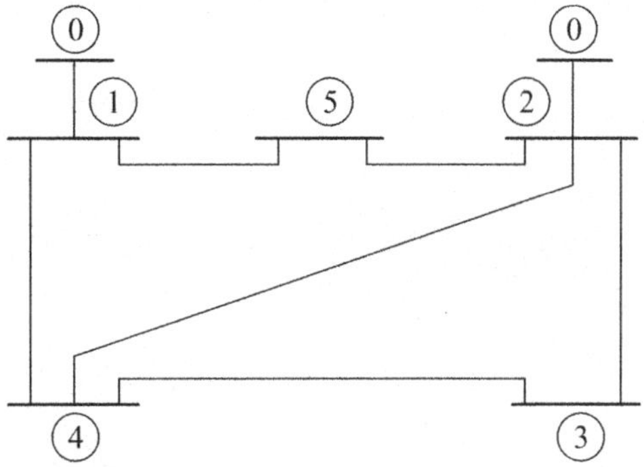

Fig. P.3.5

3.8 The system impedance data for the system in Fig. P.3.5 is given in the Table P.3.1. Lines 1-0 and 2-0 represent the transient reactances of the generators connected at buses **1** and **2**, respectively.

Table P 3.1

From Bus	*To Bus*	*R*	*X*
1	4	0.15	0.60
1	5	0.05	0.20
2	3	0.05	0.20
2	4	0.10	0.40
2	5	0.05	0.20
3	4	0.10	0.40
1	0	—	0.25
2	0	—	1.25

(a) Write a general purpose computer program for eliminating buses from the given $\boldsymbol{Y}_{\text{bus}}$.

(b) Implement the program for the above system to eliminate all the buses except **7** and **8**, and obtain a reduced bus admittance matrix at buses **7** and **8**.

3.9 For the system in Problem 3.8, neglect all series resistance and also omit the line 2-4.

(a) Compute $\boldsymbol{Z}_{\text{bus}}$ with ground as the reference bus.

(b) Compute $\boldsymbol{Y}_{\text{bus}}$ with ground as the reference.

(c) Verify that $\boldsymbol{Z}_{\text{bus}}\boldsymbol{Y}_{\text{bus}} = \mathbf{U}$.

4 BUS IMPEDANCE ALGORITHM

4.1 INTRODUCTION

As discussed in Chapter 3, the bus impedance matrix of a transmission network is a mathematical equivalent through which the bus voltages are explicitly expressed in terms of the bus currents, giving rise to a multiport representation of the transmission network. The primitive impedance matrix and the network topology are required to construct the bus impedance matrix Z_{bus}. The elements of the bus impedance matrix are the open-circuit impedance parameters of the multiport representation. The topological formula for the computation of Z_{bus}, however, requires the inversion of a matrix whose order is equal to the number of independent loops in the network and hence, the method cannot be used to compute Z_{bus} for large networks. An alternate method of constructing the bus impedance matrix by adding one element at a time has been very popular with power engineers, and is known as the Z_{bus} building algorithm. In this chapter, we shall develop the algorithm from topological considerations and illustrate it by means of examples.

4.2 PARTIAL NETWORK

A partial network is a connected network consisting of some or all of the buses, and some of the lines of a given transmission network. Let z_0, A_0, and $Z_{bus(a)}$ be the primitive impedance matrix, bus incidence matrix and the bus impedance matrix, respectively, of the partial network shown in Fig. 4.1. When certain modifications such as adding of new line(s) or removal of certain line(s) in the network are made, we would like to know the modified bus impedance matrix Z_{bus} of the new network in terms of $Z_{bus(0)}$.

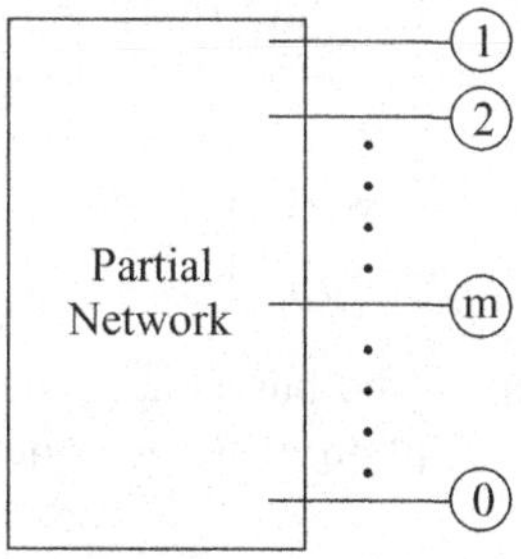

Fig. 4.1 *Partial network (interconnections are not shown)*

For the partial network, using the expression for the bus admittance matrix (3.73), we have

$$Z_{\text{bus}(0)} = \left[Y_{\text{bus}(0)}\right]^{-1} = \left[\mathbf{A}_0\, \mathbf{z}_0^{-1}\, \mathbf{A}_0^T\right]^{-1} \tag{4.1}$$

Suppose an element α having self-impedance $Z_{\alpha\alpha}$ and mutual impedance $z_{0\alpha}$ with the elements of the partial network, is added to the partial network. Then for the new network, the element voltage and current vectors respectively are

$$\boldsymbol{v} = \begin{bmatrix} \boldsymbol{v}_0 \\ \boldsymbol{v}_\alpha \end{bmatrix} \text{ and } \boldsymbol{i} = \begin{bmatrix} \boldsymbol{i}_0 \\ \boldsymbol{i}_\alpha \end{bmatrix}$$

where $\boldsymbol{v}_0$ and $\boldsymbol{i}_0$ are the element voltage and current vectors respectively in the partial network, and $\boldsymbol{v}_\alpha$ and $\boldsymbol{i}_\alpha$ are the voltage and current variables respectively of the added element.

The primitive equations in impedance form are

$$\begin{bmatrix} v_0 \\ v_\alpha \end{bmatrix} = \begin{bmatrix} z_0 & z_{0\alpha} \\ z_{\alpha 0} & z_{\alpha\alpha} \end{bmatrix} \begin{bmatrix} i_0 \\ i_\alpha \end{bmatrix} \tag{4.2}$$

and in the admittance form they are given by

$$\begin{bmatrix} i_0 \\ i_\alpha \end{bmatrix} = \begin{bmatrix} y_{00} & y_{0\alpha} \\ y_{\alpha 0} & y_{\alpha\alpha} \end{bmatrix} \begin{bmatrix} v_0 \\ v_\alpha \end{bmatrix} \tag{4.3}$$

where

$$z_0^{-1} = y_{00} - y_{0\alpha} y_{\alpha\alpha}^{-1} y_{\alpha 0} \tag{4.4}$$

Note that $y_{00} \neq z_0^{-1}$ unless $y_{0\alpha} = y_{\alpha 0}$, i.e., the added element is not coupled to any of the elements of the partial network.

Combining (4.2), (4.3), and (4.4), the primitive equations can be written in a mixed form as

$$\begin{bmatrix} \boldsymbol{i}_0 \\ \boldsymbol{v}_\alpha \end{bmatrix} = \begin{bmatrix} z_0^{-1} & \boldsymbol{y}_{0\alpha}\, \boldsymbol{y}_{\alpha\alpha}^{-1} \\ -\boldsymbol{y}_{\alpha\alpha}^{-1}\, \boldsymbol{y}_{\alpha 0} & \boldsymbol{y}_{\alpha\alpha}^{-1} \end{bmatrix} \begin{bmatrix} \boldsymbol{v}_0 \\ \boldsymbol{i}_\alpha \end{bmatrix} \tag{4.5}$$

If $\boldsymbol{V}_{\text{bus}}$ and $\boldsymbol{I}_{\text{bus}}$ denote the bus voltage and the bus current vectors corresponding to the multiport description of the new network, then from Eqs. (3.59) and (3.71) (Chapter 3), we get

$$\boldsymbol{v} = \begin{bmatrix} \boldsymbol{v}_0 \\ \boldsymbol{v}_\alpha \end{bmatrix} = \boldsymbol{A}^T \boldsymbol{V}_{\text{bus}} \tag{4.6}$$

$$\boldsymbol{I}_{\text{bus}} = \boldsymbol{A}\boldsymbol{i} = \boldsymbol{A} \begin{bmatrix} \boldsymbol{i}_0 \\ \boldsymbol{i}_\alpha \end{bmatrix} \tag{4.7}$$

where $\boldsymbol{A}$ is the bus incidence matrix of the *new network*. With these preliminaries, we now proceed to specific cases.

4.3 ADDITION OF LINK(S)

4.3.1 Addition of a Link

Let the element α be added between the buses p and q, with the orientation from p to q as shown in Fig. 4.2. Since the addition of an element connecting two nodes creates a closed path in the partial network, we designate such an element as a link. This is slightly different from the concept of a link we defined in Chapter 3 where it constituted an element of the co-tree. We retain the term link to be consistent with the prevailing terminology. When the link is added, no new bus is created. The bus incidence matrix $\boldsymbol{A}$ for the new network can be partitioned as

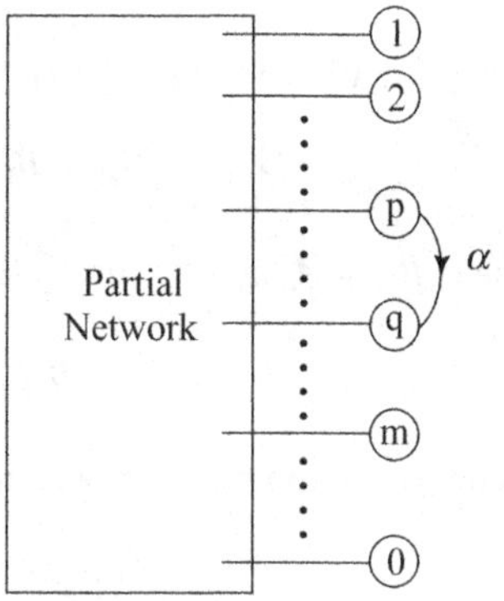

Fig. 4.2 *New network showing addition of a link*

$$\boldsymbol{A} = [\boldsymbol{A}_0 \quad \boldsymbol{K}_1] \tag{4.8}$$

where,

$\boldsymbol{K}_1 = [0 \ \cdots \ 1 \ 0 \ \cdots \ 0 \ -1 \ 0 \ \cdots \ 0]^T$

$\boldsymbol{K}_1$ is the sub-matrix of the bus incidence matrix corresponding to the added element. It essentially indicates how the new element is incident to the partial network. Hence, $\boldsymbol{K}_1$ has only two non-zero entries, +1 and -1 in the pth row and the qth row, respectively. Making use of Eq. (4.8) in Eqs. (4.6) and (4.7), the bus variables and primitive variables for the new network are related as

$$\boldsymbol{v}_0 = \boldsymbol{A}_0^T \boldsymbol{V}_{\text{bus}} \tag{4.9}$$

$$\boldsymbol{v}_\alpha = \boldsymbol{K}_1^T \boldsymbol{V}_{\text{bus}} \tag{4.10}$$

$$\boldsymbol{A}_0 \boldsymbol{i}_0 + \boldsymbol{K}_1 \boldsymbol{i}_\alpha = \boldsymbol{I}_{\text{bus}} \tag{4.11}$$

Now eliminating the variables v_0, i_0, v_α, and i_α from the set of equations (4.5), (4.9), (4.10), and (4.11), an explicit relation between V_{bus} and I_{bus} is obtained as

$$\left[\boldsymbol{B}_{11} - \boldsymbol{B}_{12}\boldsymbol{B}_{22}^{-1}\boldsymbol{B}_{21}\right]\boldsymbol{V}_{\text{bus}} = \boldsymbol{I}_{\text{bus}} \tag{4.12}$$

where
$$\begin{bmatrix} \boldsymbol{B}_{11} & \boldsymbol{B}_{12} \\ \boldsymbol{B}_{21} & \boldsymbol{B}_{22} \end{bmatrix} = \begin{bmatrix} \boldsymbol{A}_0 z_0^{-1} \boldsymbol{A}_0^T & \boldsymbol{K}_1 + \boldsymbol{A}_0 \boldsymbol{y}_{0\alpha} \boldsymbol{y}_{\alpha\alpha}^{-1} \\ -\left(\boldsymbol{K}_1^T + \boldsymbol{y}_{\alpha\alpha}^{-1} \boldsymbol{y}_{\alpha 0} \boldsymbol{A}_0^T\right) & \boldsymbol{y}_{\alpha\alpha}^{-1} \end{bmatrix} \tag{4.13}$$

Hence, the bus impedance matrix (Z_{bus}) of the new network can be written as

$$\boldsymbol{Z}_{\text{bus}} = \left[\boldsymbol{B}_{11} - \boldsymbol{B}_{12}\, \boldsymbol{B}_{22}^{-1}\, \boldsymbol{B}_{21}\right]^{-1} \tag{4.14}$$

To evaluate the inverse on the right hand side of Eq. (4.14), the following special matrix inversion formula is used.

$$[F+GHK]^{-1}=-F^{-1}G\left[H^{-1}+KF^{-1}G\right]^{-1}KF^{-1}+F^{-1} \tag{4.15}$$

Using Eq. (4.15) in Eq. (4.14), we get

$$Z_{\text{bus}}=B_{11}^{-1}+B_{11}^{-1}B_{12}\left[-B_{21}B_{11}^{-1}B_{12}+B_{22}\right]^{-1}B_{21}B_{11}^{-1} \tag{4.16}$$

Since $B_{11}=A_0\,z_0^{-1}\,A_0^T$, using Eq. (4.1),

$$B_{11}^{-1}=\left[A_0\,z_0^{-1}\,A_0^T\right]^{-1}=Z_{\text{bus(0)}} \tag{4.17}$$

Hence, the expression for Z_{bus} in Eq. (4.16) reduces to

$$Z_{\text{bus}}=Z_{\text{bus(0)}}-Z_{\text{bus(0)}}C_{12}D^{-1}C_{21}\,Z_{\text{bus(0)}} \tag{4.18}$$

where

$$C_{12}=K_1+\frac{A_0y_{0\alpha}}{y_{\alpha\alpha}} \tag{4.19}$$

$$C_{21}=K_1^T+\frac{y_{\alpha 0}A_0^T}{y_{\alpha\alpha}} \tag{4.20}$$

$$D=C_{21}Z_{bus(0)}C_{12}+\frac{1}{y_{\alpha\alpha}} \tag{4.21}$$

When there is no mutual coupling between the added element and the elements in the partial network, both $y_{0\alpha}$ and $y_{\alpha 0}$ become null vectors and hence, the bus impedance matrix of the new network is given by

$$Z_{\text{bus}}=Z_{\text{bus(0)}}-\frac{Z_{\text{bus(0)}}\,K_1\,K_1^T\,Z_{\text{bus(0)}}}{K_1^T\,Z_{\text{bus(0)}}\,K_1+z_{\alpha\alpha}} \tag{4.22}$$

where $Z_{\alpha\alpha}$ is the primitive impedance of the added element.

It may be noted that C_{12} and C_{21} in Eqs. (4.19) and (4.20) need not be the transpose of each other. If a phase shifting component is the added element, then $y_{0\alpha}\neq y_{\alpha 0}^T$ and hence $C_{12}\neq C_{21}^T$. Phase shifting components such as the tap changing transformer with a complex turns ratio have a primitive admittance matrix of order 2 × 2 which is asymmetric as well as singular. The presence of such asymmetry in the primitive admittance matrix will make the bus impedance matrix asymmetric. So far, no assumption has been made regarding the symmetry of the primitive impedance and the bus impedance matrices. Hence, Eqs. (4.18) and (4.22) are applicable even when the added element is a phase-shifting component. However, the singular nature of the primitive admittance matrix calls for a special computational procedure. For further discussion, unless otherwise stated, it is assumed that the added element is not a phase-shifting component.

Addition of a Link With Mutual Coupling Since the added element is not a phase-shifting component, therefore,

$$y_{0\alpha} = y_{\alpha 0}^T \text{ and } C_{12} = C_{21}^T \tag{4.23}$$

Hence Eq. (4.18) gets simplified to

$$Z_{\text{bus}} = Z_{\text{bus}(0)} - \frac{Z_{\text{bus}(0)} C_1 C_1^T Z_{\text{bus}(0)}}{C_1^T Z_{\text{bus}(0)} C_1 + (1/y_{\alpha\alpha})} \tag{4.24}$$

where

$$C_1 = K_1 + \frac{A_0 y_{0\alpha}}{y_{\alpha\alpha}} \tag{4.25}$$

Addition of a Link with no Mutual Coupling When there is no mutual coupling between the added link and the elements in the partial network, Eq. (4.23) reduces to Eq. (4.22).

As a special case, if the added link is between the reference bus and the bus q, then the vector K_1 reduces to

$$K_1 = [0 \quad \cdots \quad 0 \quad -1 \quad 0 \quad \cdots \quad 0]^T \tag{4.26}$$

This vector K_1 has to be used in the appropriate equations.

Addition of More Than one Link (With Mutual Coupling) When a number of links are added at a time, following the procedure similar to that of the addition of a single link, we get

$$Z_{\text{bus}} = Z_{\text{bus}(0)} - Z_{\text{bus}(0)} C_1 \hat{z}^{-1} C_1^T Z_{\text{bus}(0)} \tag{4.27}$$

where

$$\hat{z} = C_1^T Z_{\text{bus}(0)} C_1 + y_{\alpha\alpha}^{-1} \tag{4.28}$$

$$C_1 = K_1 + A_0 y_{0\alpha} y_{\alpha\alpha}^{-1} \tag{4.29}$$

K_1 is the bus incidence matrix corresponding to the added links. $y_{0\alpha}$ and $y_{\alpha\alpha}$ are also matrices here.

Addition of More Than One Link (With no Mutual Coupling) When the added links are not mutually coupled with any element in the partial network, Eq. (4.27) reduces to

$$Z_{\text{bus}} = Z_{\text{bus}(0)} - Z_{\text{bus}(0)} K_1 \hat{z}^{-1} K_1^T Z_{\text{bus}(0)} \tag{4.30}$$

where

$$\hat{z} = K_1^T Z_{\text{bus}(0)} K_1 + Z_{\alpha\alpha} \tag{4.31}$$

in which $z_{\alpha\alpha}$ is the primitive impedance matrix of the added elements.

4.4 ADDITION OF BRANCH(ES)

4.4.1 Addition of a Branch

Let the element α be added at the bus p creating a new bus q with the orientation from p to q, as shown in Fig. 4.3. This is termed as the addition of a branch. The word "branch" used here has a slightly different connotation from that used in Chapter 3 where the elements of a tree were defined as branches. We, however, stick to the terminology because of its widespread use in the literature while discussing bus impedance algorithm.

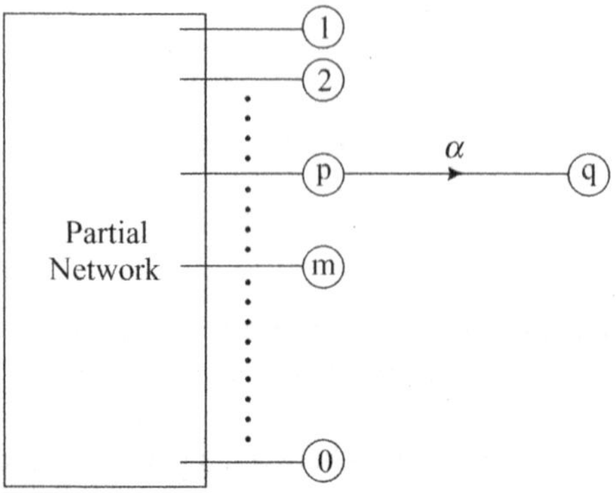

Fig. 4.3 *New network owing to the addition of a branch*

The bus incidence matrix for the new network is given by

$$A = \begin{bmatrix} & A_0 & & K_2 \\ & & & \\ 0 & \cdots & 0 & -1 \end{bmatrix} \tag{4.32}$$

where,
$$K_2 = [0 \;\; \cdots \;\; 0 \;\; 1 \;\; 0 \;\; \cdots \;\; 0]^T \tag{4.33}$$

For the new network, let the vectors V_{bus} and I_{bus} in the multiport description be partitioned as follows: Let V'_{bus}, and I'_{bus} be the variables corresponding to all the buses except the newly added bus q, whose variables are denoted by V_q and I_q. Thus

$$V_{bus} = \left[V'_{bus} \;\; V_q\right]^T ; \; I_{bus} = \left[I'_{bus} \;\; I_q\right]^T ; \tag{4.34}$$

Substituting (4.32) and Eq. (4.34) in Eqs. (4.6) and (4.7), we get

$$v_0 = A_0^T V'_{bus} \tag{4.35}$$

$$v_\alpha = K_2^T V'_{bus} - V_q \tag{4.36}$$

$$A_0 \, i_0 + K_2 i_\alpha = I'_{bus} \tag{4.37}$$

$$i_\alpha = -I_q \tag{4.38}$$

The primitive equations for the new network are given by Eq. (4.5). To obtain the relation between V_{bus} and I_{bus}, the other variables v_0, i_0, v_α, and i_α are to be eliminated from Eq. (4.5), and Eqs. (4.35) – (4.38).

Elimination of these variables yields (details omitted)

$$\begin{bmatrix} A_0 z_0^{-1} A_0^T & -\left(K_2 + A_0 y_{0\alpha} y_{\alpha\alpha}^{-1}\right) \\ -\left(K_2^T + y_{\alpha\alpha}^{-1} y_{\alpha 0} A_0^T\right) & -y_{\alpha\alpha}^{-1} \end{bmatrix} \begin{bmatrix} V'_{\text{bus}} \\ I_q \end{bmatrix} = \begin{bmatrix} I'_{\text{bus}} \\ -V_q \end{bmatrix} \tag{4.39}$$

Rearranging Eqs. (4.39) and using the relation $A_0 z_0^{-1} A_0^T = Z_{\text{bus}(0)}^{-1}$, we get

$$\begin{bmatrix} V'_{\text{bus}} \\ V_q \end{bmatrix} = \begin{bmatrix} Z_{\text{bus}(0)} & Z_{\text{bus}(0)} D_{12} \\ D_{21} Z_{\text{bus}(0)} & D_{21} Z_{\text{bus}(0)} D_{12} + (1/y_{\alpha\alpha}) \end{bmatrix} \begin{bmatrix} I'_{\text{bus}} \\ I_q \end{bmatrix} \tag{4.40}$$

where

$$D_{12} = K_2 + \frac{A_0 y_{0\alpha}}{y_{\alpha\alpha}} \tag{4.41}$$

$$D_{21} = K_2^T + \frac{y_{\alpha 0} A_0^T}{y_{\alpha\alpha}} \tag{4.42}$$

Hence, the bus impedance of the new network is given by

$$Z_{\text{bus}} = \begin{bmatrix} Z_{\text{bus}(0)} & Z_{\text{bus}(0)} D_{12} \\ D_{21} Z_{\text{bus}(0)} & D_{21} Z_{\text{bus}(0)} D_{12} + (1/y_{\alpha\alpha}) \end{bmatrix} \tag{4.43}$$

It should be noted that D_{12} and D_{21} in Eqs. (4.41) and (4.42) need not be the transpose of each other. Therefore, Eqs. (4.40) and (4.43) are very general and can be used even if the added element is a phase-shifting component. However, to simplify matters for subsequent discussions, it is assumed that the added element is not a phase-shifting component, so that $y_{0\alpha} = y_{\alpha 0}^T$.

Addition of a Branch with Mutual Coupling ***The assumption of the additional element being not a phase-shifting component leads to the relation*** $D_{12} = D_{21}^T = C_2$ (say). Hence, Eq. (4.43) reduces to

$$Z_{\text{bus}} = \begin{bmatrix} Z_{\text{bus}(0)} & Z_{\text{bus}(0)} C_2 \\ C_2^T Z_{\text{bus}(0)} & C_2^T Z_{\text{bus}(0)} C_2 + (1/y_{\alpha\alpha}) \end{bmatrix} \tag{4.44}$$

where

$$C_2 = K_2 + \frac{A_0 y_{0\alpha}}{y_{\alpha\alpha}} \tag{4.45}$$

Addition of a Branch with no Mutual Coupling When there is no mutual coupling between the added element and the elements in the partial network, Eq. (4.44) simplifies to

$$Z_{\text{bus}} = \begin{bmatrix} Z_{\text{bus}(0)} & Z_{\text{bus}(0)} K_2 \\ K_2^T Z_{\text{bus}(0)} & K_2^T Z_{\text{bus}(0)} K_2 + z_{\alpha\alpha} \end{bmatrix} \tag{4.46}$$

where $z_{\alpha\alpha}$ is the primitive impedance of the added element.

As a special case, the added branch may be from the reference bus. In such a case, the vector K_2 reduces to a null vector. The null vector has to be used in the appropriate equations.

Addition of More Than One Branch with Mutual Coupling A similar procedure can be followed when a number of branches are added at a time. In such a case,

$$Z_{\text{bus}} = \begin{bmatrix} Z_{\text{bus}(0)} & Z_{\text{bus}(0)}C_2 \\ C_2^T Z_{\text{bus}(0)} & C_2^T Z_{\text{bus}(0)}C_2 + y_{\alpha\alpha}^{-1} \end{bmatrix} \tag{4.47}$$

where the matrix

$$C_2 = K_2 + A_0 y_{0\alpha} y_{\alpha\alpha}^{-1} \tag{4.48}$$

in which K_2 is the sub-matrix of the bus incidence matrix corresponding to the added branches.

Addition of More Than one Branch with no Mutual Coupling When the added branches are not mutually coupled with any element in the partial network, Eq. (4.47) reduces to

$$Z_{\text{bus}} = \begin{bmatrix} Z_{\text{bus}(0)} & Z_{\text{bus}(0)}K_2 \\ K_2^T Z_{\text{bus}(0)} & K_2^T Z_{\text{bus}(0)}K_2 + z_{\alpha\alpha} \end{bmatrix} \tag{4.49}$$

in which $z_{\alpha\alpha}$ is the primitive impedance matrix of the added elements.

4.5 REMOVAL OF LINK(S)

Removal of a link from a network means the removal of the self and mutual admittances corresponding to the element from the primitive admittance matrix of the network. The effect can be realized by adding a fictitious link at exactly the same location of the network (as the link being removed). The fictitious link should have the self and mutual admittances of values exactly equal to those of the link to be removed, but with a negative sign; and its mutual admittance with the link to be removed should be zero. Now the same algorithm for the addition of a link can be used to include this fictitious link (thus effectively removing the original link). Because of the zero mutual coupling between the fictitious link and the link to be removed, it may be noticed that the product of the bus incidence matrix of the network and the mutual admittance vector corresponding to the fictitious element is equal to $-A_0 y_{0\alpha}$, where A_0 is the bus incidence matrix of the network with element α removed, and $y_{0\alpha}$ is the vector showing the mutual admittance between the element α and the other elements of the network. Also, the self admittance of the fictitious element is $-y_{\alpha\alpha}$. Therefore, the vector C_1 can still be computed from Eq. (4.25). Thus, when a link α is removed from the network whose bus impedance matrix is $Z_{\text{bus}(0)}$, the modified bus impedance matrix Z_{bus} is given by

$$Z_{\text{bus}} = Z_{\text{bus}(0)} - \frac{Z_{\text{bus}(0)} C_1 C_1^T Z_{\text{bus}(0)}}{C_1^T Z_{bus(0)} C_1 - (1/y_{\alpha\alpha})} \tag{4.50}$$

where
$$C_1 = K_1 + \frac{A_0 y_{0\alpha}}{y_{\alpha\alpha}} \tag{4.51}$$

Removal of a Link with no Mutual Coupling If the link to be removed is not mutually coupled with any other element in the network, Eq. (4.50) reduce to

$$Z_{bus} = Z_{bus(0)} - \frac{Z_{bus(0)} K_1 K_1^T Z_{bus(0)}}{K_1^T Z_{bus(0)} K_1 - z_{\alpha\alpha}} \tag{4.52}$$

where $z_{\alpha\alpha}$ is the primitive impedance of the element to be removed.

Removal of More Than One Link with Mutual Coupling Extending the method of removal of a single link (Eq. (4.50)), the bus impedance matrix for the removal of more than one link is obtained as

$$Z_{bus} = Z_{bus(0)} - Z_{bus(0)} C_1 \hat{z}^{-1} C_1^T Z_{bus(0)} \tag{4.53}$$

where
$$\hat{z} = C_1^T Z_{bus(0)} C_1 - y_{\alpha\alpha}^{-1} \tag{4.54}$$

$$C_1 = K_1 + A_0 y_{0\alpha} y_{\alpha\alpha}^{-1} \tag{4.55}$$

It may be noted that K_1, $y_{\alpha\alpha}$, and $y_{0\alpha}$ in Eqs. (4.53) – (4.55) are no longer vectors, but matrices.

Removal of More Than One Link with no Mutual Coupling When the links to be removed are not mutually coupled with any element in the partial network, Eq. (4.53) reduces to

$$Z_{bus} = Z_{bus(0)} - Z_{bus(0)} K_1 \hat{z}^{-1} K_1^T Z_{bus(0)} \tag{4.56}$$

where
$$\hat{z} = K_1^T Z_{bus(0)} K_1 - z_{\alpha\alpha} \tag{4.57}$$

in which $z_{\alpha\alpha}$ is the primitive impedance matrix of the links to be removed.

4.6 IMPLEMENTATION OF THE ALGORITHM

Consider the case of the addition of a link. If the added element has no mutual coupling, the vector $y_{0\alpha}$ becomes a null vector. In such a case, while computing $A_0\, y_{0\alpha}$ in Eq. (4.24), the bus incidence matrix A_0 need not be considered at all. If the added element is mutually coupled, it is enough to consider the elements in the coupled group alone, since the components of $y_{0\alpha}$ corresponding to the elements that are not in the couple group, are zero. While computing $A_0\, y_{0\alpha}$, the reduced vector $y_{0\alpha}$ denoted by $y_{0\alpha}$, and the reduced matrix A_0 denoted by A_0, can be used. The rows of $y_{0\alpha}$ and the columns of A_0 correspond to the elements having mutual coupling with the added element α. The order of the matrix to be inverted to find $y_{0\alpha}$ and $y_{\alpha\alpha}$ is now limited to the number of elements in the coupled group. All these remarks are applicable not only in the case of the addition of a link but also for all other cases discussed.

A word of caution is necessary regarding the orientations of the elements and the sign of the mutual impedance in the primitive impedance matrix of the coupled group. In circuit theory, the proper dot convention is normally followed to specify the sign of mutual impedances. But in power systems, mutual coupling arises when two transmission lines run parallel for a considerable distance. The sign of mutual impedance depends on the orientations assigned to them in the oriented graph. If the elements have like orientations when emanating from a bus, the mutual impedance will have a positive value; and if the element shave opposite orientations, the corresponding mutual impedance will have a negative value. All these points are brought out in the solved numerical examples at the end of the chapter.

4.7 SPECIAL CASES

4.7.1 Removal of Radial Line

When an element corresponding to a radial line is removed, one bus gets isolated and the number of buses in the network is reduced by one. From Eq. (4.44) it can be inferred that the modified bus impedance matrix due to the removal of the radial line can be obtained from the original bus impedance matrix by simply deleting the row and the column corresponding to the isolated bus. If the isolated bus is the reference bus itself, then the bus impedance matrix of the new network is indefinite.

4.7.2 Parameter Changes

When the parameter of an element is changed, the bus impedance matrix can be modified by simultaneously removing the element with the old parameter and adding an element with the revised parameter. This particular case has been illustrated through a numerical example (Example 4.3).

4.8 CONCLUSION

In this chapter, a generalized bus impedance algorithm has been developed based on certain fundamental concepts of the graph theory. The bus incidence matrix which shows how the various elements are interconnected, is essentially used at various stages. In other words, this development exploits the topology of the network. The merits of this algorithm are summarized below.

1. The algorithm for the addition of a link is a one-step procedure instead of a two-step procedure as in other algorithms.
2. Mutual couplings between the elements of the network do not pose any problem either at the addition or the removal stage.
3. The algorithm is amenable for the addition of the link and the branch either one at a time or in groups.
4. The algorithm is comprehensive so as to include the effects of phase-shifting components, changes in network parameters, etc.

Example 4.1 Compute the bus impedance matrix for the network shown in Fig. 4.4. Bus 1 can be taken as the reference bus, since there are no shunt connections to the ground in this case.

Solution: The oriented linear graph for the network is shown in Fig 4.5

Step 1: Element a is added. It is a branch from the reference bus. Then,

$$\mathbf{Z}_{\text{bus}} = \begin{matrix} & a \\ \mathbf{2} & [0.1] \end{matrix}$$

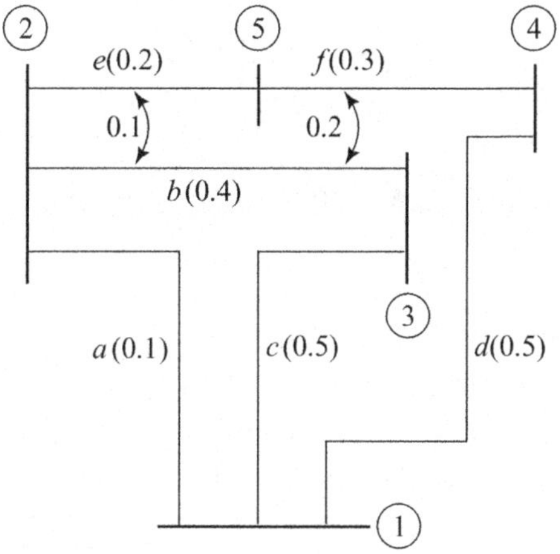

Fig. 4.4 *A five-bus example*

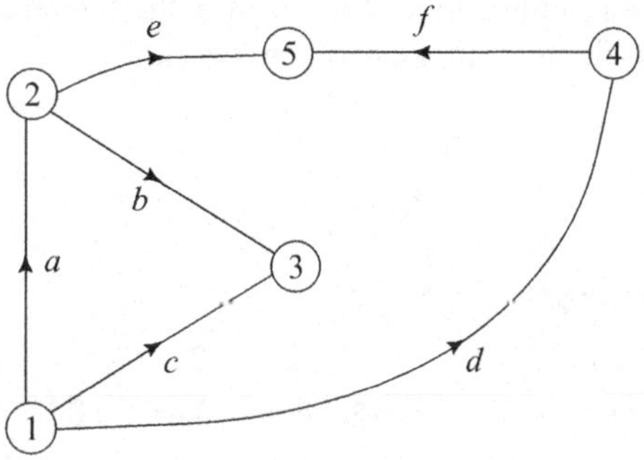

Fig. 4.5 *Oriented graph for the network in Fig 4.4*

Step 2: Element b is added. It is a branch from bus 2 to bus 3 and has no mutual coupling with the element in the partial network.

$$\mathbf{K}_2 = \begin{matrix} & b \\ \mathbf{2} & [1] \end{matrix}; \; z_{\alpha\alpha} = 0.4$$

Using Eq. (4.46), we get

$$\mathbf{Z}_{\text{bus}} = \begin{matrix} & \mathbf{2} & \mathbf{3} \\ \mathbf{2} & 0.1 & 0.1 \\ \mathbf{3} & 0.1 & 0.5 \end{matrix}$$

Step 3: Element c is added. It is a link between the reference buses 1 and 3, and has no mutual coupling with elements in the partial network.

Thus,

$$K_1 = \begin{matrix} \\ 2 \\ 3 \end{matrix}\overset{c}{\begin{bmatrix} 0 \\ -1 \end{bmatrix}}; \; z_{\alpha\alpha} = 0.5$$

$$Z_{\text{bus}(0)} K_1 = \begin{bmatrix} 0.1 & 0.1 \\ 0.1 & 0.5 \end{bmatrix} \begin{bmatrix} 0 \\ -1 \end{bmatrix} = \begin{bmatrix} -0.1 \\ -0.5 \end{bmatrix}$$

$$K_1^T Z_{\text{bus}(0)} K_1 = [0 \;\; -1] \begin{bmatrix} -0.1 \\ -0.5 \end{bmatrix} = 0.5$$

From Eq. (4.22)

$$Z_{\text{bus}} = \begin{bmatrix} 0.1 & 0.1 \\ 0.1 & 0.5 \end{bmatrix} - \left(\begin{bmatrix} -0.1 \\ -0.5 \end{bmatrix} [-0.1 \;\; -0.5] \right) \Big/ (0.5 + 0.5)$$

$$= \begin{bmatrix} 0.1 & 0.1 \\ 0.1 & 0.5 \end{bmatrix} - \begin{bmatrix} 0.01 & 0.05 \\ 0.05 & 0.025 \end{bmatrix}$$

$$= \begin{matrix} & 2 & 3 \\ 2 & 0.09 & 0.05 \\ 3 & 0.05 & 0.25 \end{matrix}$$

Step 4: Element d is added. It is a branch from the reference bus and has no mutual coupling with the elements in the partial network.

$$K_2 = \begin{matrix} \\ 2 \\ 3 \end{matrix}\overset{d}{\begin{bmatrix} 0 \\ 0 \end{bmatrix}}; \; z_{\alpha\alpha} = 0.5$$

Using Eq. (4.46), we get

$$Z_{\text{bus}} = \begin{matrix} & 2 & 3 & 4 \\ 2 & 0.09 & 0.05 & 0.0 \\ 3 & 0.05 & 0.25 & 0.0 \\ 4 & 0.0 & 0.0 & 0.5 \end{matrix}$$

Step 5: Element e is added. It is a branch from bus 2 creating a new bus 5. It has mutual coupling with the existing element b.

$$K_2 = \begin{matrix} \\ 2 \\ 3 \\ 4 \end{matrix}\overset{e}{\begin{bmatrix} 1 \\ 0 \\ 0 \end{bmatrix}}$$

Let z_c and y_c be the primitive impedance and the admittance matrices of the coupled group. Then,

$$z_c = \begin{matrix} & b & e \\ b & 0.4 & 0.1 \\ e & 0.1 & 0.2 \end{matrix}$$

Inverting z_c we get,

$$y_c = \begin{matrix} & b & e \\ b & 2.8570 & -1.4285 \\ e & -1.4285 & 5.7140 \end{matrix}$$

This gives

$$y_{0\alpha} = \overset{\quad a \quad c \quad d \quad b}{[0 \quad 0 \quad 0 \quad -1.4285]^T}\ ; \quad y_{\alpha\alpha} = 5.7140;$$

A_0 is given by

$$A_0 = \begin{matrix} & a & c & d & b \\ \mathbf{2} & -1 & 0 & 0 & 1 \\ \mathbf{3} & 0 & -1 & 0 & -1 \\ \mathbf{4} & 0 & 0 & -1 & 0 \end{matrix}$$

Clearly we can see that for calculating the product $A_0 y_{0\alpha}$, the inversion of the complete primitive impedance matrix and the use of the complete A_0 matrix is not necessary. Only the column of A_0 corresponding to the branch b (denoted by $\bar{A}_0$) and the element of $y_{0\alpha}$ corresponding to the branch b (denoted by $y_{0\alpha}$) are sufficient. Using A_0 and $\bar{y}_{0\alpha}$ in Eq. (4.45), we get

$$\boldsymbol{C}_2 = \begin{bmatrix} 1 \\ 0 \\ 0 \end{bmatrix} + \begin{bmatrix} 1 \\ -1 \\ 0 \end{bmatrix} [-1.4285] \frac{1}{5.7140} = \begin{bmatrix} 0.75 \\ 0.25 \\ 0.00 \end{bmatrix}$$

$$\boldsymbol{Z}_{\text{bus}(0)} \boldsymbol{C}_2 = \begin{bmatrix} 0.09 & 0.05 & 0.00 \\ 0.05 & 0.25 & 0.00 \\ 0.00 & 0.00 & 0.50 \end{bmatrix} \begin{bmatrix} 0.75 \\ 0.25 \\ 0.00 \end{bmatrix} = \begin{bmatrix} 0.08 \\ 0.10 \\ 0.00 \end{bmatrix}$$

$$\boldsymbol{C}_2^T \boldsymbol{Z}_{\text{bus}(0)} \boldsymbol{C}_2 = [0.75 \quad 0.25 \quad 0.00] \begin{bmatrix} 0.08 \\ 0.10 \\ 0.00 \end{bmatrix} = 0.085$$

$$\boldsymbol{C}_2^T \boldsymbol{Z}_{\text{bus}(0)} \boldsymbol{C}_2 + (1/y_{\alpha\alpha}) = 0.085 + (1/5.7140) = 0.26$$

Hence, using Eq. (4.44),

$$Z_{\text{bus}} = \begin{matrix} a & c & d & b \\ 0.09 & 0.05 & 0.00 & 0.08 \\ 0.05 & 0.25 & 0.00 & 0.10 \\ 0.00 & 0.00 & 0.50 & 0.00 \\ 0.08 & 0.10 & 0.00 & 0.26 \end{matrix}$$

Step 6: Element f is added. It is a link between buses **4** and **5** and has mutual coupling with element b.

$$K_1 = \begin{matrix} & f \\ 2 & \\ 3 & \\ 4 & \\ 5 & \end{matrix}\begin{bmatrix} 0 \\ 0 \\ 1 \\ -1 \end{bmatrix}; \quad z_c = \begin{matrix} & b & f \\ b & 0.4 & 0.2 \\ f & 0.2 & 0.3 \end{matrix}; \quad y_c = \begin{matrix} & b & f \\ b & 3.75 & -2.5 \\ f & -2.5 & 5.00 \end{matrix}$$

$$\bar{A}_0 = \begin{matrix} & b \\ 2 & 1 \\ 3 & -1 \\ 4 & 0 \\ 5 & 0 \end{matrix}; \quad \bar{y}_{0\alpha} = \overset{b}{[-2.5]}; \quad y_{\alpha\alpha} = 5.00;$$

Hence, using Eq. (4.25)

$$C_1 = K_1 + \frac{\bar{A}_0 \bar{y}_{0\alpha}}{y_{\alpha\alpha}} = \begin{bmatrix} -0.50 \\ 0.50 \\ 1.00 \\ -1.00 \end{bmatrix}$$

Using Eq. (4.24), we get,

$$Z_{\text{bus}} = Z_{\text{bus}(0)} - \frac{Z_{\text{bus}(0)} C_1 C_1^T Z_{\text{bus}(0)}}{C_1^T Z_{\text{bus}(0)} C_1 + (1/y_{\alpha\alpha})} = \begin{bmatrix} 0.080 & 0.050 & 0.050 & 0.055 \\ 0.050 & 0.250 & 0.000 & 0.100 \\ 0.050 & 0.000 & 0.250 & 0.125 \\ 0.055 & 0.100 & 0.125 & 0.1975 \end{bmatrix}$$

Example 4.2 For the network given in Example 4.1, calculate the modified bus impedance matrix when the self impedance of the line f is changed from 0.3 to 0.24, and also when the mutual impedance between the line b and f is changed from 0.2 to 0.0.

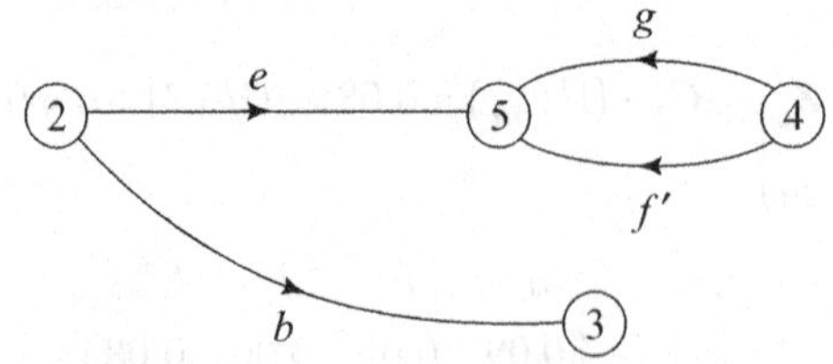

Fig. 4.6 *Oriented graph of Example 4.2*

Solution: This case can be treated as the simultaneous removal of the line f with the old impedance value (which can be viewed as the addition of a fictitious link f') and the addition of a line g with the new impedance value. Thus this case belongs to the case of addition of links, with mutual coupling. The fictitious link f' has the

self and the mutual admittances as the negative of the link to be removed, and has no mutual admittance with the link to be removed. Link g is not mutually coupled to the other elements. Fig. 4.6 shows the oriented graph of the coupled group.
The primitive admittance matrix of the coupled group is obtained as

$$\begin{array}{c} \\ b \\ e \\ f' \\ g \end{array}\begin{array}{c} \begin{array}{cccc} \quad b & \qquad e & \qquad f' & \qquad g \end{array} \\ \begin{bmatrix} 4.6153 & -2.3077 & -3.0769 & 0.0 \\ -2.3077 & 6.1538 & 1.5385 & 0.0 \\ -3.0769 & 1.5385 & -5.3846 & 0.0 \\ 0.0 & 0.0 & 0.0 & 4.1667 \end{bmatrix} \end{array}$$

$$\bar{y}_{0\alpha} = \begin{bmatrix} -3.0769 & 0.0 \\ 1.5385 & 0.0 \end{bmatrix}; \; y_{\alpha\alpha} = \begin{bmatrix} -5.3846 & 0.0 \\ 0.0 & 4.1667 \end{bmatrix}$$

Also, from the oriented graph in Fig. 4.6,

$$\bar{A}_0 = \begin{array}{c} \\ 2 \\ 3 \\ 4 \\ 5 \end{array}\begin{array}{c} \begin{array}{cc} b & \quad e \end{array} \\ \begin{bmatrix} 1 & 1 \\ -1 & 0 \\ 0 & 0 \\ 0 & -1 \end{bmatrix} \end{array}; \; K_1 = \begin{array}{c} \\ 2 \\ 3 \\ 4 \\ 5 \end{array}\begin{array}{c} \begin{array}{cc} f' & \quad g \end{array} \\ \begin{bmatrix} 0 & 0 \\ 0 & 0 \\ 1 & 1 \\ -1 & -1 \end{bmatrix} \end{array}$$

For this case,

$$y_{\alpha\alpha}^{-1} = \begin{array}{c} \\ f' \\ g \end{array}\begin{array}{c} \begin{array}{cc} f' & \qquad g \end{array} \\ \begin{bmatrix} -0.186 & 0.0 \\ 0.0 & 0.240 \end{bmatrix} \end{array}$$

Substituting the values of $\bar{y}_{0\alpha}$, $y_{\alpha\alpha}^{-1}$, $\bar{A}_0$, , and K_1 in Eq. (4.29), C_1 is obtained as

$$C_1 = \begin{bmatrix} 0.2855 & 0.0 \\ -0.5715 & 0.0 \\ 1.0 & 1.0 \\ -0.7140 & -1.0 \end{bmatrix}$$

Knowing $Z_{bus(0)}$ in Example 4.1, C_1 and $y_{\alpha\alpha}^{-1}$, $\hat{z}$ is obtained from Eq. (4.28) as

$$\hat{z} = \begin{bmatrix} -0.0369 & 0.1428 \\ 0.1428 & 0.4485 \end{bmatrix}$$

Hence,

$$\hat{z}^{-1} = \begin{bmatrix} 12.230 & 3.890 \\ 3.890 & -1.007 \end{bmatrix}$$

Finally, substituting the values of $\boldsymbol{Z}_{\text{bus}(0)}$ and the calculated $\boldsymbol{C}_1$ and $\hat{z}^{-1}$ in Eq. (4.27), the modified bus impedance matrix is obtained as

$$\boldsymbol{Z}_{\text{bus}} = \begin{array}{c} \\ \mathbf{2} \\ \mathbf{3} \\ \mathbf{4} \\ \mathbf{5} \end{array} \begin{array}{c} \begin{array}{cccc} \mathbf{2} & \mathbf{3} & \mathbf{4} & \mathbf{5} \end{array} \\ \begin{bmatrix} 0.0836 & 0.0420 & 0.0400 & 0.0592 \\ 0.0420 & 0.2400 & 0.5000 & 0.0740 \\ 0.0400 & 0.5000 & 0.2500 & 0.1300 \\ 0.0592 & 0.0740 & 0.1300 & 0.1920 \end{bmatrix} \end{array}$$

Problems

4.1 Using the building algorithm, construct the bus impedance matrix $\boldsymbol{Z}_{\text{bus}}$ for the network in Fig. P.4.1. Choose bus **3** as the reference bus. The system data is given below:

From bus	*To bus*	*X*
1	2	0.5
1	3	0.5
2	3	0.6
2	4(1)	0.4
2	4(2)	0.4
3	4	0.5

Mutual coupling between 2-4(1) and 2-4(2) may be taken as 0.1

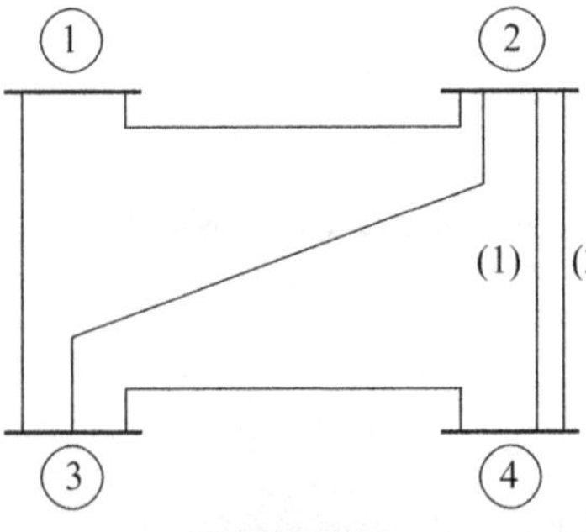

Fig. P4.1

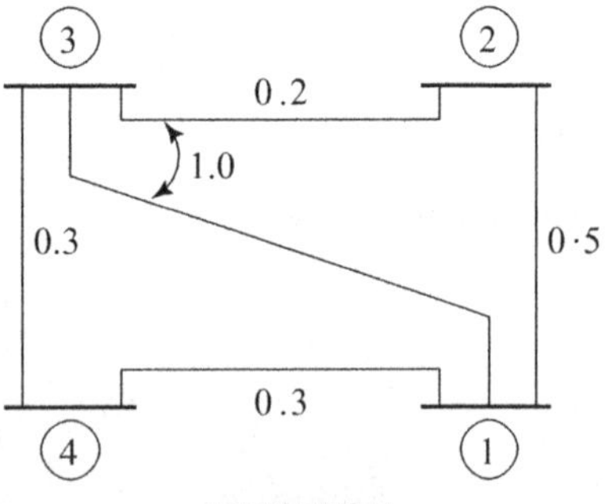

Fig. P4.2

4.2 Using the building algorithm, construct $\boldsymbol{Z}_{\text{bus}}$ for the system in Fig. P.4.2. Choose 4 as the reference bus.

4.3 Modify the $\boldsymbol{Z}_{\text{bus}}$ of Problem 4.1 when
  (a) line 1-2 is removed,
  (b) line 2-4(1) is removed,
  (c) mutual impedance between lines 2-4(1) and 2-4(2) is changed to 0.15, and
  (d) another line is added between buses **1** and **2** having self-impedance of $j0.4$.

4.4 Modify $\boldsymbol{Z}_{\text{bus}}$ of Problem 4.2 if the mutual reactance between lines 1-3 and 2-3 is absent.

5 SHORT CIRCUIT STUDIES

5.1 INTRODUCTION

When an abnormal condition arises in a power system such as a fault, an insulator flashover, or a lightning stroke to the transmission tower, high currents flow in the system depending on the nature and location of the fault. These currents are sensed through relays that take about half a cycle or so, and with proper coordination among the relays, faulty sections of the power system are isolated quickly by actuation of the circuit breakers in another three to four cycles. Thus, the time scale of interest is of the order of 100 milliseconds. Any delay in the operation of the circuit breaker action might result in creating instability in the system. When the circuit breakers operate under abnormal conditions, high currents have to be interrupted and the circuit breaker must be able to withstand such a high MVA interruption. In the planning and design stage, studies are done to determine the short-circuit currents throughout the network for various types of faults. Based on these studies, the short-circuit MVA at various points in the network can be calculated. These short-circuit levels provide the basis for specifying the interrupting capacities of the circuit breakers. Constant changes in the configuration of the transmission network alter both the short-circuit levels and the short-circuit currents at various points in the system. Hence, when any major modifications to the power system are made, these computations must be repeated to determine the adequacy of the protective equipment. The short-circuit studies cover both the planning, and to a certain extent, the operation phase. In this chapter, techniques for the determination of bus voltages and line currents under short-circuit conditions are developed for various types of faults, e.g., three-phase, line-to-ground, line-to-line, line-line-to-ground, and three-phase to ground. Based on these studies, the planning engineer can specify both the circuit breaker rating and the relay settings.

Analysis is done on the basis of a steady state AC model by representing the generators as constant voltage sources in series with appropriate machine reactances, transient or sub-transient. The transmission network is also assumed to be in steady state. The analysis can be carried out either in terms of the phase quantities or in terms of the sequence quantities. In arriving at a mathematical model for the short-circuit studies, a number of assumptions are made which simplify the formulation of the problem, and in addition, facilitate the solution without introducing significant inaccuracies in the results. The main assumptions are as follows:

1. The normal loads, line-charging capacitances, and other shunt connections to the ground are neglected. This is based on the fact that the fault circuit

has predominantly lower impedance than the shunt impedance. The saving in computational effort as a result of this assumption justifies the slight loss in accuracy.

2. The generator is represented by a voltage source in series with a reactance that is taken as the sub-transient or transient reactance. Such a representation is adequate to compute the magnitudes of currents in the first 3-4 cycles after the occurrence of the fault.
3. All the voltages of the voltage sources representing the generators are assumed to be equal. This is based on the fact that all the currents flowing in the network can be considered as zero due to assumption (1).
4. All the transformers are considered to be at their nominal taps.
5. If the resistances of the transmission lines are smaller than the reactances by a factor of six or more, the resistances are neglected. This obviates the need for complex arithmetic and the system will then consist of voltage sources and reactances only.

5.2 THREE-PHASE BALANCED NETWORKS AND FAULTS

The short-circuit problem essentially consists of determining the steady-state solution of a linear network with a balanced three-phase sinusoidal excitation. The linear network comprises three main sub-networks: (1) the transmission network, (2) the generator network, and (3) the fault. The transmission network and the generator network consist of an interconnection of three-phase balanced components. For example, if the lines are assumed to be completely transposed, the impedance matrix for a three-phase transmission line is of the form

$$\begin{bmatrix} z_s & z_m & z_m \\ z_m & z_s & z_m \\ z_m & z_m & z_s \end{bmatrix} \tag{5.1}$$

This type of symmetry is called 'complete' symmetry. The three-phase transformer also has a similar type of symmetry. The synchronous generator is represented by a three-phase balanced source in series with the impedances. For a cylindrical rotor machine, the impedance matrix is of the same form as shown in Eq. (5.1), except for the fact that the mutual inductance has a negative value.

If we restrict our attention to shunt type faults, the fault can be one of the following:

1. Symmetrical three-phase-to-ground fault
2. Symmetrical three-phase fault
3. Single-line-to-ground (LG) fault
4. Line-to-line (LL) fault
5. Double-line-to-ground (LLG) fault

Faults of the type (1) and (2) have a symmetric impedance and/or admittance representation. Hence, for the faults of these two types, i.e., three-phase to ground and three-phase fault, we have a three-phase network with balanced three-phase components only, and balanced excitation. Analysis of such balanced three-phase networks can be carried out on the basis of a per phase equivalent, except that the impedances are now replaced by what are called the 'positive sequence' impedances. This fact will be demonstrated mathematically in the next section. In the case of unsymmetrical faults of the type (3), (4), and (5), such a simplified analysis is not possible and the analysis is either carried out on a three-phase basis itself or through sequence components.

5.3 SYMMETRICAL COMPONENTS

The voltage-current relation of the network can be written as:

$$\begin{bmatrix} V^a \\ V^b \\ V^c \end{bmatrix} = \begin{bmatrix} z_s & z_m & z_m \\ z_m & z_s & z_m \\ z_m & z_m & z_s \end{bmatrix} \begin{bmatrix} I^a \\ I^b \\ I^c \end{bmatrix} \tag{5.2}$$

In symbolic notation,

$$V^{a,b,c} = z^{a,b,c} \, I^{a,b,c} \tag{5.3}$$

Let us define a transformation of variables as follows

$$V^S = S_3 V^{a,b,c}; \qquad I^S = S_3 I^{a,b,c}; \tag{5.4}$$

where

$$S_3 = \frac{1}{\sqrt{3}} \begin{bmatrix} 1 & 1 & 1 \\ 1 & a & a^2 \\ 1 & a^2 & a \end{bmatrix}; \qquad a = 1\angle 120° = -0.5 + j0.866 \tag{5.5}$$

This is known as the symmetric component transformation. S_3 has the interesting property that

$$S_3^{-1} = \left(S_3^*\right)^T = \frac{1}{\sqrt{3}} \begin{bmatrix} 1 & 1 & 1 \\ 1 & a^2 & a \\ 1 & a & a^2 \end{bmatrix} \tag{5.6}$$

Such a matrix is termed as a unitary matrix.
Pre-multiplying both sides of the Eq. (5.4) by S_3, we can re-arrange it as

$$S_3 V^{a,b,c} = S_3 \, z^{a,b,c} S_3^{-1} S_3 \, I^{a,b,c}$$

More compactly,

$$V^s = z^s I^s \tag{5.7}$$

where

$$z^S = S_3 z^{a,b,c} S_3^{-1};$$

Simplifying, z^s turns out to be

$$z^S = \begin{bmatrix} z_s + 2z_m & 0 & 0 \\ 0 & z_s - z_m & 0 \\ 0 & 0 & z_s - z_m \end{bmatrix} = \begin{bmatrix} z^0 & 0 & 0 \\ 0 & z^1 & 0 \\ 0 & 0 & z^2 \end{bmatrix} \tag{5.8}$$

z^0, z^1, and z^2 are termed the zero sequence, the positive sequence and the negative impedance respectively, and we denote z^s by the symbol $z^{0,1,2}$. We thus see that the symmetric component transformation diagonalizes the matrix $z^{a,b,c}$. Fortescue (in the classic paper of 1918) was the first to propose the symmetrical component transformation.

The transformed voltage and current variables $\boldsymbol{V}^s$ and $\boldsymbol{I}^s$ are denoted by $\boldsymbol{V}^{0,1,2}$ and $\boldsymbol{I}^{0,1,2}$ respectively, and are given by

$$\boldsymbol{V}^{0,1,2} = \begin{bmatrix} V^0 \\ V^1 \\ V^2 \end{bmatrix} = \frac{1}{\sqrt{3}} \begin{bmatrix} V^a + V^b + V^c \\ V^a + aV^b + a^2V^c \\ V^a + a^2V^b + aV^c \end{bmatrix} \tag{5.9}$$

Similarly,

$$\boldsymbol{I}^{0,1,2} = \begin{bmatrix} I^0 \\ I^1 \\ I^2 \end{bmatrix} = \frac{1}{\sqrt{3}} \begin{bmatrix} I^a + I^b + I^c \\ I^a + aI^b + a^2I^c \\ I^a + a^2I^b + aI^c \end{bmatrix} \tag{5.10}$$

In terms of the new variables, Eq. (5.3) becomes

$$V^0 = z^0 I^0; \quad V^1 = z^1 I^1; \quad V^2 = z^2 I^2; \tag{5.11}$$

i.e., we have three uncoupled equations. If we interpret them in terms of networks, we have the three sequence networks: the zero, the positive, and the negative sequence networks which have no coupling between them (Fig. 5.1).

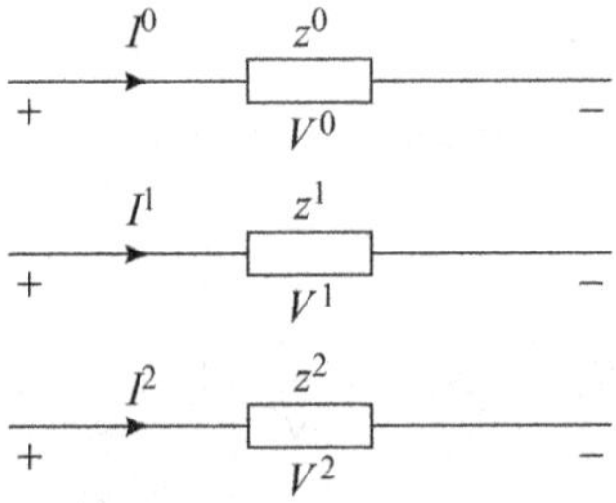

Fig. 5.1 *Three-phase components after decoupling into sequence components*

It should be noted that if the self impedances of the three phases are different and the mutual impedances are also different from each other, then this decoupling does not take place. In such a case, the impedance matrix takes the form

$$z^{a,b,c} = \begin{bmatrix} z_{saa} & z_{mab} & z_{mac} \\ z_{mab} & z_{sbb} & z_{mbc} \\ z_{mac} & z_{mbc} & z_{scc} \end{bmatrix} \tag{5.12}$$

The matrix is still symmetric. However, when subjected to sequence component transformation as in Eq. (5.7), it takes the following form:

$$z^S = S_3 z^{a,b,c} S_3^{-1}$$

$$= \frac{1}{3}\begin{bmatrix} \begin{matrix}(z_{saa}+z_{Sbb}+z_{Scc}) \\ +2(z_{mab}+z_{mbc}+z_{mac})\end{matrix} & \begin{matrix}(z_{saa}+a^2 z_{sbb}+az_{scc}) \\ -(z_{mbc}+az_{mab}+a^2 z_{mac})\end{matrix} & \begin{matrix}(z_{saa}+az_{sbb}+a^2 z_{scc}) \\ -(z_{mbc}+az_{mac}+a^2 z_{mab})\end{matrix} \\ \begin{matrix}(z_{saa}+az_{sbb}+a^2 z_{scc}) \\ -(z_{mbc}+az_{mac}+a^2 z_{mab})\end{matrix} & \begin{matrix}(z_{saa}+z_{sbb}+z_{scc}) \\ -(z_{mab}+z_{mbc}+z_{mac})\end{matrix} & \begin{matrix}(z_{saa}+a^2 z_{sbb}+az_{scc}) \\ +2(z_{mbc}+az_{mab}+a^2 z_{mac})\end{matrix} \\ \begin{matrix}(z_{saa}+a^2 z_{sbb}+az_{scc}) \\ -(z_{mbc}+az_{mab}+a^2 z_{mac})\end{matrix} & \begin{matrix}(z_{saa}+az_{sbb}+a^2 z_{scc}) \\ +2(z_{mbc}+az_{mac}+a^2 z_{mab})\end{matrix} & \begin{matrix}(z_{saa}+z_{sbb}+z_{scc}) \\ -(z_{mab}+z_{mbc}+z_{mac})\end{matrix} \end{bmatrix} \quad (5.13)$$

Quite clearly, the decoupling takes place only if

$$z_{saa} = z_{sbb} = z_{scc} \text{ and } z_{mab} = z_{mac} = z_{mbc}$$

In the admittance form, if the admittance matrix in the phase quantities is $y^{a,b,c}$, then in the symmetric component variables, it is given by

$$\boldsymbol{y}^{0,1,2} = \boldsymbol{S}_3\, \boldsymbol{y}^{a,b,c} \boldsymbol{S}_3^{-1} = \begin{bmatrix} y^0 & 0 & 0 \\ 0 & y^1 & 0 \\ 0 & 0 & y^2 \end{bmatrix} \quad (5.14)$$

where $\quad y^0 = \frac{1}{z^1},\ y^1 = \frac{1}{z^1},\ \text{and } y^2 = \frac{1}{z^2}.$

Therefore, the application of the sequence component transformation to a three-phase network consisting of balanced three-phase passive components results in three uncoupled networks, namely the positive sequence, the negative sequence, and the zero sequence networks. The topology of all three networks is the same as the per phase equivalent, except that the impedances correspond to the respective sequence quantities.

Let us now consider the three-phase voltage source to be balanced taking V^a as the reference phasor and the phase sequence as a-b-c.

Then,

$$V^a = |V^a|\angle 0^\circ;\ V^b = |V^a|\angle -120^\circ;\ V^c = |V^a|\angle -240^\circ \quad (5.15)$$

If θ is the phase angle between $\boldsymbol{V}^a$ and $\boldsymbol{I}^a$, then for a balanced network,

$$I^a = |I^a|\angle\theta;\ I^b = |I^a|\angle(\theta-120)^\circ;\ I^c = |I^a|\angle(\theta-240)^\circ \quad (5.16)$$

The sequence voltages and currents then become

$$V^0 = 0;\ V^1 = \sqrt{3}\,V^a;\ V^2 = 0 \tag{5.17}$$

$$I^0 = 0;\ I^1 = \sqrt{3}\,I^a;\ I^2 = 0$$

i.e., only positive sequence currents and voltages exist. There is no source in the negative and zero sequence network. Hence, we need to consider only the positive sequence network and its equations. Now for the positive sequence network, we have

$$V^1 = z^1 I^1 \tag{5.18}$$

Since

$$V^1 = \sqrt{3}V^a;\quad I^1 = \sqrt{3}I^a;\quad I^a = \frac{V^a}{z^1} \tag{5.19}$$

This means that a balanced three-phase network with balanced excitation can be analyzed by considering only the positive sequence network and applying the '*a*' phase voltage which will yield the corresponding '*a*' phase current. Other phase currents can be obtained using Eq. (5.16).

5.4 SEQUENCE IMPEDANCES FOR SYMMETRICAL FAULTS

5.4.1 Symmetrical Three-phase to Ground Fault

The general fault impedance diagram is shown in Fig. 5.2(a). There are five nodes in all — 'a', 'b', 'c', '0', and 'n'—the node connecting the three fault impedances (z_F) to the ground impedance z_g. We first eliminate this last node 'n', to get a three-port impedance description of Fig. 5.2(b). This can be obtained by writing the three-loop equations in Fig. 5.2(c) noting that $I^a + I^b + I^c$ flows through z_g.

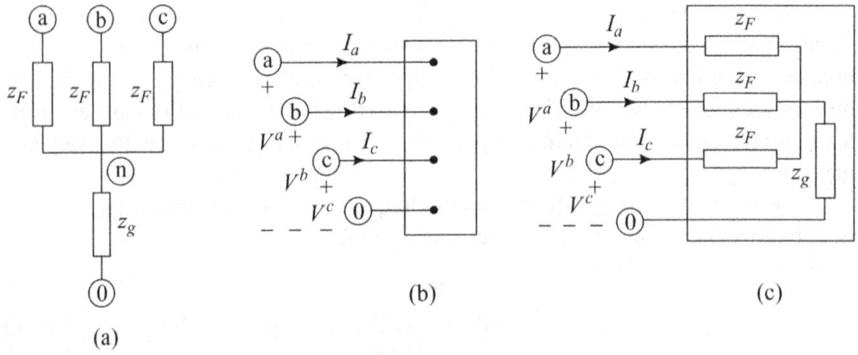

Fig. 5.2 *Representation of symmetrical three-phase to ground fault*

Hence,

$$\begin{bmatrix} V^a \\ V^b \\ V^c \end{bmatrix} = \begin{bmatrix} z_F + z_g & z_g & z_g \\ z_g & z_F + z_g & z_g \\ z_g & z_g & z_F + z_g \end{bmatrix} \begin{bmatrix} I^a \\ I^b \\ I^c \end{bmatrix} \tag{5.20}$$

$$\boldsymbol{V}^{a,b,c} = \boldsymbol{Z}^{a,b,c} \boldsymbol{I}^{a,b,c}$$

Application of the symmetrical component transformation to $\boldsymbol{Z}^{a,b,c}$ yields

$$\boldsymbol{Z}^{0,1,2} = \begin{bmatrix} z^0 & 0 & 0 \\ 0 & z^1 & 0 \\ 0 & 0 & z^2 \end{bmatrix} = \begin{bmatrix} z_F + 3z_g & 0 & 0 \\ 0 & z_F & 0 \\ 0 & 0 & z_F \end{bmatrix} \tag{5.21}$$

Thus the positive sequence impedance is only z_F and is independent of z_g. $\boldsymbol{Y}^{0,1,2}$ which is the inverse of $\boldsymbol{Z}^{0,1,2}$, is given by

$$\boldsymbol{Y}^{0,1,2} = \begin{bmatrix} y^0 & 0 & 0 \\ 0 & y^1 & 0 \\ 0 & 0 & y^2 \end{bmatrix} = \begin{bmatrix} (z_F + 3z_g)^{-1} & 0 & 0 \\ 0 & (z_F^{-1}) & 0 \\ 0 & 0 & (z_F^{-1}) \end{bmatrix} \tag{5.22}$$

The positive sequence admittance is thus $y^1 = \dfrac{1}{z_F} = y_F$.

5.4.2 Sequence Impedances of Symmetrical Three-phase Fault (Not Involving Ground)

The fault impedance diagram is given in Fig. 5.3(a). Since the ground bus is isolated, the impedance matrix corresponding to the three-port representation of Fig. 5.3(b) does not exist. However, the admittance form does exist. This is derived by considering the limiting case of $z_g \to \infty$. In phase quantities, from Eq. (5.20)

$$\boldsymbol{I}^{a,b,c} = \boldsymbol{Y}^{a,b,c} \boldsymbol{V}^{a,b,c} \tag{5.23}$$

where

$$\boldsymbol{Y}^{a,b,c} = \left(\boldsymbol{Z}^{a,b,c}\right)^{-1} = \frac{1}{3} \begin{bmatrix} y_t + 2y_F & y_t - y_F & y_t - y_F \\ y_t - y_F & y_t + 2y_F & y_t - y_F \\ y_t - y_F & y_t - y_F & y_t + 2y_F \end{bmatrix} \tag{5.24}$$

with

$$y_t = \frac{1}{z_F + 3z_g}$$

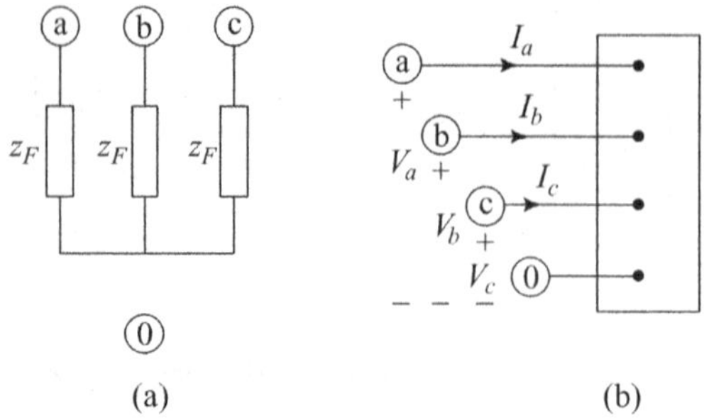

Fig. 5.3 *Representation of symmetrical three-phase fault*

$Y_F^{a,b,c}$ can be derived from Fig. 5.2(a), by writing node equations at the buses 'a', 'b', 'c', and 'n'. Since there is no injection at bus 'n', it can be eliminated using Kron reduction. If we now let $z_g = \infty$, then $\boldsymbol{Y}^{a,b,c}$ becomes

$$\boldsymbol{Y}^{a,b,c} = \frac{1}{3}\begin{bmatrix} 2y_F & -y_F & -y_F \\ -y_F & 2y_F & -y_F \\ -y_F & -y_F & 2y_F \end{bmatrix} \tag{5.25}$$

Application of the symmetrical component transformation to Eq. (5.24) yields

$$\boldsymbol{Y}^{0,1,2} = \begin{bmatrix} 0 & 0 & 0 \\ 0 & y_F & 0 \\ 0 & 0 & y_F \end{bmatrix} = \begin{bmatrix} y^0 & 0 & 0 \\ 0 & y^1 & 0 \\ 0 & 0 & y^2 \end{bmatrix} \tag{5.26}$$

We could have also obtained the above expression by letting $Z_g = \infty$ in Eq. (5.21). The positive sequence admittance is again y_F.

We have now completed the representation of the power system in terms of its sub-networks, viz. the generator network, the transmission network, and the fault representation for symmetrical fault analysis. To summarize,

1. The transmission network is replaced by its positive sequence network
2. The phase 'a' voltage is taken as the excitation to this network
3. At the point of fault, we connect the fault impedance z_F if it is a three-phase to ground fault, and fault admittance y_F if it is a three-phase fault not involving ground (as the impedance matrix does not exist in this case). Thus both the negative and the zero sequence networks are not needed for symmetrical fault calculations.

5.5 SYMMETRICAL FAULT CALCULATIONS

Let the transmission network consist of $(n + 1)$ buses including the ground bus denoted by 0. Since the ground bus is isolated, $\boldsymbol{Z}_{\text{bus}}$ with the ground as the reference, does not exist. The transmission network consists of positive sequence impedances only. There is no loss of generality if we assume that the first m buses are the buses where generators are connected. Since the generator representation consists of a constant voltage source corresponding to phase 'a' in series with the machine reactance, and since by assumption, all these voltage sources are assumed to be equal, we may augment the network by the generator representation. This is shown in Fig. 5.4(a) where the first m buses are augmented by the machine positive sequence reactances connected to a common fictitious node 0′, and the common generator voltage source corresponding to phase 'a' is connected between 0′ and 0.

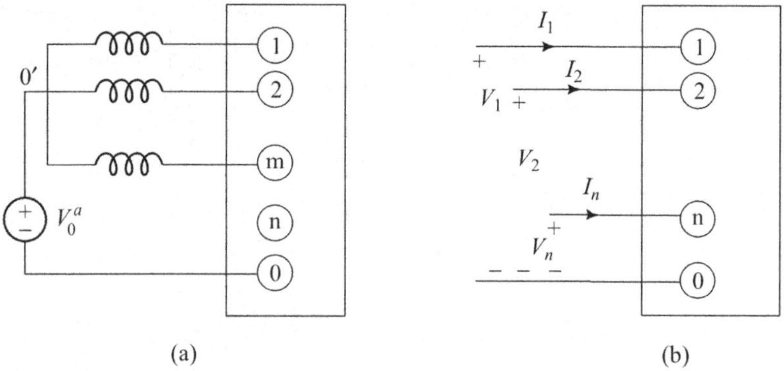

Fig. 5.4 *Power system representation for short-circuit studies*

As an intermediate step in our procedure of fault calculations, we shall get an n-port description of the transmission-generator network as shown in Fig. 5.4(b). It is obtained in two stages as described below:

1. First, consider that the voltage source V_0^a is shorted, i.e., nodes 0 and 0′ merge. Then for the resulting passive network, $\boldsymbol{Z}_{\text{bus}}$ can be obtained with the ground bus as the reference, as

$$V_{\text{bus}} = Z_{\text{bus}} I_{\text{bus}} \tag{5.27}$$

2. Now introduce the voltage source between 0 and 0′. This results in raising all the bus voltages by V_0^a. Hence, the modified n-port description is obtained adding V_0^a for all the equations in Eq. (5.27), i.e.,

$$\boldsymbol{V}_{\text{bus}} = \boldsymbol{Z}_{\text{bus}} \boldsymbol{I}_{\text{bus}} + \boldsymbol{b} V_0^a \quad \text{where } \boldsymbol{b} = [1 \;\; 1 \;\; \cdots \;\; \cdots \;\; 1]^T \tag{5.28}$$

To this representation we shall now apply the symmetrical faults.

5.5.1 Symmetrical Three-phase to Ground Fault

Calculations in Impedance Form Suppose the fault occurs at the pth bus. Let the fault network description (Fig. 5.5(a)) in the positive sequence impedance form be

$$V_F = z_F I_F \tag{5.29}$$

The pth port is terminated with this impedance (Fig. 5.5(b)).

The constraints between the variables of the fault impedance and the port variables are written from inspection as

$$I_{p(F)} = -I_F;\ V_{p(F)} = V_F;\ I_{i(F)} = 0;\ V_{i(F)} = \text{unknown for } i = 1, 2, \cdots n \text{ and } i \neq p \quad (5.30)$$

where $I_{p(F)}$ and $V_{p(F)}$ are the port currents and voltages respectively, under faulted conditions.

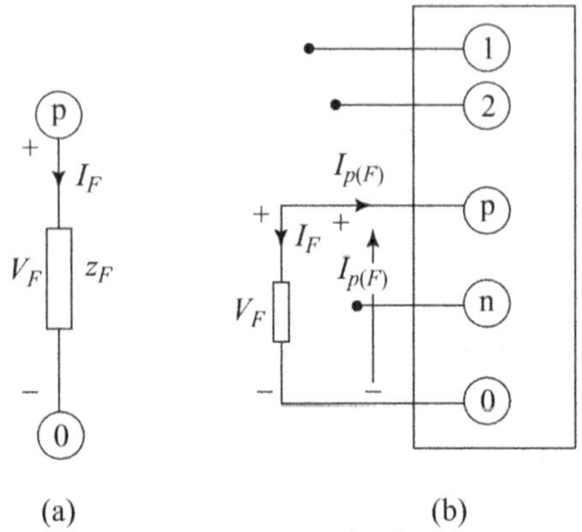

Fig. 5.5 *Fault network representation and the power system*

The n-port description form Eq. (5.28) is

$$\begin{aligned}
V_{1(F)} &= Z_{11}I_{1(F)} + \cdots + Z_{1p}I_{p(F)} + \cdots + Z_{1n}I_{n(F)} + V_0^a \\
V_{2(F)} &= Z_{21}I_{1(F)} + \cdots + Z_{2p}I_{p(F)} + \cdots + Z_{2n}I_{n(F)} + V_0^a \\
&\ \ \vdots \\
V_{p(F)} &= Z_{p1}I_{1(F)} + \cdots + Z_{pp}I_{p(F)} + \cdots + Z_{pn}I_{n(F)} + V_0^a \\
&\ \ \vdots \\
V_{n(F)} &= Z_{n1}I_{1(F)} + \cdots + Z_{np}I_{p(F)} + \cdots + Z_{nn}I_{n(F)} + V_0^a
\end{aligned} \quad (5.31)$$

Using the relations in Eqs (5.29) and (5.30) in the pth row of Eq. (5.31), we obtain

$$z_F I_F = -Z_{pp}I_F + V_0^a \quad \Rightarrow I_F = \frac{V_0^a}{Z_{pp} + z_F} \quad (5.32)$$

Also,

$$V_F = V_{p(F)} = z_F I_F = z_F \left(Z_{pp} + z_F\right)^{-1} V_0^a. \quad (5.33)$$

The other bus voltages are obtained using Eqs (5.32) and (5.33) in Eq. (5.31) as

$$V_{iF} = V_0^a - Z_{ip}\left(Z_{pp} + z_F\right)^{-1} V_0^a \quad i = 1,\ 2, \ldots,\ n \quad i \neq p \quad (5.34)$$

This determines all the bus voltages in the system that in turn will determine the line currents in all the lines by the elementary application of the Ohm's law.

Calculations in Admittance Form If the fault is in the admittance form, the fault description is

$$I_F = y_F V_F \tag{5.35}$$

The port constraints are as given earlier in Eq. (5.30). Instead of substituting for V_F, we now substitute for I_F in the pth row of Eq. (5.30) using Eq. (5.35), and get

$$V_{p(F)} = -Z_{pp} y_F V_{p(F)} + V_0^a \tag{5.36}$$

Hence,

$$V_{p(F)} = \left(1 + Z_{pp} y_F\right)^{-1} V_0^a \tag{5.37}$$

and

$$I_F = y_F \left(1 + Z_{pp} y_F\right)^{-1} V_0^a \tag{5.38}$$

Voltages at the other buses (except the pth bus) can be obtained as

$$V_{i(F)} = V_0^a + Z_{ip} I_{p(F)} = V_0^a - Z_{ip} I_F = V_0^a - Z_{ip} y_F \left(1 + Z_{pp} y_F\right)^{-1} V_0^a \tag{5.39}$$

5.5.2 Symmetrical Three-phase Fault (Not Involving Ground)

Since there is no impedance description for the fault, we only have the positive sequence admittance description form Eq. (5.26), i.e.,

$$I_F = y_F V_F$$

Since it is the same as Eq. (5.35), the currents and voltages at the different buses can be obtained using the Eqs (5.37)–(5.39). For symmetrical fault, the symbolic formulation for the admittance description of a fault has been included because the procedure will be useful and valid when we discuss unsymmetrical faults not involving the ground.

5.5.3 Some Comments on Z_{bus} and Y_{bus} Formulation for Short-Circuit Studies

The use of $\boldsymbol{Z}_{\text{bus}}$ for short-circuit calculations is seen to involve only trivial computations. Most of the effort will be expended in constructing the $\boldsymbol{Z}_{\text{bus}}$ using the building algorithm. Once this is done, the rest of the computations are straightforward. Prior to the advent of the building algorithm, the method using $\boldsymbol{Y}_{\text{bus}}$ was used. But this requires the solution of simultaneous linear equations as the following development shows.

For $\boldsymbol{Y}_{\text{bus}}$ formulation, we augment the transmission network by a Norton equivalent of the generator representation, i.e., at each generator bus, we have a current source V_0^a / z_j in parallel with an impedance $z_j = (j = 1, 2, \ldots, m)$ corresponding to the transient or the subtransient reactance of the generator. The admittance description of the n-port transmission generator network is now given by

$$\boldsymbol{I}_{\text{bus}} = \boldsymbol{Y}_{\text{bus}} \boldsymbol{V}_{\text{bus}} - \boldsymbol{I}_0^a \quad \text{where,} \quad \boldsymbol{I}_0^a = \begin{bmatrix} \dfrac{V_0^a}{z_1} & \dfrac{V_0^a}{z_2} & \cdots & \dfrac{V_0^a}{z_m} & 0 & 0 & \cdots & 0 \end{bmatrix}^T \tag{5.40}$$

Now if we apply the fault at a bus other than the generator bus in either the impedance or the admittance form, and impose the resulting constraints of Eq. (5.30) in Eq. (5.40), we have the following set of unknowns:

(a) Fault in impedance form: $V_{iF}\,(i = 1, 2, \ldots, n)\,(i \neq p)$ and $I_{p(F)}$
(b) Fault in admittance form: $V_{iF}\,(i = 1, 2, \ldots, n)$

Thus in either case, these are n unknowns and n simultaneous linear equations. To solve these unknowns for a large network on a computer, some kind of an iterative solution will have to be used. This is quite unattractive as compared to the $\boldsymbol{Z}_{\text{bus}}$ method where, having constructed $\boldsymbol{Z}_{\text{bus}}$, the remaining computations are trivial.

With the popularity of the sparse matrix techniques however, the $\boldsymbol{Y}_{\text{bus}}$ can be stored in the form of triangularized factors from which the elements of the $\boldsymbol{Z}_{\text{bus}}$ matrix can be retrieved easily. Thus the advantages of both the $\boldsymbol{Z}_{\text{bus}}$ and $\boldsymbol{Y}_{\text{bus}}$ formulation can be retained.

Example 5.1 The positive sequence network of a three-bus power system is shown in Fig. 5.6. The three-port description of the generator–transmission network is given to be

$$\begin{bmatrix} V_1 \\ V_2 \\ V_3 \end{bmatrix} = j \begin{bmatrix} 0.1274 & 0.1061 & 0.0981 \\ 0.1061 & 0.1345 & 0.1151 \\ 0.0981 & 0.1151 & 0.1215 \end{bmatrix} \begin{bmatrix} I_1 \\ I_2 \\ I_3 \end{bmatrix} + \begin{bmatrix} 1 \\ 1 \\ 1 \end{bmatrix} V_0^a$$

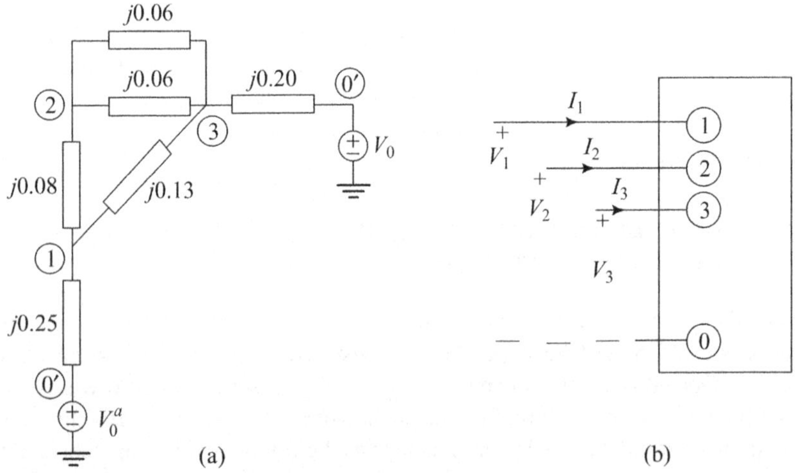

Fig. 5.6 *Example for three-phase fault study*

For a symmetrical three-phase to ground fault with $z_F = j\,0.1$ p.u., find the fault currents for the faults at the buses 1, 2, and 3. For the fault at bus 1, find all the bus voltages and line currents. Assume $V_0^a = 1 + j0$.

Solution: For fault at bus 1, the fault current is given by

$$I_F = \frac{V_0^a}{Z_{11} + z_F} = \frac{1}{j(0.1274+0.1)}$$

$$= \frac{1}{j0.2274} = -j4.3975 \text{ p.u.}$$

Similarly, for fault at bus 2,

$$I_F = \frac{V_0^a}{Z_{22} + z_F} = \frac{1}{j0.2345} = -j4.2644 \text{ p.u.}$$

and for fault at bus 3,

$$I_F = \frac{V_0^a}{Z_{33} + z_F}$$

$$= \frac{1}{j0.2215} = -j4.5146 \text{ p.u.}$$

For fault at bus 1, the voltage at bus 1, using Eq. (5.32) is obtained as

$$V_{1(F)} = z_F I_F = 0.43975 \text{ p.u.}$$

Using Eq. (5.33), the voltages at the other buses are

$$V_{2(F)} = V_0^a - Z_{21} I_F$$

$$= 1 - j0.1061(-j4.3975)$$

$$= 0.53342 \text{ p.u.}$$

and

$$V_{3(F)} = V_0^a - Z_{31} I_F$$

$$= 1 - j0.0981(-j4.3975)$$

$$= 0.5686 \text{ p.u.}$$

The currents in the lines are

$$I_{12(F)} = \left(V_{1(F)} - V_{2(F)}\right)\frac{1}{j0.08} = (0.43975 - 0.53342)\frac{1}{j0.08}$$

$$= -j1.1709 \text{ p.u.}$$

and

$$I_{23(1)(F)} = I_{23(2)(F)}$$

$$= (V_{2(F)} - V_{3(F)})\frac{1}{j0.06} = (0.53342 - 0.5686)\frac{1}{j0.06}$$

$$= -j0.58633 \text{ p.u.}$$

$$I_{0'3(F)} = (V_0^a - V_{3(F)}) \frac{1}{j0.02} = (1 - 0.5686) \frac{1}{j0.02} = -j\,2.157 \text{ p.u.}$$

$$I_{0'3(F)} = (V_0^a - V_{1(F)}) \frac{1}{j0.25} = (1 - 0.439795) \frac{1}{j0.25} = -j\,2.241 \text{ p.u.}$$

$$I_{13(F)} = (V_{1(F)} - V_{3(F)}) \frac{1}{j0.13} = (0.43975 - 0.5686) \frac{1}{j0.13} = -j\,0.99115 \text{ p.u.}$$

5.6 THREE-PHASE NETWORKS

Although the analysis of unsymmetrical faults by symmetrical components has been popular for the past 50 years or so, the analysis of three-phase networks in its phase impedance form has become necessary owing to the introduction of partially transposed or untransposed lines, dc links, etc. Consider the transmission system (single-line diagram) in Fig. 5.7(a) whose system graph is shown in Fig. 5.7(b). In the system graph, each element represents a three-phase component as shown in Fig. 5.7(c) for element 1. The primitive impedance matrix of this three-phase element is denoted by the 3×3 symmetric matrix $z^{a,b,c}$. However, the symmetry can be either as in (5.1) or as in (5.12).

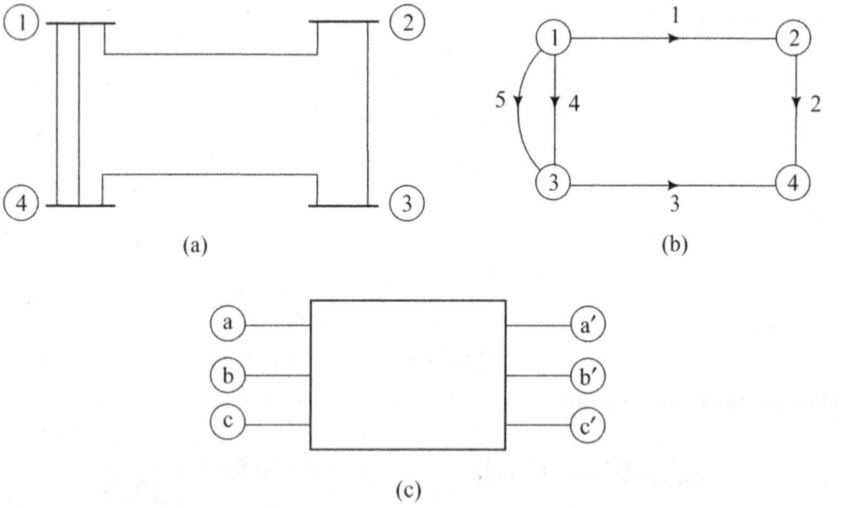

Fig. 5.7 *Three-phase network representation*

The system graph in terms of the three-phase element representation is identical to the system graph on a per phase basis. This enables us to generalize the results of the analysis from a per phase basis to three-phase variables easily. Take for example the fundamental cut-set and circuit equations.

For the graph of Fig. 5.7(b) and for the tree $T(1, 2, 3)$, the circuit equations are

$$\begin{bmatrix} -U & -U & -U & \vdots & U & 0 \\ -U & -U & -U & \vdots & 0 & U \end{bmatrix} \begin{bmatrix} V_1^{a,b,c} \\ V_2^{a,b,c} \\ V_3^{a,b,c} \\ \cdots\cdots \\ V_4^{a,b,c} \\ V_5^{a,b,c} \end{bmatrix} = 0 \tag{5.41}$$

where U is a 3×3 unit matrix and $V_i^{a,b,c}$ is the three-element column vector corresponding to the three-phase voltages $V_i^a, V_i^b, and\ V_i^c$, of element $i(i = 1, 2, \ldots, 5)$. It is easy to see that Eq. (5.41) can be readily written as an extension of the circuit equations for the single-phase case which is

$$\begin{bmatrix} -1 & -1 & -1 & \vdots & 1 & 0 \\ -1 & -1 & -1 & \vdots & 0 & 1 \end{bmatrix} \begin{bmatrix} V_1 \\ V_2 \\ V_3 \\ \cdots \\ V_4 \\ V_5 \end{bmatrix} = 0 \tag{5.42}$$

The cutset and the primitive equations, respectively, can likewise be written as

$$\begin{bmatrix} U & 0 & 0 & \vdots & U & U \\ 0 & U & 0 & \vdots & U & U \\ 0 & 0 & U & \vdots & U & U \end{bmatrix} \begin{bmatrix} I_1^{a,b,c} \\ I_2^{a,b,c} \\ I_3^{a,b,c} \\ \cdots\cdots \\ I_4^{a,b,c} \\ I_5^{a,b,c} \end{bmatrix} = 0 \tag{5.43}$$

and

$$\begin{bmatrix} V_1^{a,b,c} \\ V_2^{a,b,c} \\ V_3^{a,b,c} \\ V_4^{a,b,c} \\ V_5^{a,b,c} \end{bmatrix} = \begin{bmatrix} z_{11}^{a,b,c} & 0 & 0 & 0 & 0 \\ 0 & z_{22}^{a,b,c} & 0 & 0 & 0 \\ 0 & 0 & z_{33}^{a,b,c} & 0 & 0 \\ 0 & 0 & 0 & z_{44}^{a,b,c} & z_{45}^{a,b,c} \\ 0 & 0 & 0 & z_{54}^{a,b,c} & z_{55}^{a,b,c} \end{bmatrix} \begin{bmatrix} I_1^{a,b,c} \\ I_2^{a,b,c} \\ I_3^{a,b,c} \\ I_4^{a,b,c} \\ I_5^{a,b,c} \end{bmatrix} \tag{5.44}$$

Mutual coupling has been assumed between the lines 4 and 5 denoted by $z_{45}^{a,b,c}$ and $z_{54}^{a,b,c}$, where

$$z_{45}^{a,b,c} = \left[z_{54}^{a,b,c}\right]^T$$

In view of the close resemblance of the structure of equations for the single and three-phase cases, much of the analysis of Chapters 3, 4, and 5 can be carried over to the three-phase case. In particular, the building algorithm of Chapter 4 is readily applicable.

If there are n buses in the network, the terminal representation will correspond to a 3-n port network. The variables of the three phases at any one bus are always treated as one 3-component variable. Thus, in the bus frame of reference with the ground as reference bus, we have

$$V_{\text{bus}}^{a,b,c} = Z_{\text{bus}}^{a,b,c} I_{\text{bus}}^{a,b,c}$$

or

$$I_{\text{bus}}^{a,b,c} = Y_{\text{bus}}^{a,b,c} V_{\text{bus}}^{a,b,c} \tag{5.45}$$

where

$$V_{\text{bus}}^{a,b,c} = \begin{bmatrix} V_1^a \\ V_1^b \\ V_1^c \\ \cdots \\ V_2^a \\ V_2^b \\ V_2^c \\ \cdots \\ \vdots \\ \cdots \\ V_n^a \\ V_n^b \\ V_n^c \end{bmatrix} = \begin{bmatrix} V_1^{a,b,c} \\ V_2^{a,b,c} \\ \vdots \\ V_n^{a,b,c} \end{bmatrix}; \quad I_{\text{bus}}^{a,b,c} = \begin{bmatrix} I_1^a \\ I_1^b \\ I_1^c \\ \cdots \\ I_2^a \\ I_2^b \\ I_2^c \\ \cdots \\ \vdots \\ \cdots \\ I_n^a \\ I_n^b \\ I_n^c \end{bmatrix} = \begin{bmatrix} I_1^{a,b,c} \\ I_2^{a,b,c} \\ \vdots \\ I_n^{a,b,c} \end{bmatrix}$$

and

$$Z_{\text{bus}}^{a,b,c} = \begin{bmatrix} Z_{11}^{a,b,c} & Z_{12}^{a,b,c} & \cdots & Z_{1n}^{a,b,c} \\ \vdots & & & \\ Z_{n1}^{a,b,c} & Z_{n2}^{a,b,c} & \cdots & Z_{nn}^{a,b,c} \end{bmatrix}$$

$Y_{\text{bus}}^{a,b,c}$ has a similar structure.

5.7 FAULT ANALYSIS IN THE PHASE IMPEDANCE FORM

Consider a three-phase transmission network with the same assumptions as in Section 5.2. Augment the 3-n port network with a three-phase representation of the generator together with the machine reactance. Since we assume all generators to have the same three-phase voltage source, the result of augmentation can be represented as in Fig. 5.8.

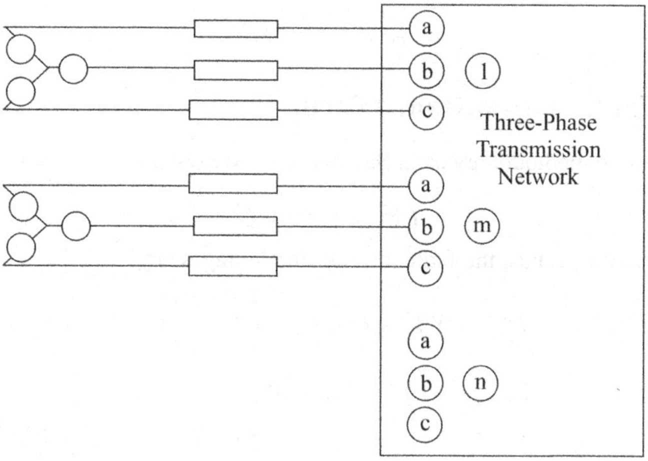

Fig. 5.8 *Three-phase power system representation for short-circuit studies*

As in the single-phase case, we can derive $Z_{\text{bus}}^{a,b,c}$ by short-circuiting the voltage sources first, and then following the procedure in Section 5.5. The result is

$$\left[V_{\text{bus}}^{a,b,c}\right]=\left[Z_{\text{bus}}^{a,b,c}\right]\left[I_{\text{bus}}^{a,b,c}\right]+[B]\left[V_0^{a,b,c}\right] \tag{5.46}$$

where

$$[B]=\begin{bmatrix} U \\ U \\ \vdots \\ U \end{bmatrix} \text{ and } V_0^{a,b,c}=\begin{bmatrix} V_0^a \\ V_0^b \\ V_0^c \end{bmatrix}$$

This is the general 3-n port description in the phase impedance form that can be used to analyze faulted conditions whether they are symmetrical or unsymmetrical.

5.7.1 Fault in Impedance Form

Let the 3-port description of the fault be

$$V_{\text{F}}^{a,b,c}=Z_F^{a,b,c}\,I_F^{a,b,c} \tag{5.47}$$

Proceeding as in Section 5.5, the fault current and voltages for a fault at the pth bus are given by

$$I_F^{a,b,c}=\left[Z_F^{a,b,c}+Z_{pp}^{a,b,c}\right]^{-1}V_0^{a,b,c} \tag{5.48}$$

$$I_{p(F)}^{a,b,c}=-I_F^{a,b,c};\; I_{i(F)}^{a,b,c}=0 \qquad i=1,2,\ldots,n \quad i\neq p \tag{5.49}$$

$$V_{i(F)}^{a,b,c}=V_0^{a,b,c}-Z_{ip}^{a,b,c}\,I_F^{a,b,c} \qquad i=1,2,\ldots,n \quad i\neq p \tag{5.50}$$

$$V_{p(F)}^{a,b,c} = Z_F^{a,b,c} I_F^{a,b,c} \tag{5.51}$$

5.7.2 Fault in Admittance Form

Using the same methodology as in Section 5.5.1, we obtain the following

$$I_F^{a,b,c} = Y_F^{a,b,c} V_F^{a,b,c} \tag{5.52}$$

For a fault at the pth bus, the fault currents and voltages are

$$I_F^{a,b,c} = Y_F^{a,b,c}\left[U + Z_{pp}^{a,b,c} Y_F^{a,b,c}\right]^{-1} V_0^{a,b,c} \tag{5.53}$$

$$I_{p(F)}^{a,b,c} = -I_F^{a,b,c};\ I_{i(F)}^{a,b,c} = 0 \quad i = 1,\ 2,\ \ldots n \quad i \neq p \tag{5.54}$$

Voltages are given by

$$V_{i(F)}^{a,b,c} = V_0^{a,b,c} - Z_{ip}^{a,b,c} I_F^{a,b,c} \qquad i = 1,\ 2,\ \ldots n \quad i \neq p \tag{5.55}$$

$$V_{p(F)}^{a,b,c} = \left[U + Z_{pp}^{a,b,c} Y_F^{a,b,c}\right]^{-1} V_0^{a,b,c} \tag{5.56}$$

For the various types of unsymmetrical faults, the appropriate $Y_F^{a,b,c}$ is substituted in the above equation. $Y_F^{a,b,c}$ for different faults is derived in the following section.

5.8 GENERAL FAULT REPRESENTATION IN PHASE QUANTITIES

We shall derive the various fault descriptions from the general diagram in Fig. 5.9(a).

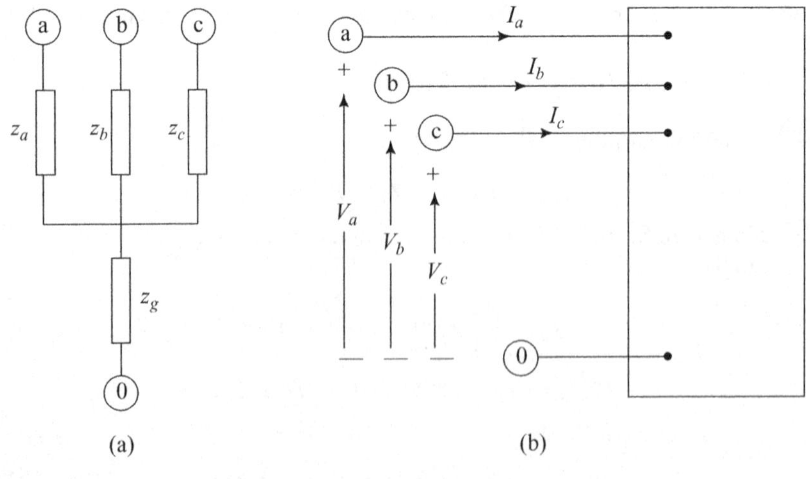

Fig. 5.9 *Three-phase unsymmetrical fault representation*

By reasoning similar to that in Section 5.4, the 3-port impedance description of Fig. 5.10(b) is obtained as

$$\begin{bmatrix} V^a \\ V^b \\ V^c \end{bmatrix} = \begin{bmatrix} z_a + z_g & z_g & z_g \\ z_g & z_b + z_g & z_g \\ z_g & z_g & z_c + z_g \end{bmatrix} \begin{bmatrix} I^a \\ I^b \\ I^c \end{bmatrix} \tag{5.57}$$

i.e.,

$$\boldsymbol{V}_F^{a,b,c} = \boldsymbol{Z}_F^{a,b,c} \boldsymbol{I}_F^{a,b,c}$$

In the admittance form it can be written as

$$\begin{bmatrix} \boldsymbol{I}^a \\ \boldsymbol{I}^b \\ \boldsymbol{I}^c \\ 0 \end{bmatrix} = \begin{bmatrix} y_a & 0 & 0 & -y_a \\ 0 & y_b & 0 & -y_b \\ 0 & 0 & y_c & -y_c \\ -y_a & -y_b & -y_c & y_a + y_b + y_c + y_g \end{bmatrix} \begin{bmatrix} \boldsymbol{V}^a \\ \boldsymbol{V}^b \\ \boldsymbol{V}^c \\ \boldsymbol{V}^n \end{bmatrix}$$

which after reduction becomes

$$\begin{bmatrix} I^a \\ I^b \\ I^c \end{bmatrix} = \frac{1}{y} \begin{bmatrix} y_a (y_b + y_c + y_g) & -y_a y_b & -y_a y_c \\ -y_a y_b & y_b (y_a + y_c + y_g) & -y_b y_c \\ -y_a y_c & -y_b y_c & y_c (y_a + y_b + y_g) \end{bmatrix} \begin{bmatrix} V^a \\ V^b \\ V^c \end{bmatrix} \tag{5.58}$$

i.e.,

$$\boldsymbol{I}_F^{a,b,c} = \boldsymbol{Y}_F^{a,b,c} \boldsymbol{V}_F^{a,b,c}$$

where z_a, z_b, z_c, and z_g are the impedances, and y_a, y_b, y_c, and y_g are the corresponding admittances, and

$$y = y_a + y_b + y_c + y_g$$

The 3-port description for other types of faults can be deduced through an appropriate modification of these two descriptions. Note that $\boldsymbol{Z}_F^{a,b,c}$ does not exist for any other fault except a three-phase to ground fault, as one or more elements of $\boldsymbol{Z}_F^{a,b,c}$ become ∞ for other faults because of the isolation of one or more of the four nodes a, b, c, and g. We therefore drive the $\boldsymbol{Y}_F^{a,b,c}$ for various unsymmetrical faults shown in Fig. 5.10.

Single-Line to Ground (L-G)Fault Here, $y_b = y_c = 0$ (Fig. 5.10(a)). So Eq. (5.58) takes the form

$$\boldsymbol{Y}_F^{a,b,c} = \begin{bmatrix} \dfrac{y_a y_g}{y_a + y_g} & 0 & 0 \\ 0 & 0 & 0 \\ 0 & 0 & 0 \end{bmatrix} \tag{5.59}$$

Line-Line to Ground (L-L-G) Fault In this case, $y_a = 0$ (Fig. 5.10(b)). Hence,

$$\mathbf{Y}_F^{a,b,c} = \frac{1}{y}\begin{bmatrix} 0 & 0 & 0 \\ 0 & y_b(y_c + y_g) & -y_b y_c \\ 0 & -y_b y_c & y_c(y_b + y_g) \end{bmatrix} \text{ with } y = y_b + y_c + y_g. \quad (5.60)$$

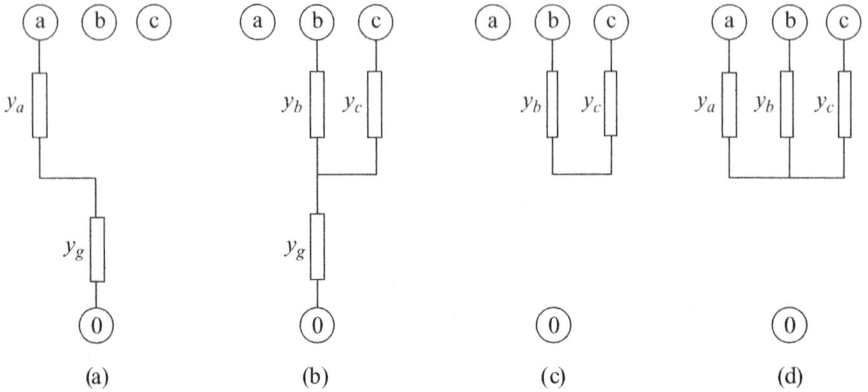

Fig. 5.10 *Representation of unsymmetrical fault*

Line-Line (L-L) Fault In this case, $y_a = y_g = 0$ (Fig. 5.10(c)). Hence,

$$\mathbf{Y}_F^{a,b,c} = \frac{1}{y_b + y_c}\begin{bmatrix} 0 & 0 & 0 \\ 0 & y_b y_c & -y_b y_c \\ 0 & -y_b y_c & y_b y_c \end{bmatrix} \quad (5.61)$$

Three-Phase Fault (Unsymmetrical) In this case, $y_g = 0$ (Fig. 5.10(d)). Hence,

$$\mathbf{Y}_F^{a,b,c} = \frac{1}{y}\begin{bmatrix} y_a(y_b + y_c) & -y_a y_b & -y_a y_c \\ -y_a y_b & y_b(y_a + y_c) & -y_b y_c \\ -y_a y_c & -y_b y_c & y_c(y_a + y_b) \end{bmatrix} \quad (5.62)$$

with $y = y_a + y_b + y_c$.

For the Balanced Three-Phase Fault Set $y_a = y_b = y_c = y_F$ in Eq. (5.62)

$$\mathbf{Y}_F^{a,b,c} = \frac{y_F}{3}\begin{bmatrix} 2 & -1 & -1 \\ -1 & 2 & -1 \\ -1 & -1 & 2 \end{bmatrix} \quad (5.63)$$

Using Eqs (5.59)–(5.62) in Eqs (5.53)–(5.56), the fault currents and voltages for any unsymmetrical fault can be computed. The summary of various fault impedance and admittance matrices in the phase component form is given in Table 5.1.

Table 5.1 *Fault impedance and admittance matrices (phase components)*

Type of Fault	*Three-Phase Components*	
	$Z_F^{a,b,c}$	$Y_F^{a,b,c}$
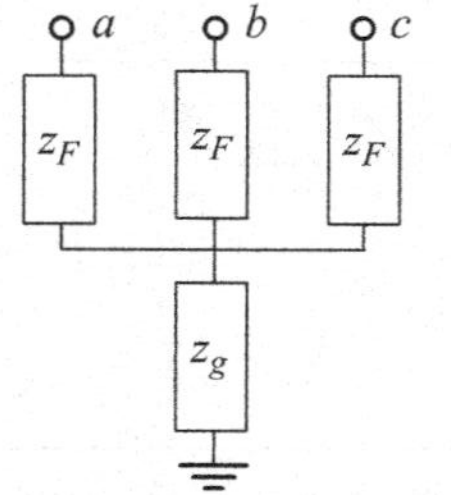 Three-Phase-to-Ground	$\begin{bmatrix} z_F+z_g & z_g & z_g \\ z_g & z_F+z_g & z_g \\ z_g & z_g & z_F+z_g \end{bmatrix}$	$\frac{1}{3}\begin{bmatrix} y_t+2y_F & y_t-y_F & y_t-y_F \\ y_t-y_F & y_t+2y_F & y_t-y_F \\ y_t-y_F & y_t-y_F & y_t+2y_F \end{bmatrix}$ where $y_t = \frac{1}{z_F+3z_g}$
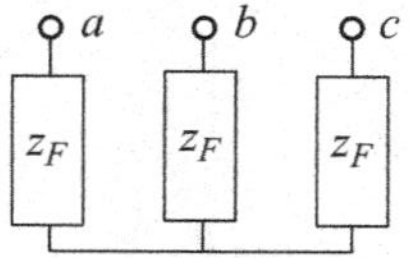 Three-Phase	Not Defined	$\frac{y_F}{3}\begin{bmatrix} 2 & -1 & -1 \\ -1 & 2 & -1 \\ -1 & -1 & 2 \end{bmatrix}$
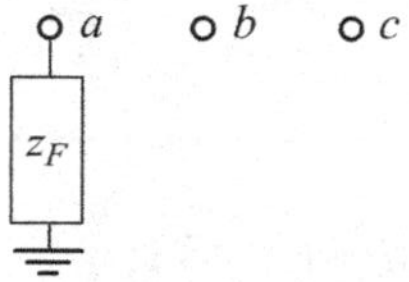 Line-to-Ground	$\begin{bmatrix} z_F & 0 & 0 \\ 0 & \infty & 0 \\ 0 & 0 & \infty \end{bmatrix}$	$\begin{bmatrix} y_F & 0 & 0 \\ 0 & 0 & 0 \\ 0 & 0 & 0 \end{bmatrix}$

Type of Fault	Three-Phase Components	
	$Z_F^{a,b,c}$	$Y_F^{a,b,c}$
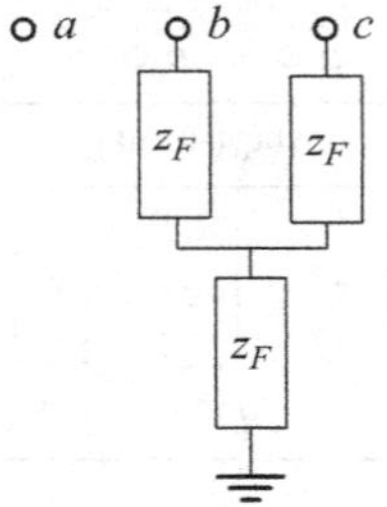 Line-Line-to-Ground	$\begin{bmatrix} \infty & 0 & 0 \\ 0 & z_F + z_g & z_g \\ 0 & z_g & z_F + z_g \end{bmatrix}$	$\begin{bmatrix} 0 & 0 & 0 \\ 0 & \dfrac{z_F + z_g}{z_F^2 + 2z_F z_g} & \dfrac{-z_g}{z_F^2 + 2z_F z_g} \\ 0 & \dfrac{-z_F}{z_F^2 + 2z_F z_g} & \dfrac{z_F + z_g}{z_F^2 + 2z_F z_g} \end{bmatrix}$
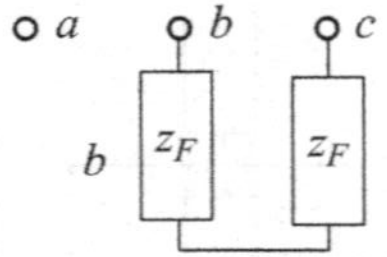 Line-to-Line	Not Defined	$\dfrac{y_F}{2}\begin{bmatrix} 0 & 0 & 0 \\ 0 & 1 & -1 \\ 0 & -1 & 1 \end{bmatrix}$

5.9 SYMMETRICAL COMPONENT ANALYSIS

The short-circuit study based on phase impedance quantities involves $Z_{\text{bus}}^{a,b,c}$, which besides being large in size, is also a nearly full matrix. Hence, a computation using such a representation may have limitations in terms of the computer size for large networks. If all components in the system excepting the fault are balanced, a symmetrical component transformation of the generators and the transmission network yields three uncoupled networks thereby simplifying the analysis. Through such an analysis, the degree of unbalance also gets clearly exhibited. We shall now demonstrate that under certain additional assumptions, the mutual coupling between the transmission lines is present only in the zero sequence network.

If two transmission lines are coupled, simplification will depend on the nature of the coupling submatrix. For example, in Fig. 5.7(a), lines 4 and 5 are coupled and the coupling submatrices $z_{45}^{a,b,c}$ and $z_{54}^{a,b,c}$ are related with the phase voltages $V_4^{a,b,c}$ and $V_5^{a,b,c}$, and currents $I_4^{a,b,c}$ and $I_5^{a,b,c}$ of the two lines as shown below:

$$\begin{bmatrix} V_4^{a,b,c} \\ V_5^{a,b,c} \end{bmatrix} = \begin{bmatrix} z_{44}^{a,b,c} & z_{45}^{a,b,c} \\ z_{54}^{a,b,c} & z_{55}^{a,b,c} \end{bmatrix} \begin{bmatrix} I_4^{a,b,c} \\ I_5^{a,b,c} \end{bmatrix} \quad \text{where, } z_{45}^{a,b,c} = \begin{bmatrix} z_{45}^{aa} & z_{45}^{ab} & z_{45}^{ac} \\ z_{45}^{ba} & z_{45}^{bb} & z_{45}^{bc} \\ z_{45}^{ca} & z_{45}^{cb} & z_{45}^{cc} \end{bmatrix} = \left[z_{54}^{a,b,c} \right]^T$$

It is to be noted that all the elements of the $z_{45}^{a,b,c}$ matrix are mutual impedances between different phases of lines 4 and 5.

The transformation to sequence components results in the following:

$$\begin{bmatrix} V_4^{0,1,2} \\ V_5^{0,1,2} \end{bmatrix} = \begin{bmatrix} S_3 & 0 \\ 0 & S_3 \end{bmatrix} \begin{bmatrix} z_{44}^{a,b,c} & z_{45}^{a,b,c} \\ z_{54}^{a,b,c} & z_{55}^{a,b,c} \end{bmatrix} \begin{bmatrix} S_3^{-1} & 0 \\ 0 & S_3^{-1} \end{bmatrix} \begin{bmatrix} I_4^{0,1,2} \\ I_5^{0,1,2} \end{bmatrix}$$

$$= \begin{bmatrix} S_3 z_{44}^{a,b,c} S_3^{-1} & S_3 z_{45}^{a,b,c} S_3^{-1} \\ S_3 z_{54}^{a,b,c} S_3^{-1} & S_3 z_{55}^{a,b,c} S_3^{-1} \end{bmatrix} \begin{bmatrix} I_4^{0,1,2} \\ I_5^{0,1,2} \end{bmatrix} = \begin{bmatrix} z_{44}^{0,1,2} & z_{45}^{0,1,2} \\ z_{54}^{0,1,2} & z_{55}^{0,1,2} \end{bmatrix} \begin{bmatrix} I_4^{0,1,2} \\ I_5^{0,1,2} \end{bmatrix}$$

Here, S_3 is the sequence component transformation matrix defined in (5.5). If complete transposition is considered, then $z_{44}^{0,1,2}$ and $z_{55}^{0,1,2}$ will have the form as shown in (5.8). $z_{45}^{0,1,2}$ can be obtained as

$$z_{45}^{0,1,2} = S_3 z_{45}^{a,b,c} S_3^{-1} = \frac{1}{3} \begin{bmatrix} 1 & 1 & 1 \\ 1 & a & a^2 \\ 1 & a^2 & a \end{bmatrix} \begin{bmatrix} z_{45}^{aa} & z_{45}^{ab} & z_{45}^{ac} \\ z_{45}^{ba} & z_{45}^{bb} & z_{45}^{bc} \\ z_{45}^{ca} & z_{45}^{cb} & z_{45}^{cc} \end{bmatrix} \begin{bmatrix} 1 & 1 & 1 \\ 1 & a^2 & a \\ 1 & a & a^2 \end{bmatrix}$$

Further simplification gives the complete form of the impedance matrix $z_{45}^{0,1,2}$ as

$$
z_{45}^{0,1,2} = \frac{1}{3}\begin{bmatrix}
\begin{array}{l}\left(z_{45}^{aa}+z_{45}^{bb}+z_{45}^{cc}\right)\\ +\left(z_{45}^{ab}+z_{45}^{ca}+z_{45}^{bc}\right)\\ +\left(z_{45}^{ba}+z_{45}^{ac}+z_{45}^{cb}\right)\end{array} &
\begin{array}{l}\left(z_{45}^{aa}+a^2 z_{45}^{bb}+a z_{45}^{cc}\right)\\ +\left(z_{45}^{ba}+z_{45}^{ca}\right)+a\left(z_{45}^{ac}+z_{45}^{bc}\right)\\ +a^2\left(z_{45}^{ab}+z_{45}^{cb}\right)\end{array} &
\begin{array}{l}\left(z_{45}^{aa}+a z_{45}^{bb}+a^2 z_{45}^{cc}\right)\\ +\left(z_{45}^{ba}+z_{45}^{ca}\right)+\\ a\left(z_{45}^{ab}+z_{45}^{cb}\right)\\ +a^2\left(z_{45}^{ac}+z_{45}^{bc}\right)\end{array} \\[2em]
\begin{array}{l}\left(z_{45}^{aa}+a z_{45}^{bb}+a^2 z_{45}^{cc}\right)\\ +\left(z_{45}^{ab}+z_{45}^{ac}\right)\\ +a\left(z_{45}^{ba}+z_{45}^{bc}\right)\\ +a^2\left(z_{45}^{ca}+z_{45}^{cc}\right)\end{array} &
\begin{array}{l}\left(z_{45}^{aa}+z_{45}^{bb}+z_{45}^{cc}\right)\\ +a\left(z_{45}^{ac}+z_{45}^{ba}+z_{45}^{cb}\right)\\ +a^2\left(z_{45}^{ab}+z_{45}^{bc}+z_{45}^{ca}\right)\end{array} &
\begin{array}{l}\left(z_{45}^{aa}+a^2 z_{45}^{bb}+a z_{45}^{cc}\right)\\ +\left(z_{45}^{bc}+z_{45}^{cb}\right)+\\ a\left(z_{45}^{ab}+z_{45}^{ba}\right)\\ +a^2\left(z_{45}^{ac}+z_{45}^{ca}\right)\end{array} \\[2em]
\begin{array}{l}\left(z_{45}^{aa}+a^2 z_{45}^{bb}+a z_{45}^{cc}\right)\\ +\left(z_{45}^{ab}+z_{45}^{ac}\right)\\ +a\left(z_{45}^{ca}+z_{45}^{cb}\right)\\ +a^2\left(z_{45}^{ba}+z_{45}^{bc}\right)\end{array} &
\begin{array}{l}\left(z_{45}^{aa}+a z_{45}^{bb}+a^2 z_{45}^{cc}\right)\\ +\left(z_{45}^{bc}+z_{45}^{cb}\right)+a\left(z_{45}^{ac}+z_{45}^{ca}\right)\\ +a^2\left(z_{45}^{ab}+z_{45}^{ba}\right)\end{array} &
\begin{array}{l}\left(z_{45}^{aa}+z_{45}^{bb}+z_{45}^{cc}\right)\\ +a\left(z_{45}^{ab}+z_{45}^{bc}+z_{45}^{ca}\right)\\ +a^2\left(z_{45}^{ac}+z_{45}^{ba}+z_{45}^{cb}\right)\end{array}
\end{bmatrix}
$$

All the elements of $z_{45}^{a,b,c}$ being mutual impedances, if the two lines are parallel or share the same right of way, it can be assumed that all the elements have equal values.

$$
z_{45}^{aa} = z_{45}^{bb} = z_{45}^{cc} = z_{45}^{ab} = z_{45}^{ac} = z_{45}^{ba} = z_{45}^{bc} = z_{45}^{ca} = z_{45}^{cb} = z_{45}^{m}
$$

In that case, the coupling terms will appear only in the zero sequence network as is evident from the above expression of $z_{45}^{0,1,2}$. However, if this assumption is not valid, then there will be mutual coupling even between the sequence component networks of lines 4 and 5. Hence, with the assumption of complete transposition and equal values of elements of $z_{45}^{a,b,c}$, the sequence component impedance matrix for the whole network of Fig. 5.7(a) becomes

$$
z^{0,1,2} = \begin{bmatrix}
z_{11}^{0,1,2} & 0 & 0 & 0 & 0 \\
0 & z_{22}^{0,1,2} & 0 & 0 & 0 \\
0 & 0 & z_{33}^{0,1,2} & 0 & 0 \\
0 & 0 & 0 & z_{44}^{0,1,2} & z_{45}^{0,1,2} \\
0 & 0 & 0 & z_{54}^{0,1,2} & z_{55}^{0,1,2}
\end{bmatrix}
$$

where

$$z_{kk}^{0,1,2} = \begin{bmatrix} z_{ks} + 2z_{km} & 0 & 0 \\ 0 & z_{ks} - z_{km} & 0 \\ 0 & 0 & z_{ks} - z_{km} \end{bmatrix}; \quad z_{45}^{0,1,2} = \begin{bmatrix} 3z_{45}^{m} & 0 & 0 \\ 0 & 0 & 0 \\ 0 & 0 & 0 \end{bmatrix}$$

z_{k_s} is the self inductance of each phase and z_{k_m} is the mutual inductance between the phases of line 'k'. z_{45}^{m} is the mutual inductance between the phases of line 4 and phases of line 5.

The voltages behind the transient reactances are also expressed in terms of the symmetrical component variables. For balanced excitation, $V_0^{\mathrm{a}} = |V_{\mathrm{a}}|\angle 0^\circ$, $V_0^{\mathrm{b}} = |V_{\mathrm{a}}|\angle -120^\circ$, $V_0^{\mathrm{c}} = |V_{\mathrm{a}}|\angle -240^\circ$. Hence,

$$\boldsymbol{V}^S = \boldsymbol{V}_0^{0,1,2} = \frac{1}{\sqrt{3}}\begin{bmatrix} 1 & 1 & 1 \\ 1 & a & a^2 \\ 1 & a^2 & a \end{bmatrix}\begin{bmatrix} V_0^a \\ V_0^b \\ V_0^c \end{bmatrix} = \frac{1}{\sqrt{3}}\begin{bmatrix} 1 & 1 & 1 \\ 1 & a & a^2 \\ 1 & a^2 & a \end{bmatrix}\begin{bmatrix} |V_a\angle 0^\circ| \\ |V_a\angle -120^\circ| \\ |V_a\angle 120^\circ| \end{bmatrix} = \begin{bmatrix} 0 \\ \sqrt{3} \\ 0 \end{bmatrix}|V_a|$$

Thus, only the positive sequence component of the voltage source is present.

Expressing the variables of the 3-n port network in terms of symmetrical components, the Z representation of the generator-transmission network can be arrived at along the same lines as in the phase impedance form as

$$\left[\boldsymbol{V}_{\text{bus}}^{0,1,2}\right] = \left[\boldsymbol{Z}_{\text{bus}}^{0,1,2}\right]\left[\boldsymbol{I}_{\text{bus}}^{0,1,2}\right] + [\boldsymbol{B}]\left[\boldsymbol{V}_0^{0,1,2}\right] \tag{5.64}$$

The fault description may be given either in the impedance form as $\boldsymbol{Z}_F^{0,1,2}$, or in the admittance form as $\boldsymbol{Y}_F^{0,1,2}$. $\boldsymbol{Z}_F^{0,1,2}$ and $\boldsymbol{Y}_{\mathrm{F}}^{0,1,2}$ for various types of faults are obtained by applying the symmetrical component transformation to the corresponding $\boldsymbol{Z}_F^{\mathrm{a,b,c}}$ or $\boldsymbol{Y}_F^{a,b,c}$. As an example, for the single-line to ground fault,

$$\begin{aligned} \boldsymbol{Y}_F^{0,1,2} &= \boldsymbol{S}_3\boldsymbol{Y}_F^{a,b,c}\boldsymbol{S}_3^{-1} \\ &= \frac{1}{\sqrt{3}}\begin{bmatrix} 1 & 1 & 1 \\ 1 & a & a^2 \\ 1 & a^2 & a \end{bmatrix}\begin{bmatrix} y_F & 0 & 0 \\ 0 & 0 & 0 \\ 0 & 0 & 0 \end{bmatrix}\frac{1}{\sqrt{3}}\begin{bmatrix} 1 & 1 & 1 \\ 1 & a^2 & a \\ 1 & a & a^2 \end{bmatrix} \\ &= \frac{1}{3}\begin{bmatrix} 1 & 1 & 1 \\ 1 & a & a^2 \\ 1 & a^2 & a \end{bmatrix}\begin{bmatrix} y_F & y_F & y_F \\ 0 & 0 & 0 \\ 0 & 0 & 0 \end{bmatrix} \\ &= \frac{1}{3}\begin{bmatrix} y_F & y_F & y_F \\ y_F & y_F & y_F \\ y_F & y_F & y_F \end{bmatrix} \end{aligned} \tag{5.65}$$

Similarly, for the symmetrical three-phase to ground fault

$$z_F^{0,1,2} = \boldsymbol{S}_3 z_F^{a,b,c}\boldsymbol{S}_3^{-1} = \begin{bmatrix} z_F + 3z_g & 0 & 0 \\ 0 & z_F & 0 \\ 0 & 0 & z_F \end{bmatrix} \tag{5.66}$$

The complete list of fault matrices in the symmetrical component quantities is given in Table 5.2.

Table 5.2 *Fault impedance and admittance matrices (symmetrical components)*

Type of Fault	Symmetrical Components	
	$Z_F^{0,1,2}$	$Y_F^{0,1,2}$
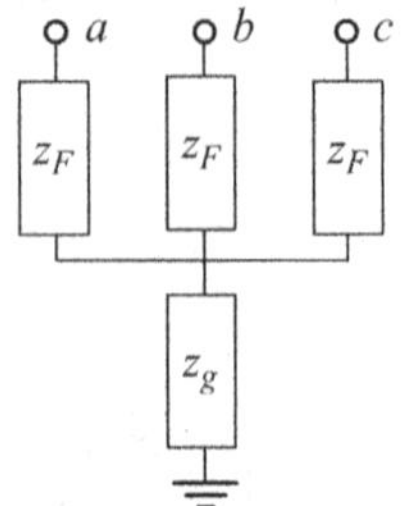 Three-Phase-to-Ground	$\begin{bmatrix} z_F+3z_g & 0 & 0 \\ 0 & z_F & 0 \\ 0 & 0 & z_F \end{bmatrix}$	$\begin{bmatrix} y_F & 0 & 0 \\ 0 & y_F & 0 \\ 0 & 0 & y_F \end{bmatrix}$ where $y_0 = \dfrac{1}{z_F+3z_g}$
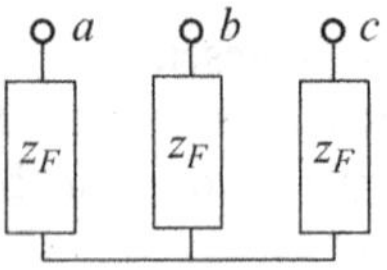 Three-Phase	$\begin{bmatrix} \infty & 0 & 0 \\ 0 & z_F & 0 \\ 0 & 0 & z_F \end{bmatrix}$	$y_F \begin{bmatrix} 0 & 0 & 0 \\ 0 & 1 & 0 \\ 0 & 0 & 1 \end{bmatrix}$
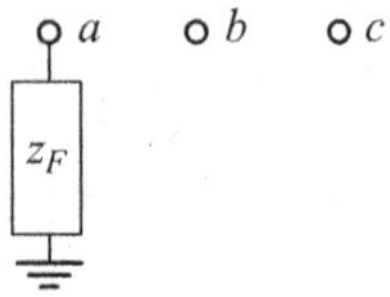 Line-to-Ground	Not Defined	$\dfrac{y_F}{3} \begin{bmatrix} 1 & 1 & 1 \\ 1 & 1 & 1 \\ 1 & 1 & 1 \end{bmatrix}$
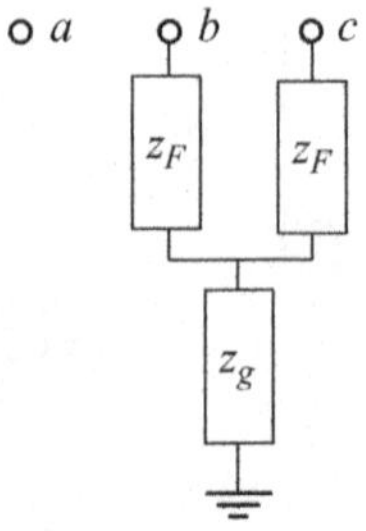 Line-Line-to-Ground	Not Defined	$\dfrac{1}{3(z_F^2+2z_F z_g)} \begin{bmatrix} 2z_F & -z_F & -z_F \\ -z_F & 2z_F+3z_g & -(z_F+3z_g) \\ -z_F & -(z_F+3z_g) & 2z_F+3z_g \end{bmatrix}$
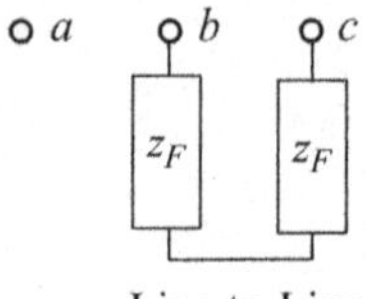 Line-to-Line	Not Defined	$\dfrac{y_F}{2} \begin{bmatrix} 0 & 0 & 0 \\ 0 & 1 & -1 \\ 0 & -1 & 1 \end{bmatrix}$

5.9.1 Computation of Currents and Voltages under Faulted Condition

The procedure is analogous to that in Section 5.7 except that the sequence quantities are involved here instead of the phase quantities.

1. When $\boldsymbol{Z}_F^{0,1,2}$ exists:

$$\boldsymbol{I}_F^{0,1,2} = \left[\boldsymbol{Z}_F^{0,1,2} + \boldsymbol{Z}_{pp}^{0,1,2}\right]^{-1} \boldsymbol{V}_0^{0,1,2} \tag{5.67}$$

$$\boldsymbol{V}_{i(F)}^{0,1,2} = \boldsymbol{V}_0^{0,1,2} - \boldsymbol{Z}_{ip}^{0,1,2} \boldsymbol{I}_F^{0,1,2} \quad i = 1,\ 2,\ \ldots, n \quad i \neq p \tag{5.68}$$

$$\boldsymbol{V}_{p(F)}^{0,1,2} = \boldsymbol{Z}_F^{0,1,2} \boldsymbol{I}_F^{0,1,2} \tag{5.69}$$

2. When $\boldsymbol{Y}_F^{0,1,2}$ exists:

$$\boldsymbol{I}_F^{0,1,2} = \boldsymbol{Y}_F^{0,1,2}\left[\boldsymbol{U} + \boldsymbol{Z}_F^{0,1,2} \boldsymbol{Y}_F^{0,1,2}\right]^{-1} \boldsymbol{V}_0^{0,1,2} \tag{5.70}$$

and

$$\boldsymbol{V}_{i(F)}^{0,1,2} = \boldsymbol{V}_0^{0,1,2} - \boldsymbol{Z}_{ip}^{0,1,2} \boldsymbol{I}_F^{0,1,2} \quad i = 1,\ 2,\ \ldots, n \quad i \neq p \tag{5.71}$$

$$\boldsymbol{V}_{p(F)}^{0,1,2} = \left[\boldsymbol{U} + \boldsymbol{Z}_{pp}^{0,1,2} \boldsymbol{Y}_F^{0,1,2}\right]^{-1} \boldsymbol{V}_0^{0,1,2} \tag{5.72}$$

5.10 CURRENTS AND VOLTAGES UNDER FAULTED CONDITION IN TERMS OF SYMMETRICAL COMPONENTS

The symbolic expression for the currents and voltages under faulted condition given in the preceding section can be simplified because of the uncoupling between the three sub-networks.

Three-Phase to Ground Fault Since $Z_F^{0,1,2}$ exists, we use Eq. (5.67) to get

$$\boldsymbol{I}_F^{0,1,2} = \left[\boldsymbol{Z}_F^{0,1,2} + \boldsymbol{Z}_{pp}^{0,1,2}\right]^{-1} \boldsymbol{V}_0^{0,1,2} = \begin{bmatrix} z_F + 3z_g + Z_{pp}^{(0)} & 0 & 0 \\ 0 & z_F + Z_{pp}^{(1)} & 0 \\ 0 & 0 & z_F + Z_{pp}^{(2)} \end{bmatrix} \begin{bmatrix} 0 \\ \sqrt{3} \\ 0 \end{bmatrix} |V_a|$$

Hence,

$$\begin{bmatrix} I_F^{(0)} \\ I_F^{(1)} \\ I_F^{(2)} \end{bmatrix} = \begin{bmatrix} 0 \\ \dfrac{\sqrt{3}}{z_F + Z_{pp}^{(1)}} \\ 0 \end{bmatrix} |V_a| \tag{5.73}$$

$$\begin{bmatrix} I_F^{(a)} \\ I_F^{(b)} \\ I_F^{(c)} \end{bmatrix} = \boldsymbol{S}_3^{-1}\boldsymbol{I}_F^{0,1,2} = \frac{1}{\sqrt{3}}\begin{bmatrix} 1 & 1 & 1 \\ 1 & a^2 & a \\ 1 & a & a^2 \end{bmatrix}\boldsymbol{I}_F^{0,1,2} = \begin{bmatrix} \dfrac{1}{z_F + Z_{pp}^{(1)}} \\ \dfrac{1}{z_F + Z_{pp}^{(1)}}\angle -120^\circ \\ \dfrac{1}{z_F + Z_{pp}^{(1)}}\angle -240^\circ \end{bmatrix}|V_a| \tag{5.74}$$

The sequence and phase voltages at the faulted bus are given by

$$\begin{bmatrix} V_F^{(0)} \\ V_F^{(1)} \\ V_F^{(2)} \end{bmatrix} = \begin{bmatrix} 0 \\ \dfrac{\sqrt{3}z_F}{z_F + Z_{pp}^{(1)}} \\ 0 \end{bmatrix}|V_a| \tag{5.75}$$

$$\begin{bmatrix} V_F^{(a)} \\ V_F^{(b)} \\ V_F^{(c)} \end{bmatrix} = \boldsymbol{S}_3^{-1}\boldsymbol{V}_F^{0,1,2} = \begin{bmatrix} \dfrac{z_F}{z_F + Z_{pp}^{(1)}} \\ \dfrac{z_F}{z_F + Z_{pp}^{(1)}}\angle -120^\circ \\ \dfrac{1}{z_F + Z_{pp}^{(1)}}\angle -240^\circ \end{bmatrix}|V_a| \tag{5.76}$$

Voltages at the other buses are given by

$$\begin{bmatrix} V_{i(F)}^{(0)} \\ V_{i(F)}^{(1)} \\ V_{i(F)}^{(2)} \end{bmatrix} = \begin{bmatrix} 0 \\ \sqrt{3} \\ 0 \end{bmatrix}|V_a| - \begin{bmatrix} Z_{ip}^{(0)} & 0 & 0 \\ 0 & Z_{ip}^{(1)} & 0 \\ 0 & 0 & Z_{ip}^{(2)} \end{bmatrix}\begin{bmatrix} 0 \\ I_F^{(1)} \\ \end{bmatrix}$$

i.e.,

$$V_{i(F)}^{(1)} = \sqrt{3}\left(1 - \frac{Z_{ip}^{(1)}}{z_F + Z_{pp}^{(1)}}\right)|V_a| \quad i = 1, 2, \ldots, n \quad i \neq p \tag{5.77}$$

Phase voltages are given by

$$\boldsymbol{V}_{1F}^{a,b,c} = \boldsymbol{S}_3^{-1}\boldsymbol{V}_{iF}^{0,1,2} \quad \text{for } i = 1, 2, \ldots, n \quad i \neq p$$

$$= \frac{1}{\sqrt{3}}\begin{bmatrix} 1 & 1 & 1 \\ 1 & a^2 & a \\ 1 & a & a^2 \end{bmatrix}\begin{bmatrix} 0 \\ V_{i(\mathrm{F})}^{(1)} \\ 0 \end{bmatrix}|V_a| = \left(1 - \frac{Z_{ip}^{(1)}}{z_F + Z_{pp}^{(1)}}\right)\begin{bmatrix} 1 \\ \angle -120^\circ \\ \angle -240^\circ \end{bmatrix}|V_a| \tag{5.78}$$

We can easily verify that the phase 'a' quantities in Eqs (5.74), (5.76), and (5.78) agree with those in Eqs (5.32), (5.33), and (5.34) respectively.

Line to Ground Fault As explained earlier, $\boldsymbol{Y}_F^{0,1,2}$ is used because for this type of fault, $\boldsymbol{Z}_F^{0,1,2}$ does not exist.

From (5.65), we get
$$\boldsymbol{Y}_F^{0,1,2} = \frac{y_F}{3}\begin{bmatrix} 1 & 1 & 1 \\ 1 & 1 & 1 \\ 1 & 1 & 1 \end{bmatrix}$$

From Eq. (5.70),

$$\boldsymbol{I}_F^{0,1,2} = \frac{y_F}{3}\begin{bmatrix} 1 & 1 & 1 \\ 1 & 1 & 1 \\ 1 & 1 & 1 \end{bmatrix}\left\{\begin{bmatrix} 1 & 0 & 0 \\ 0 & 0 & 0 \\ 0 & 0 & 1 \end{bmatrix} + \begin{bmatrix} Z_{pp}^{(0)} & 0 & 0 \\ 0 & Z_{pp}^{(1)} & 0 \\ 0 & 0 & Z_{pp}^{(2)} \end{bmatrix} \times \frac{y_F}{3}\begin{bmatrix} 1 & 1 & 1 \\ 1 & 1 & 1 \\ 1 & 1 & 1 \end{bmatrix}\right\}^{-1}\begin{bmatrix} 0 \\ \sqrt{3} \\ 0 \end{bmatrix}|V_a|$$

For short-circuit studies, it is generally assumed that for rotating elements, the positive and negative sequence impedances are equal. This implies that the primitive impedance matrix for the generator phase impedances is completely symmetric. With this assumption, $\boldsymbol{Z}_{pp}^{(1)} = \boldsymbol{Z}_{pp}^{(2)}$, since for the transmission network, the positive and negative sequence networks are identical. This assumption is, however, not necessary as the subsequent analysis shows.

$$\boldsymbol{I}_F^{0,1,2} = \frac{y_F}{3}\begin{bmatrix} 1 & 1 & 1 \\ 1 & 1 & 1 \\ 1 & 1 & 1 \end{bmatrix}\begin{bmatrix} 1+\boldsymbol{Z}_{pp}^{(0)}\frac{y_F}{3} & \boldsymbol{Z}_{pp}^{(0)}\frac{y_F}{3} & \boldsymbol{Z}_{pp}^{(0)}\frac{y_F}{3} \\ \boldsymbol{Z}_{pp}^{(1)}\frac{y_F}{3} & 1+\boldsymbol{Z}_{pp}^{(1)}\frac{y_F}{3} & \boldsymbol{Z}_{pp}^{(1)}\frac{y_F}{3} \\ \boldsymbol{Z}_{pp}^{(2)}\frac{y_F}{3} & \boldsymbol{Z}_{pp}^{(2)}\frac{y_F}{3} & 1+\boldsymbol{Z}_{pp}^{(2)}\frac{y_F}{3} \end{bmatrix}^{-1}\begin{bmatrix} 0 \\ \sqrt{3} \\ 0 \end{bmatrix}|\boldsymbol{V}_a| \tag{5.79}$$

$$= \frac{\sqrt{3}}{Z_{pp}^{(0)} + 3z_F + Z_{pp}^{(1)} + Z_{pp}^{(2)}}\begin{bmatrix} 1 \\ 1 \\ 1 \end{bmatrix}|\boldsymbol{V}_a|$$

In arriving at this simplification, the following interesting matrix identity could be used

$$(\boldsymbol{U}+\boldsymbol{B})^{-1} = \boldsymbol{U} - \frac{\boldsymbol{B}}{(1+\alpha)}$$

if $\boldsymbol{B}$ is expressible as $\boldsymbol{u}\boldsymbol{v}^T$, where $\boldsymbol{u}$ and $\boldsymbol{v}$ are vectors, and $\alpha = \sum_i \text{diag}\,(b_{ii}) = \boldsymbol{v}^T\boldsymbol{u}$.

This can be verified by simply taking the product of $(\boldsymbol{U}+\boldsymbol{B})$ with $\left(\boldsymbol{U} - \frac{\boldsymbol{B}}{(1+\alpha)}\right)$ and noting that $\boldsymbol{v}^T\boldsymbol{u} = \alpha$.

Phase currents corresponding to the fault current are

$$\begin{bmatrix} I_{(F)}^{(a)} \\ I_{(F)}^{(b)} \\ I_{(F)}^{(c)} \end{bmatrix} = \begin{bmatrix} \dfrac{3}{Z_{pp}^{(0)} + Z_{pp}^{(1)} + Z_{pp}^{(2)} + 3z_F} \\ 0 \\ 0 \end{bmatrix} |V_a| \tag{5.80}$$

Voltage at the faulted bus is

$$\begin{aligned} \begin{bmatrix} V_{(F)}^{(0)} \\ V_{(F)}^{(1)} \\ V_{(F)}^{(2)} \end{bmatrix} &= \left[\boldsymbol{U} + \boldsymbol{Z}_{pp}^{0,1,2} Y_F^{0,1,2} \right]^{-1} \begin{bmatrix} 0 \\ \sqrt{3} \\ 0 \end{bmatrix} Z_{pp}^{(2)} \\ &= \frac{\sqrt{3}}{\boldsymbol{Z}_{pp}^{(0)} + \boldsymbol{Z}_{pp}^{(1)} + \boldsymbol{Z}_{pp}^{(2)} + 3z_F} \begin{bmatrix} -Z_{pp}^{(0)} \\ \boldsymbol{Z}_{pp}^{(0)} + Z_{pp}^{(2)} + 3z_F \\ -Z_{pp}^{(2)} \end{bmatrix} |\boldsymbol{V}_a| \end{aligned} \tag{5.81}$$

Phase components of the faulted voltage are obtained from the symmetric component transformation. For conciseness, define $Z_{pp}^{(0)} + Z_{pp}^{(1)} + Z_{pp}^{(2)} + 3z_F = Z_{1g}$. Then the phase voltages at the faulted bus are

$$\begin{bmatrix} V_{(F)}^{(a)} \\ V_{(F)}^{(b)} \\ V_{(F)}^{(c)} \end{bmatrix} = \begin{bmatrix} \dfrac{3z_F}{Z_{1g}} \\ a^2 - \dfrac{Z_{pp}^{(0)} + a^2 Z_{pp}^{(1)} + aZ_{pp}^{(2)}}{Z_{1g}} \\ a - \dfrac{Z_{pp}^{(0)} + aZ_{pp}^{(1)} + a^2 Z_{pp}^{(2)}}{Z_{1g}} \end{bmatrix} |V_a| \tag{5.82}$$

The sequence voltages at the other buses are

$$\begin{aligned} \begin{bmatrix} V_{i(F)}^{(0)} \\ V_{i(F)}^{(1)} \\ V_{i(F)}^{(2)} \end{bmatrix} &= \begin{bmatrix} 0 \\ \sqrt{3} \\ 0 \end{bmatrix} |V_a| - \sqrt{3} \begin{bmatrix} \dfrac{Z_{ip}^{(0)}}{Z_{1g}} \\ \dfrac{Z_{ip}^{(1)}}{Z_{1g}} \\ \dfrac{Z_{ip}^{(2)}}{Z_{1g}} \end{bmatrix} |V_a| \quad \begin{matrix} i = 1, 2, \ldots, n \\ i \neq p \end{matrix} \\ &= \sqrt{3} \begin{bmatrix} -\dfrac{Z_{ip}^{(0)}}{Z_{1g}} \\ \left(1 - \dfrac{Z_{ip}^{(1)}}{Z_{1g}}\right) \\ -\dfrac{Z_{ip}^{(2)}}{Z_{1g}} \end{bmatrix} |V_a| \quad \begin{matrix} i = 1, 2, \ldots, n \\ i \neq p \end{matrix} \end{aligned} \tag{5.83}$$

In phase components

$$\begin{bmatrix} V_{i(F)}^{a} \\ V_{i(F)}^{b} \\ V_{i(F)}^{c} \end{bmatrix} = \begin{bmatrix} 1 - \dfrac{Z_{ip}^{(0)} + Z_{ip}^{(1)} + Z_{ip}^{(2)}}{Z_{1g}} \\ a^2 - \dfrac{Z_{ip}^{(0)} + a^2 Z_{ip}^{(1)} + a Z_{ip}^{(2)}}{Z_{1g}} \\ a - \dfrac{Z_{ip}^{(0)} + a Z_{ip}^{(1)} + a^2 Z_{ip}^{(2)}}{Z_{1g}} \end{bmatrix} |V_a| \quad \begin{matrix} i = 1, 2, \ldots, n \\ i \neq p \end{matrix} \tag{5.84}$$

Example 5.2 For the sample system shown in Fig. 5.11, the following data is given

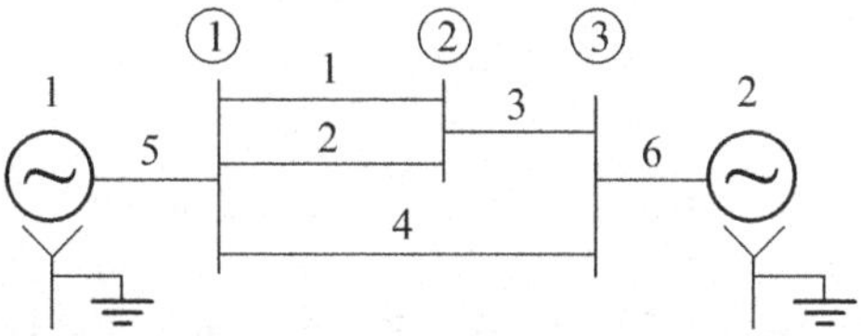

Fig. 5.11 *Example for short-circuit studies*

Line Reactances

Bus code	Self-impedance			Mutual impedance	
	+ve seq.	– ve seq.	Zero seq	Zero seq	Coupling element
1-2(1)	0.05	0.05	0.10	0.05	1-2(2)
1-2(2)	0.05	0.05	0.12	0.05	1-2(1)
2-3	0.06	0.06	0.12	—	—
1-3	0.1	0.1	0.15	—	—

Generator Reactances

Generator	Impedance		
	+ve seq.	-ve seq.	Zero seq
1	0.25	0.15	0.04
2	0.20	0.12	0.02

Assume that $V_0^a = 1 + j0$

1. Using symmetrical components, calculate the following for a symmetrical three-phase to ground fault at bus 3.
   (a) fault currents, b) line currents, and c) bus voltages during fault.
2. Using symmetrical components, calculate the following for a single-phase to ground fault at bus 3.
   (a) fault currents, b) line currents, and c) bus voltages during fault.

Solution: The positive, negative, and zero sequence bus impedance matrices are obtained by using the building algorithm.

Combining the elements of the three sequence impedance matrices, the bus impedance matrix is

$$Z_{bus}^{0,1,2} = \begin{array}{cc} & \begin{array}{ccccccccc} & \mathbf{1} & & & \mathbf{2} & & & \mathbf{3} & \\ 0 & 1 & 2 & 0 & 1 & 2 & 0 & 1 & 2 \end{array} \\ \begin{array}{c} 0 \\ \mathbf{1}\ 1 \\ 2 \\ 0 \\ \mathbf{2}\ 1 \\ 2 \\ 0 \\ \mathbf{3}\ 1 \\ 2 \end{array} & \left[\begin{array}{ccc:ccc:ccc} 0.0291 & 0 & 0 & 0.0198 & 0 & 0 & 0.0054 & 0 & 0 \\ 0 & 0.124 & 0 & 0 & 0.118 & 0 & 0 & 0.101 & 0 \\ 0 & 0 & 0.07905 & 0 & 0 & 0.07262 & 0 & 0 & 0.0556 \\ \hdashline 0.0198 & 0 & 0 & 0.0639 & 0 & 0 & 0.0105 & 0 & 0 \\ 0 & 0.118 & 0 & 0 & 0.1316 & 0 & 0 & 0.1065 & 0 \\ 0 & 0 & 0.07262 & 0 & 0 & 0.08776 & 0 & 0 & 0.06442 \\ \hdashline 0.0054 & 0 & 0 & 0.0105 & 0 & 0 & 0.017 & 0 & 0 \\ 0 & 0.101 & 0 & 0 & 0.1065 & 0 & 0 & 0.1188 & 0 \\ 0 & 0 & 0.0556 & 0 & 0 & 0.06442 & 0 & 0 & 0.0745 \end{array}\right] \end{array}$$

(A) *Three-Phase to Ground Fault Bus 3* Assuming the fault impedance to be zero, the total fault current at bus 3 using Eq. (5.73) is

$$I_F^{0,1,2} = \begin{bmatrix} 0 \\ \dfrac{\sqrt{3}}{Z_{33}^{(1)} + z_F} \\ 0 \end{bmatrix} = \begin{bmatrix} 0 \\ \dfrac{\sqrt{3}}{0.1188} \\ 0 \end{bmatrix} = \begin{bmatrix} 0 \\ 14.58 \\ 0 \end{bmatrix}$$

Using Eq. (5.74), the phase components of the fault current are

$$I_F^{a,b,c} = S_3^{-1} I_F^{0,1,2} = 8.417 \begin{bmatrix} 1 \\ a^2 \\ a \end{bmatrix}$$

From Eqs (5.75) and (5.76), the sequence and phase voltages at the faulted bus are given by

$$V_{3(F)}^{0,1,2} = \begin{bmatrix} 0 \\ \dfrac{\sqrt{3} z_F}{z_F + Z_{33}^{(1)}} \\ 0 \end{bmatrix} = \begin{bmatrix} 0 \\ 0 \\ 0 \end{bmatrix}$$

and

$$V_{3F}^{a,b,c} = S_3^{-1} V_{3(F)}^{0,1,2} = \begin{bmatrix} 0 \\ 0 \\ 0 \end{bmatrix}$$

From Eq. (5.77), voltages at the other buses are given by

$$V_{1(F)}^{0,1,2} = \begin{bmatrix} 0 \\ \sqrt{3}\left(1-\dfrac{Z_{13}^{(1)}}{z_F + Z_{33}^{(1)}}\right) \\ 0 \end{bmatrix} = \begin{bmatrix} 0 \\ 0.259 \\ 0 \end{bmatrix}$$

$$V_{2(F)}^{0,1,2} = \begin{bmatrix} 0 \\ \sqrt{3}\left(1-\dfrac{Z_{23}^{(1)}}{z_F + Z_{33}^{(1)}}\right) \\ 0 \end{bmatrix} = \begin{bmatrix} 0 \\ 0.1793 \\ 0 \end{bmatrix}$$

Using Eq. (5.78), the phase voltages at these buses are

$$V_{1(F)}^{a,b,c} = S_3^{-1} V_{1(F)}^{0,1,2} = 0.1485 \begin{bmatrix} 1 \\ a^2 \\ a \end{bmatrix}$$

$$V_{2(F)}^{a,b,c} = S_3^{-1} V_{2(F)}^{0,1,2} = 0.1035 \begin{bmatrix} 1 \\ a^2 \\ a \end{bmatrix}$$

Since the bus voltages in the zero and the negative sequence networks are zero, the short-circuit currents in the lines connected to the faulted bus are

$$I_{23(F)}^{0,1,2} = \begin{bmatrix} 0 \\ y_{23}^{(1)}\left(V_{2(F)}^{(1)} - V_{3(F)}^{(1)}\right) \\ 0 \end{bmatrix} = \begin{bmatrix} 0 \\ \dfrac{0.1793}{0.06} \\ 0 \end{bmatrix} = \begin{bmatrix} 0 \\ 2.9883 \\ 0 \end{bmatrix}$$

$$I_{13(F)}^{0,1,2} = \begin{bmatrix} 0 \\ y_{13}^{(1)}\left(V_{1(F)}^{(1)} - V_{3(F)}^{(1)}\right) \\ 0 \end{bmatrix} = \begin{bmatrix} 0 \\ \dfrac{0.259}{0.1} \\ 0 \end{bmatrix} = \begin{bmatrix} 0 \\ 2.59 \\ 0 \end{bmatrix}$$

The currents in both the lines between bus 1 and bus 2 will be the same since their positive sequence impedances are equal, i.e.,

$$I_{12(1)(F)}^{0,1,2} = I_{12(2)(F)}^{0,1,2}$$

and

$$I_{12(1)(F)}^{a,b,c} = I_{12(2)(F)}^{a,b,c}$$

$$I_{12(1)(F)}^{0,1,2} = \begin{bmatrix} 0 \\ y_{12}^{(1)}\left(V_{1(F)}^{(1)} - V_{2(F)}^{(1)}\right) \\ 0 \end{bmatrix} = \begin{bmatrix} 0 \\ \dfrac{0.0797}{0.05} \\ 0 \end{bmatrix} = \begin{bmatrix} 0 \\ 1.594 \\ 0 \end{bmatrix}$$

The phase currents are

$$I_{23(F)}^{a,b,c} = S_3^{-1} I_{23(F)}^{0,1,2} = 1.725 \begin{bmatrix} 1 \\ a^2 \\ a \end{bmatrix}$$

$$I_{13(F)}^{a,b,c} = S_3^{-1} I_{13(F)}^{0,1,2} = 1.495 \begin{bmatrix} 1 \\ a^2 \\ a \end{bmatrix}$$

$$I_{12(1)(F)}^{a,b,c} = S_3^{-1} I_{12(F)}^{0,1,2} = 0.9203 \begin{bmatrix} 1 \\ a^2 \\ a \end{bmatrix}$$

(B) *Single-Phase to Ground Fault at Bus 3* For the single-phase to ground fault at bus 3, the fault current at bus 3, assuming zero fault impedance, is given by Eq. (5.79) as

$$I_{(F)}^{0,1,2} = \frac{\sqrt{3}}{Z_{33}^{(0)} + Z_{33}^{(1)} + Z_{33}^{(2)} + 3z_F} \begin{bmatrix} 1 \\ 1 \\ 1 \end{bmatrix} = \frac{\sqrt{3}}{0.017 + 0.1188 + 0.0745} \begin{bmatrix} 1 \\ 1 \\ 1 \end{bmatrix} = \begin{bmatrix} 8.236 \\ 8.236 \\ 8.236 \end{bmatrix}$$

The phase components of the total fault current are given by Eq. (5.80) as

$$I_{(F)}^{a,b,c} = S_3^{-1} I_{(F)}^{0,1,2} = \begin{bmatrix} 14.265 \\ 0 \\ 0 \end{bmatrix}$$

The bus voltages during the fault are given by Eq. (5.81) as

$$V_{(F)}^{0,1,2} = V_{3(F)}^{0,1,2} = \frac{\sqrt{3}}{Z_{33}^{(0)} + Z_{33}^{(1)} + Z_{33}^{(2)} + 3z_F} \begin{bmatrix} -Z_{33}^{(0)} \\ Z_{33}^{(0)} + Z_{33}^{(2)} + 3z_F \\ -Z_{33}^{(2)} \end{bmatrix}$$

$$= \frac{\sqrt{3}}{0.017 + 0.1188 + 0.0745} \begin{bmatrix} -0.0170 \\ 0.0170 + 0.0745 \\ -0.0745 \end{bmatrix}$$

$$= \begin{bmatrix} -0.1400 \\ 0.7536 \\ -0.6136 \end{bmatrix}$$

The other bus voltages are given by Eq. (5.83) as

$$V_{1(F)}^{0,1,2} = \begin{bmatrix} 0 \\ \sqrt{3} \\ 0 \end{bmatrix} - \frac{\sqrt{3}}{Z_{33}^{(0)} + Z_{33}^{(1)} + Z_{33}^{(2)} + 3z_F} \begin{bmatrix} Z_{13}^{(0)} \\ Z_{13}^{(1)} \\ Z_{13}^{(2)} \end{bmatrix}$$

$$= \begin{bmatrix} 0 \\ \sqrt{3} \\ 0 \end{bmatrix} - \frac{\sqrt{3}}{0.017+0.1188+0.0745} \begin{bmatrix} 0.0054 \\ 0.101 \\ 0.0556 \end{bmatrix} = \begin{bmatrix} -0.0444 \\ 0.9002 \\ -0.4579 \end{bmatrix}$$

$$V_{2(F)}^{0,1,2} = \begin{bmatrix} 0 \\ \sqrt{3} \\ 0 \end{bmatrix} - \frac{\sqrt{3}}{Z_{33}^{(0)} + Z_{33}^{(1)} + Z_{33}^{(2)} + 3z_F} \begin{bmatrix} Z_{23}^{(0)} \\ Z_{23}^{(1)} \\ Z_{23}^{(2)} \end{bmatrix}$$

$$= \begin{bmatrix} 0 \\ \sqrt{3} \\ 0 \end{bmatrix} - \frac{\sqrt{3}}{0.0170+0.1188+0.0745} \begin{bmatrix} 0.0105 \\ 0.1065 \\ 0.06442 \end{bmatrix} = \begin{bmatrix} -0.0865 \\ 0.8549 \\ -0.5306 \end{bmatrix}$$

The phase components of the bus voltages are

$$V_{3(F)}^{a,b,c} = S_3^{-1} V_{3(F)}^{0,1,2} = \begin{bmatrix} 0 \\ -0.1212 - j0.6836 \\ -0.1212 + j0.6836 \end{bmatrix}$$

$$V_{1(F)}^{a,b,c} = S_3^{-1} V_{1(F)}^{0,1,2} = \begin{bmatrix} 0.2297 \\ -0.1533 - j0.67902 \\ -0.1533 + j0.67902 \end{bmatrix}$$

$$V_{2(F)}^{a,b,c} = S_3^{-1} V_{2(F)}^{0,1,2} = \begin{bmatrix} 0.1372 \\ -0.1436 - j0.6972 \\ 0.1436 + j0.6972 \end{bmatrix}$$

The short-circuit currents in the lines are

$$I_{13(F)}^{0,1,2} = \begin{bmatrix} \left(V_{1(F)}^{(0)} - V_{3(F)}^{(0)}\right) & y_{13}^{(0)} \\ \left(V_{1(F)}^{(1)} - V_{3(F)}^{(1)}\right) & y_{13}^{(1)} \\ \left(V_{1(F)}^{(2)} - V_{3(F)}^{(2)}\right) & y_{13}^{(2)} \end{bmatrix} = \begin{bmatrix} \frac{0.0956}{0.15} \\ \frac{0.1466}{0.1} \\ \frac{0.1557}{0.1} \end{bmatrix} = \begin{bmatrix} 0.6373 \\ 1.466 \\ 1.557 \end{bmatrix}$$

$$I_{12(1)(F)}^{0,1,2} = \begin{bmatrix} \left(V_{1(F)}^{(0)} - V_{2(F)}^{(0)}\right) & \left(y_{12(1)}^{(0)} + y_{m12}^{(0)}\right) \\ \left(V_{1(F)}^{(1)} - V_{2(F)}^{(1)}\right) & y_{12(1)}^{(1)} \\ \left(V_{1(F)}^{(2)} - V_{2(F)}^{(2)}\right) & y_{12(1)}^{(2)} \end{bmatrix} = \begin{bmatrix} 0.3102 \\ 0.906 \\ 1.454 \end{bmatrix}$$

$$I_{12(2)(F)}^{0,1,2} = \begin{bmatrix} \left(V_{1(F)}^{(0)} - V_{2(F)}^{(0)}\right) & \left(y_{12(2)}^{(0)} + y_{m12}^{(0)}\right) \\ \left(V_{1(F)}^{(1)} - V_{2(F)}^{(1)}\right) & y_{12(2)}^{(1)} \\ \left(V_{1(F)}^{(1)} - V_{2(F)}^{(1)}\right) & y_{12(2)}^{(2)} \end{bmatrix} = \begin{bmatrix} 0.2216 \\ 0.906 \\ 1.454 \end{bmatrix}$$

$$I_{23(F)}^{0,1,2} = \begin{bmatrix} \left(V_{2(F)}^{(0)} - V_{3(F)}^{(0)}\right) & y_{23}^{(0)} \\ \left(V_{2(F)}^{(1)} - V_{3(F)}^{(1)}\right) & y_{23}^{(1)} \\ \left(V_{2(F)}^{(1)} - V_{3(F)}^{(1)}\right) & y_{23}^{(2)} \end{bmatrix} = \begin{bmatrix} 0.4458 \\ 1.688 \\ 1.3833 \end{bmatrix}$$

The phase components of the line currents are

$$I_{13(F)}^{a,b,c} = S_3^{-1} I_{13(F)}^{0,1,2} = \begin{bmatrix} 2.133 \\ -0.5047 + j0.0456 \\ -0.5047 - j0.0456 \end{bmatrix}$$

$$I_{12(1)(F)}^{a,b,c} = S_3^{-1} I_{12(1)(F)}^{0,1,2} = \begin{bmatrix} 1.542 \\ -0.502 + j0.274 \\ -0.502 - j0.274 \end{bmatrix}$$

$$I_{12(2)(F)}^{a,b,c} = S_3^{-1} I_{12(2)(F)}^{0,1,2} = \begin{bmatrix} 1.4905 \\ -0.553 + j0.274 \\ -0.553 - j0.274 \end{bmatrix}$$

and

$$I_{23(F)}^{a,b,c} = S_3^{-1} I_{23(F)}^{0,1,2} = \begin{bmatrix} 2.031 \\ -0.6291 - j0.1528 \\ -0.6291 + j0.1528 \end{bmatrix}$$

5.11 CONCLUSION

In this chapter we have discussed methods for carrying out short-circuit studies for both symmetrical and unsymmetrical faults. The procedure has been outlined both in terms of the phase quantities and the symmetrical components.

Problems

5.1 The positive sequence network impedances are as shown in Fig. P.5.1. Assume the prefault voltages to be $1\angle 0°$. For a fault at bus 3, compute the fault current, currents in the transmission lines, currents supplied by the two generators and the voltages at the buses. Assume the fault impedance as zero and follow the procedure of Section 5.6.

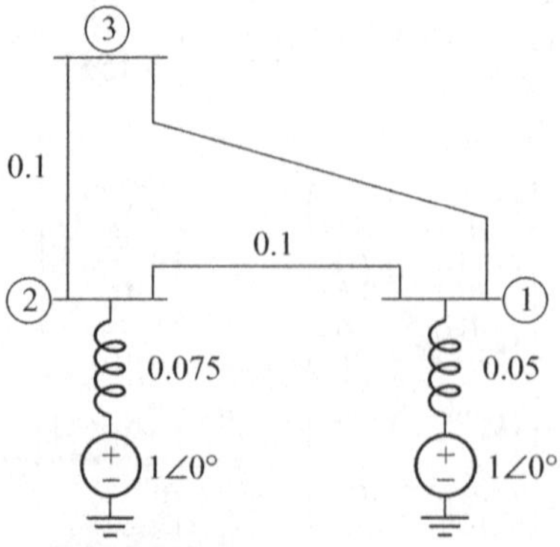

Fig. P.5.1

5.2 Consider the network in Fig. P.5.2.

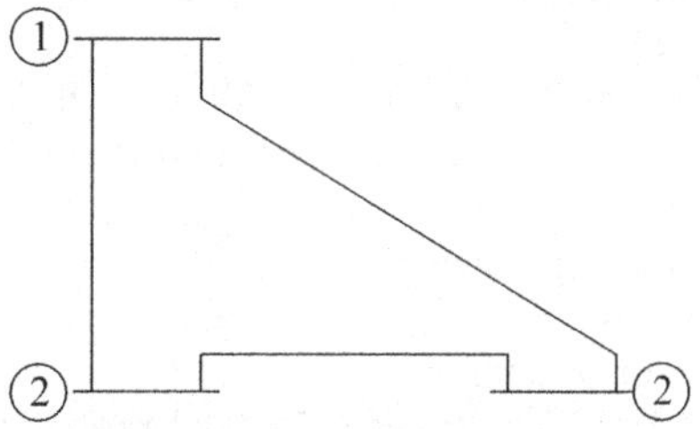

Fig. P. 5.2

For both the generators connected at buses 2 and 3, the primitive impedance matrix in sequence variables is identical and is given by

$$\begin{array}{c} \\ 0 \\ 1 \\ 2 \end{array}\begin{array}{c} \begin{array}{ccc} 0 & 1 & 2 \end{array} \\ \begin{bmatrix} 0.05 & 0 & 0 \\ 0 & 0.10 & 0 \\ 0 & 0 & 0.10 \end{bmatrix} \end{array}$$

The zero, the positive, and the negative sequence impedances for the lines are

$$\text{Line } \mathbf{1-2} \qquad \text{Line } \mathbf{1-3} \qquad \text{Line } \mathbf{2-3}$$

$$\begin{array}{c} \\ 0 \\ 1 \\ 2 \end{array}\begin{array}{c} \begin{array}{ccc} 0 & 1 & 2 \end{array} \\ \begin{bmatrix} 0.1 & 0 & 0 \\ 0 & 0.5 & 0 \\ 0 & 0 & 0.5 \end{bmatrix} \end{array} \quad \begin{array}{c} \\ 0 \\ 1 \\ 2 \end{array}\begin{array}{c} \begin{array}{ccc} 0 & 1 & 2 \end{array} \\ \begin{bmatrix} 0.2 & 0 & 0 \\ 0 & 0.5 & 0 \\ 0 & 0 & 0.5 \end{bmatrix} \end{array} \quad \begin{array}{c} \\ 0 \\ 1 \\ 2 \end{array}\begin{array}{c} \begin{array}{ccc} 0 & 1 & 2 \end{array} \\ \begin{bmatrix} 0.5 & 0 & 0 \\ 0 & 0.25 & 0 \\ 0 & 0 & 0.25 \end{bmatrix} \end{array}$$

The coupling submatrix between lines 1-3 and 1-2 is given by

$$\begin{array}{c} \\ 0 \\ 1 \\ 2 \end{array}\begin{array}{c} \begin{array}{ccc} 0 & 1 & 2 \end{array} \\ \begin{bmatrix} 0.15 & 0 & 0 \\ 0 & 0 & 0 \\ 0 & 0 & 0 \end{bmatrix} \end{array}$$

1. Construct the three sequence networks and obtain the respective $\boldsymbol{Z}_{\text{bus}}$ for each one of them. From these, construct the $\boldsymbol{Z}_{\text{bus}}^{0,1,2,}$.
2. Calculate the fault currents, line currents, bus voltages, and currents delivered by the generator for
   (a) symmetrical three-phase to ground fault at bus e,
   (b) line to ground fault at bus 3,
   (c) line to line fault at bus e, and assume the fault impedance as zero.

5.3 Derive $\boldsymbol{Z}_F^{a,b,c}$ in Eq. (5.57) using the bus impedance algorithm.

(*Hint*: Denote the node connecting $\boldsymbol{Z}_a, \boldsymbol{Z}_b, \boldsymbol{Z}_c$, to $\boldsymbol{Z}_g$, as $0'$ and after obtaining the 4th order $\boldsymbol{Z}_{\text{bus}}$, impose the restriction that the bus current at $0'$ is zero and hence, bus $0'$ can be eliminated.)

5.4 For Fig. 5.10, write the node equations taking node **0** as the reference and then eliminate the node connecting y_a, y_b, and y_c, to obtain $Y_F^{a,b,c}$, and verify this with Eq. (5.58).

5.5 Derive the equations for the total fault current in terms of symmetrical components and phase quantities for the following faults at bus **p** for a general system:
- (a) three-phase (not grounded),
- (b) line-to-line, and
- (c) line-to-line to ground.

5.6 Use expressions derived in Problem 5.4 for Example 5.2 to compute the short-circuit fault current for each of the cases.

5.7 For Example 5.2, compute the fault current and voltage at the faulted bus, and the currents delivered by the generators for fault at bus **2**. Consider both 3 phase and *L-G* fault.

5.8 From the expressions for the fault current for the line to ground fault, give an interpretation in terms of connecting the three sequence networks. Repeat this for other types of unsymmetrical faults and verify your results from a standard text on power systems analysis.

5.9 Derive the expression (5.79) using the matrix identity given.

6 POWER FLOW STUDIES

6.1 INTRODUCTION

Power flow constitutes perhaps the most important study in the planning and expansion of power systems. It is also the study frequently carried out by the utilities for planning the operations. The purpose of the study is to compute the steady state operating condition of the system, i.e., voltage magnitudes and phase angles at the buses. From these, other quantities like the real and reactive power supplied by the generators, the power flow through the lines, and the loading of the transformers can be calculated. This helps to detect overloaded conditions as well as poor voltages existing in parts of the system. The angle separation between the generator buses or between any two buses gives a qualitative idea of the steady state stability of the power system. The planning engineer performs load flows for different configurations and loading conditions before deciding on a final configuration. The mathematical model for the study of power flow is a set of nonlinear algebraic equations that basically express power (real and reactive) conservation at a bus. These equations can be expressed as a set of real equations with the voltages either in the rectangular or the polar form. The network matrix used in deriving the above equations is generally the Y_{bus} matrix although there are techniques that use the Z_{bus} matrix. Y_{bus} is preferred because of its sparse structure and efficient programming techniques that enable fast solutions using only the nonzero entries. Hence, we confine to formulations that use the Y_{bus} only.

The mathematical techniques used to solve the nonlinear set of algebraic equations are either the Gauss–Seidel or the Newton's method. The solution techniques are iterative starting from an initial guess. The Gauss–Seidel method takes a large number of iterations depending on the system size, while the Newton's method converges in three to four iterations. Modern day power flow programs invariably use Newton's method.

Shunt and series compensation is often required for a stable operation with system variables within the acceptable range. To start with, fixed capacitors/inductors were used. However, the inclusion of power electronics in the form of flexible AC transmission system (FACTS) devices has made the compensation more effective and helped to operate the system close to its limits. High voltage direct current (HVDC) transmission forms a considerable portion of the present day power system. The increasing requirement of long distance transmission and integration of renewable power sources located at remote regions will certainly necessitate the installation of more number of HVDC lines. Therefore, it is important to know the methodology for including the effects of HVDC and FACTS devices in the studies involving load flow, which will be discussed here.

6.2 MATHEMATICAL MODEL

6.2.1 Classification of Buses

A balanced three-phase system is assumed so that considering only the positive sequence network is sufficient. In the positive sequence network, the element values are those obtained on a per phase basis and there is no mutual coupling between the network elements. To start with, we assume a transmission network with no phase-shifters or tap changing transformers. The Y_{bus} can hence be written by inspection. The n-port representation shown in Fig. 6.1 is given by

$$I_{\text{bus}} = Y_{\text{bus}} V_{\text{bus}} \tag{6.1}$$

When loads and/or generators are connected at the buses, these give rise to port-constraints. We can specify the constraints in terms of P_i, Q_i, the net injected real and reactive power, respectively, or $|V_i|$, the magnitude of voltages. The voltage magnitude at a bus can be kept constant through a voltage regulator as in the case of a generator bus or by means of a tap changing transformer in the case of a load bus. In addition, we need to have a reference bus in the system whose phase angle is zero. At any bus, there are therefore four variables, P_i, Q_i, $|V_i|$, and θ_i (the phase angle of the voltage at the bus). Two of these four quantities are specified giving rise to the following classification.

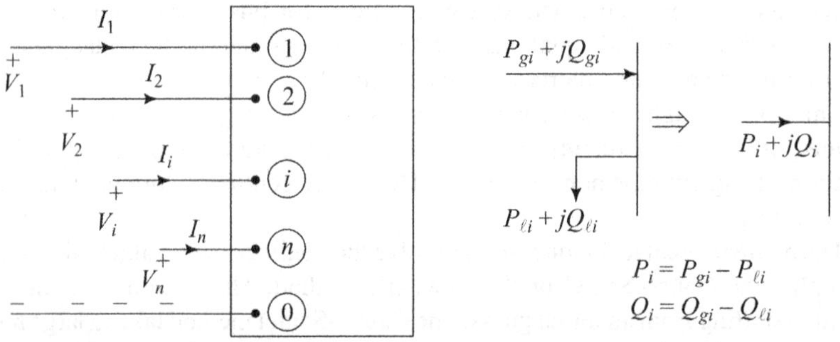

Fig. 6.1 *n-port description of a transmission network*

***P-Q* bus** A P-Q bus is one where the total injected power (real and reactive) is specified. The unknowns are $|V_i|$ and θ_i. At such a bus (say the ith bus), suppose there is both generation and load, then the port constraint is

$$V_i I_i^* = S_i^{sp} = P_i^{sp} + jQ_i^{sp} \tag{6.2}$$

where $P_i^{sp} = P_{gi}^{sp} - P_{\ell i}^{sp}$ and $Q_i^{sp} = Q_{gi}^{sp} - Q_{\ell i}^{sp}$.

From Eq. (6.2), the constraints are

$$P_i^{sp} = Re\left(V_i I_i^*\right) \tag{6.3}$$

$$Q_i^{sp} = Im\left(V_i I_i^*\right) \tag{6.4}$$

The subscripts g and ℓ refer to generation and load respectively, and the superscript 'sp' refers to specified value.

***P-V* bus** A *P-V* bus is one where the real power injection is specified as P_i^{sp} and the voltage magnitude is held constant as $|V_i|^{sp}$. The unknowns are Q_i and θ_i. Thus the constraints are

$$P_i^{sp} = Re\left(V_i I_i^*\right) \text{ where } P_i^{sp} = P_{gi}^{sp} - P_{\ell i}^{sp} \tag{6.5}$$

and

$$|V_i| = |V_i|^{sp} \tag{6.6}$$

***V-θ* bus** At this bus, the voltage magnitude is specified as $|V_s|^{sp}$ and the angle θ_s is specified, generally as 0°. In the steady state AC analysis of networks, we generally have a reference phasor. There is one more reason for specifying both the voltage magnitude and the angle at one bus in the system. In the practical sense, the system I^2R real power loss and I^2X reactive power loss are not known in advance and hence, it is not possible to fix the injected real and reactive powers at all the buses. At one of the buses, generally the bus having the largest generation, the injected powers P_i and Q_i are not specified but calculated at the end of the computation. Thus, effectively in the systems, I^2R and I^2X losses are absorbed at this bus. It is called the *swing* or the *slack* bus and hence the subscript 's' for quantities at this bus. Usually, $|V_s|^{sp}$ is set at a value between 1.0 and 1.05 p.u. so that the other bus voltages converge at values close to unity. At the end of the computation, we can compute the injected powers P_s and Q_s by summing the powers over the lines connected to bus s. As we know,

$$P_s = P_{gs} - P_{\ell s}^{sp} \tag{6.7}$$

$$Q_s = P_{gs} - Q_{\ell s}^{sp} \tag{6.8}$$

the generated powers P_{gs} and Q_{gs} at the slack bus can be computed since $P_{\ell s}^{sp}$ and $Q_{\ell s}^{sp}$ are the specified load powers. Table 6.1 summarizes the known and the unknown quantities at the various buses.

Table 6.1 *Bus information*

Type of Bus	*Specified*	*Unknowns*	*Number of Buses*
P-Q (Load Bus)	P_i^{sp}, Q_i^{sp}	$\lvert V_i\rvert$, θ_i	n_1
P-V (Voltage-Controlled Bus)	P_i^{sp}, $\lvert V_i\rvert^{sp}$	Q_i, θ_i	n_2
V-θ (Swing-Bus)	$\lvert V_s\rvert^{sp}$, θ_s^{sp} ($\theta_s^{sp} \equiv 0$)	P_s, Q_s	1

The total number of buses is equal to $n_1 + n_2 + 1 = n$.

Notation

We use the following notation

$$Y_{ij} = i - j \text{ element of } \boldsymbol{Y}_{\text{bus}} = G_{ij} + jB_{ij} = Y_{ij}\angle\phi_{ij}$$

$$V_i = e_i + jf_i = V_i\angle\theta_i$$

$$\theta_{ij} = \theta_i - \theta_j$$

In summation, notation (6.1) can be written as

$$I_i = \sum_{j=1}^{n} Y_{ij}V_j \quad i = 1, 2, \ldots, n \tag{6.9}$$

and the expression for the injected power is

$$V_i I_i^* = V_i \sum_{j=1}^{n} Y_{ij}^* V_j^* \tag{6.10}$$

6.2.2 Convergence Criteria

Since P_i^{sp} and Q_i^{sp} are nonlinear constraints in terms of V_i and θ_i, the load flow equations will be nonlinear. A solution is said to have been reached when the constraints (6.3)–(6.6) are exactly satisfied. Nearly all the load flow methods are devised to satisfy Eq. (6.6) exactly. Then Eq. (6.9) is substituted in Eqs. (6.3)–(6.5), and thus expressed in terms of V_i only ($i \neq s$). There are two popular criteria for the convergence of iterative schemes: (1) in terms of the bus mismatches, and (2) in terms of voltages.

Bus Mismatch Criteria

Substituting Eq. (6.9) in the right-hand sides of Eqs. (6.3)–(6.5), we get, in terms of the polar coordinates for V_i and the rectangular notation of Y_{ij}, the following

P-Q Bus

$$P_i^{sp} = Re\left[V_i \sum_{j=1}^{n} Y_{ij}^* V_j^*\right] \tag{6.11}$$

$$= Re\left[|V_i|\angle\theta_i \sum_{j=1}^{n}(G_{ij} - jB_{ij})|V_j|\angle -\theta_j\right]$$

$$= |V_i|\left[\sum_{j=1}^{n}(G_{ij}\cos\theta_{ij} + B_{ij}\sin\theta_{ij})|V_j|\right] \tag{6.12}$$

Similarly,

$$Q_i^{sp} = |V_i|\left[\sum_{j=1}^{n}(G_{ij}\sin\theta_{ij} - B_{ij}\cos\theta_{ij})|V_j|\right] \tag{6.13}$$

P-V Bus

$$P_i^{sp} = |V_i| \left[\sum_{j=1}^{n} \left(G_{ij} \cos \theta_{ij} + B_{ij} \sin \theta_{ij} \right) |V_i| \right] \tag{6.14}$$

At any stage of the iteration, the difference between the left and right-hand sides of Eqs. (6.12)–(6.14) is known as the bus power mismatch and is denoted by ΔP_i and ΔQ_i for the real and reactive powers, respectively. The most common criteria for convergence are

$$\Delta P_i \le C_p \text{ for all } P\text{-}Q \text{ and } P\text{-}V \text{ buses}$$

$$\Delta Q_i \le C_q \text{ for all } P\text{-}Q \text{ buses} \tag{6.15}$$

The values of C_p and C_q are typically of the order of 0.01 p.u. which translated in terms of a base 100 MVA becomes 1 MW or 1 MVAR.

Voltage Magnitude Criteria

Let the iteration count be denoted by k. The difference of voltage between the kth iteration and the previous $(k-1)^{\text{th}}$ iteration is given by

$$|\Delta V_i| = \left| V_i^{(k)} - V_i^{(k-1)} \right| \tag{6.16}$$

Convergence is said to have been reached when $|\Delta V_i| \le C_v$ for all P-Q buses where C_v is typically in the range of .01 to .0001 p.u.

We are now in a position to discuss the two specific methods of solving the load flow equations.

6.3 GAUSS–SEIDEL METHOD

6.3.1 No *P-V* Buses Present

We first discuss this method when there are no P-V buses in the system and then introduce the P-V buses to explain the changes in the algorithm. Let the swing bus be denoted by s where the voltage is assumed to be $V_s = |V_s| \angle \theta_s$.

From Eq. (6.2), the real and reactive power equations together are

$$S_i^{sp} = P_i^{sp} + jQ_i^{sp} = V_i \sum_{j=1} Y_{ij}^* V_j^* \qquad \begin{matrix} i = 1, 2, \ldots, n \\ i \ne s \end{matrix} \tag{6.17}$$

Dividing by V_i and conjugating,

$$\frac{\left(S_i^{sp}\right)^*}{V_i^*} = \sum_{j-1}^{n} Y_{ij} V_j \qquad \begin{matrix} i = 1, 2, \ldots, n \\ i \ne s \end{matrix} \tag{6.18}$$

We can express V_i on the right hand side of Eq. (6.18) in terms of the rest of the variables as

$$V_i = \frac{1}{Y_{ii}} \left[\frac{\left(S_i^{sp}\right)^*}{V_i^*} - \sum_{\substack{j=1 \\ \ne i}}^{n} Y_{ij} V_j \right]_{\substack{i=1, 2, \ldots, n \\ i \ne s}} \tag{6.19}$$

Note that in the above equations, V_s assumes the known value $|V_s|\angle\ \theta_s$, and hence there are only $n - 1$ complex equations. In an expanded form, the Gauss–Seidel iterates corresponding to Eq. (6.19) can be written as

$$V_1^{(k)} = \frac{1}{Y_{11}}\left[\frac{P_1^{sp} - jQ_1^{sp}}{V_1^{(k-1)*}} - Y_{12}V_2^{(k-1)} - Y_{13}V_3^{(k-1)} \cdots - Y_{1s}V_s \cdots - Y_{1n}V_nV_n^{(k-1)}\right]$$

$$V_2^{(k)} = \frac{1}{Y_{22}}\left[\frac{P_2^{sp} - jQ_2^{sp}}{V_2^{(k-1)*}} - Y_{21}V_1^{(k)} - Y_{23}V_3^{(k-1)} \ldots - Y_{2s}V_s \ldots - Y_{2n}V_n^{(k-1)}\right]$$

$$V_i^{(k)} = \frac{1}{Y_{ii}}\left[\frac{P_i^{sp} - jQ_i^{sp}}{V_i^{(k-1)*}} - Y_{i1}V_1^{(k)} \cdots - Y_{i(i-1)}V_{i-1}^{(k)} - Y_{ii}V_i^{k-1} \cdots\cdots - Y_{is}V_s \cdots - Y_{in}V_n^{(k-1)}\right]$$

$$V_n^{(k)} = \frac{1}{Y_{nn}}\left[\frac{P_n^{sp} - jQ_n^{sp}}{V_n^{(k-1)*}} - Y_{n1}V_1^{(k)} \cdots - Y_{ns}V_s \cdots - Y_{n(n-1)}V_n^{(k)}{}_{(n-1)}\right]$$

$$k = 1, 2, \ldots \tag{6.20}$$

A good guess at the initial value $V_i^{(0)}$ $(i = 1, 2, \ldots, n, i \neq s)$ is made. Generally a flat start namely $V_i^{(0)} = 1\angle 0°$ is made. There is no equation for iterating with respect to $V_i(i = 5)$ since the slack bus voltage is known. The iterations $k = 1, 2, \ldots$ are carried on until the test for convergence, either in terms of bus mismatches or voltage, is satisfied. With the Gauss–Seidel method, voltage magnitude criterion is generally specified as

$$\underset{i}{\text{Max}} \left|V_i^{(k)} - V_i^{(k-1)}\right| < C_v \tag{6.21}$$

where $.01 < C_v < .0001$. A flat start namely $V_i^{(0)} = 1\angle 0°(i \neq s)$ is generally made.

6.3.2 *P-V* Buses Present

If there are *P-V* buses in the system, the following changes in the algorithm are made. At a *P-V* bus, the injected Q_i is not specified but $|V_i|$ is specified as $|V_i|^{sp}$. Hence, a different set of computations will have to be carried out for that bus. Generally, we have limitations on Q_{gi}, i.e., $Q_{gi}^{\min} < Q_{gi} < Q_{gi}^{\max}$. When a *P-V* bus is encountered during the computation, we perform the following steps for that bus.

1. We first calculate the injected Q_i^{calc} based on the already calculated voltages as

$$Q_i^{\text{calc}} = |V_i|\left[\sum_{j=1}^{n}\left(G_{ij}\ \sin\ \theta_{ij} - B_{ij}\ \cos\ \theta_{ij}\right)|V_j|\right] \tag{6.22}$$

2. Check if $Q_{gi} = Q_i^{\text{calc}} + Q_{\ell i}$ satisfies the limits

$$Q_{gi}^{\min} < Q_{gi} < Q_{gi}^{\max}$$

If the limits are violated, go to step 5. If Q_{gi} is within the limits, go to step 3.

3. V_i is calculated as

$$V_i^{(k)\text{calc}} = \frac{1}{Y_{ii}}\left[\frac{P_i^{sp} - jQ_i^{\text{calc}}}{V_i^{(k-1)*}} - Y_{i1}V_i^{(k)} \cdots - Y_{i(i-1)}V_{i-1}^{(k)} \cdots - Y_{is}V_s \cdots - Y_{in}V_n^{(k-1)}\right] \tag{6.23}$$

$$= e_i^{(k)\text{calc}} + j\, f_i^{(k)\text{calc}}$$

4. Adjust $\left|V_i^{(k)\text{calc}}\right| = |V_i|^{sp}$ as follows:

From Eq. (6.23), compute

$$\theta_i^{(k)\text{calc}} = \tan^{-1}\left(f_i^{(k)\text{calc}} / e_i^{(k)\text{calc}}\right) \tag{6.24}$$

$V_i^{(k)}$ is adjusted as

$$V_i^{(k)\text{new}} = |V_i|^{sp}\left[\cos\,\theta_i^{(k)\text{calc}} + j\,\sin\,\theta_i^{(k)\text{calc}}\right] \tag{6.25}$$

This value of $V_i^{(k)}$ is used until the next iteration for that P-V bus.

5. If $Q_{gi} < Q_{gi}^{\min}$, set $Q_{gi} = Q_{gi}^{\min}$

If $Q_{gi} > Q_{gi}^{\max}$, set $Q_{gi} = Q_{gi}^{\max}$

Then do the computations as for a *P-Q* bus, i.e., step 3 only and use the value of $V_i^{(k)\text{calc}}$ until the next iteration for that *P-V* bus.

Note: Steps 1-5 have to be done for each iteration at a *P-V* bus since the *P-V* bus may become a *P-Q* bus or vice versa as iterations proceed.

6.3.3 Computation of Line Flows, Bus Powers, and Slack Bus Powers

After the algorithm has converged, the following computations are done to find other quantities of interest. Let $y_{ij} = -Y_{ij}$ be the series admittance of the line between the buses *i* and *j*. Let its line charging (total susceptance) be y'_{ij} so that the π representation is given as in Fig. 6.2(a). The complex power from the bus *i* to *j* is computed as

$$S_{ij} = V_i I_{ij}^* = V_i\left[\left(V_i - V_j\right)^* y_{ij}^* + V_i^* \frac{y'^*_{ij}}{2}\right] = P_{ij} + jQ_{ij} \tag{6.26}$$

Similarly, the line flow from bus *j* to bus *i* is given by

$$S_{ji} = V_j\left[\left(V_j - V_i\right)^* y_{ij}^* + V_j^* \frac{y'^*_{ij}}{2}\right] \tag{6.27}$$

The total line loss in line *i-j* is given by Fig. 6.2(b).

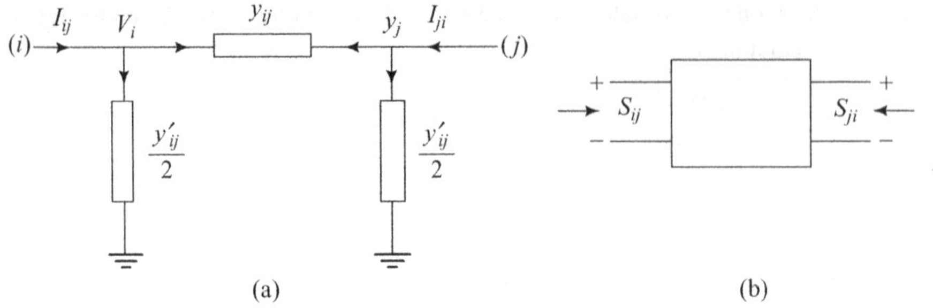

Fig. 6.2 *Calculation of line flows*

$$S_{ij} + S_{ji} = \left(|V_i|^2 + |V_j|^2\right)\left(y_{ij}^* + \frac{y_{ij}'^*}{2}\right) - \left(V_i V_j^* + V_j V_i^*\right) y_{ij}^* \tag{6.28}$$

$$= P_{ij,\text{loss}} + jQ_{ij,\text{loss}}$$

If $y_{ij} = g_{ij} + jb_{ij}$ and $\frac{y_{ij}}{2} = jb_{ii}$, i.e., pure capacitive then (6.28) can be simplified as

$$P_{ij,\text{loss}} = g_{ij}\,|V_i - V_j|^2$$

and
$$Q_{ij,\text{loss}} = -b_{ii}\left(|V_i|^2 + |V_j|^2\right) - b_{ij}\,|V_i - V_j|^2$$

Since $|V_i - V_j|$ = the magnitude of voltage drop across the line, $g_{ij}\,|V_i - V_j|^2$ represents the real power loss on the line. The reactive power loss consists of two components. The term associated with b_{ii} is in the shunt elements and the term associated with b_{ij} is the reactive power loss in the series portion.

6.3.4 Bus Powers and Slack Bus Power

P-V Bus

At a generator bus or a bus where reactive power is supplied, let there be a load of $S_{\ell i} = P_{\ell i} + jQ_{\ell i}$ and for the sake of illustration, let there be three lines going out from that bus (Fig. 6.3). Then,

$$Q_{gi} = Q_{\ell i} + Q_1 + Q_2 + Q_3 \tag{6.29}$$

where Q_i $(i = 1, 2, 3)$ are the reactive powers away from bus i through each of the lines and calculated as in Eq. (6.26).

Slack Bus

If i = s, the slack bus power is

$$S_{gs} = S_{\ell i} + S_1 + S_2 + S_3 \tag{6.30}$$

where S_i $(i = 1, 2, 3)$ are the complex powers away from bus s through each of the lines, and $S_{\ell s}$ is the complex power to a local load.

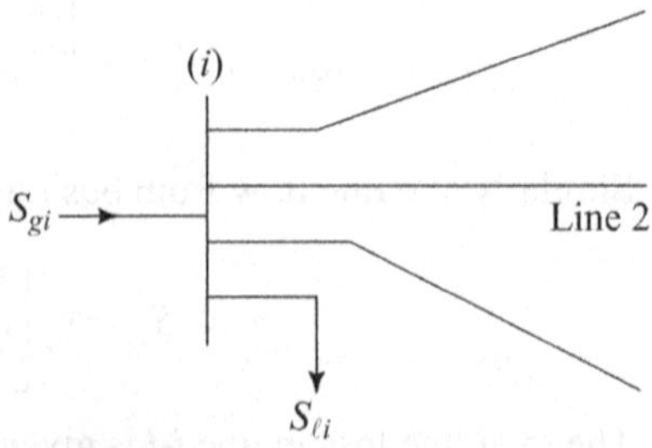

Fig. 6.3 *Bus power calculations*

6.3.5 Acceleration of Convergence

The convergence of the Gauss–Seidel method using Y_{bus} is slow since in a large network, each bus is connected generally at most to three or four buses. Hence, when one voltage is updated, it directly affects the buses connected to it. Acceleration methods are therefore used to improve the convergence. The standard method for acceleration is as follows:

1. After each iteration at each P-Q bus, compute the difference $\Delta V_i^{(k)} = \left(V_i^{(k)} - V_i^{(k-1)}\right)$ and multiply it by a number α, so that

$$\alpha\,\Delta V_i^{(k)} = \alpha\left(V_i^{(k)} - V_i^{(k-1)}\right) \tag{6.31}$$

2. For the next iteration, the voltages at P-Q buses are updated as

$$V_i^{(k)} = V_i^{(k-1)} + \alpha\,\Delta V_i^{(k)}$$

Typical value of α will range from 0.75 to 2.0 and is largely dependent on the particular system. A typical value of α seems to lie between 1.4 to 1.6. α can also be chosen as a complex number, or the real and imaginary parts of $V_i^{(k)}$ may be accelerated by separate amounts. In general, therefore,

$$\Delta V_i^{(k)} = \beta Re\left[\Delta V_i^{(k)}\right] + j\gamma Im\left[\Delta V_i^{(k)}\right] \tag{6.32}$$

and

$$V_i^{(k)} = V_i^{(k-1)} + \Delta V_i^{(k)} \tag{6.33}$$

where both β and γ are real.

The Gauss–Seidel method using Y_{bus} is easy to program. If only the nonzero elements are stored, there is a lot of saving in the computer memory requirements. For large systems, it is difficult to predict in advance, the number of iterations or the tolerance and acceleration factor required for convergence. Experience is therefore necessary. Under certain conditions such as high R/X ratios in transmission lines (e.g., due to the introduction of series capacitor), convergence difficulty can be considerable. In view of all these uncertain factors, Newton's method has gained in popularity and forms the basis of the present day production grade programs.

In the 1960s, there was interest for a brief time in the load flow techniques, to use Z_{bus} instead of Y_{bus}. Z_{bus} being full, there was a speed up in convergence. But the computational effort at computing and storing Z_{bus} entries far outweighed the advantages in improved convergence. Hence, load flow using Z_{bus} is not discussed.

6.4 NEWTON'S METHOD

For the n-dimensional case, the system of nonlinear equations is of the form

$$F(x) = 0 \tag{6.34}$$

Where F is a N-dimensional vector of the variables $(x_1, x_2, \ldots x_n)$. Eq. (6.34) can be written as n equations.

$$f_i(x_1, x_2, \ldots, x_n) = 0,\ i = 1, 2, \ldots, n \tag{6.35}$$

The Newton iterates are

$$x^{(k)} = x^{(k-1)} - \left[J\left(x^{(k-1)}\right)\right]^{-1} F\left(x^{(k-1)}\right); \quad k = 1, 2, \ldots \tag{6.36}$$

where the Jacobian $\boldsymbol{J}(\boldsymbol{x})$ of $\boldsymbol{F}(\boldsymbol{x})$ is defined as

$$J(x) = \begin{bmatrix} \frac{\partial f_1}{\partial x_1} & \frac{\partial f_1}{\partial x_2} & \cdots & \frac{\partial f_1}{\partial x_n} \\ \frac{\partial f_2}{\partial x_1} & \frac{\partial f_2}{\partial x_2} & \cdots & \frac{\partial f_2}{\partial x_n} \\ \cdots & \cdots & \cdots & \cdots \\ \frac{\partial f_n}{\partial x_1} & \frac{\partial f_n}{\partial x_2} & \cdots & \frac{\partial f_n}{\partial x_n} \end{bmatrix} \tag{6.37}$$

Equation (6.36) can be split up into two parts: one consisting of solving a set of algebraic equations of the form $\mathbf{A}\Delta\boldsymbol{x} = \boldsymbol{b}$, and the other, to update $\boldsymbol{x}$ as follows: We define $\boldsymbol{x}^{(k)} - \boldsymbol{x}^{(k-1)} = \Delta\boldsymbol{x}^{(k)}$. Then Eq. (6.36) is equivalent to the following two equations

$$\boldsymbol{F}\left(\boldsymbol{x}^{(k-1)}\right) = -\boldsymbol{J}^{(k-1)}\Delta\boldsymbol{x}^{(k)} \tag{6.38}$$

$$\boldsymbol{x}^{(k)} = \boldsymbol{x}^{(k-1)} + \Delta\boldsymbol{x}^{(k)} \tag{6.39}$$

Equation (6.38) is a set of linearized simultaneous equations to be solved for $\Delta\boldsymbol{x}^{(k)}$ and Eq. (6.39) is the updating equation. Newton's method possesses quadratic convergence as opposed to linear convergence by the $\boldsymbol{Y}_{\text{bus}}$ based Gauss–Seidel method. However, the initial guess $\boldsymbol{x}^{(0)}$, i.e., for $k = 1$, must be a good one for the method to converge. In power system problems, the voltages are near 1 p.u. in magnitude and phase angles are in a 0 – 30° range. Sometimes, the converged results of a previous run is saved and used as the initial value for running a load flow for the same configuration with minor changes in loading conditions. Hence, a good initial guess is available.

6.4.1 Power Flow Equations—Polar Version

We need to put the load flow equations in the format of Eq. (6.34) where $x_i (i = 1, 2, \ldots, n)$ are real variables. We take due note of the n_1 P-Q buses, n_2 P-V buses, and one slack bus so that $n_1 + n_2 + 1 = n$. The network equations are

$$I_i = \sum_{J=1}^{n} Y_{ij} V_{ij} \qquad i = 1, 2, \ldots, n \tag{6.40}$$

The expressions for injected powers are

$$P_i = Re\left[V_i \sum_{j=1}^{n} Y_{ij}^* V_j^*\right] \quad i = 1, 2, \ldots, n \tag{6.41}$$

$$Q_i = Im\left[V_i \sum_{j=1}^{n} Y_{ij}^* V_j^*\right] \quad i = 1, 2, \ldots, n \tag{6.42}$$

The widely used polar version of Newton's method actually uses polar coordinates for the voltages and rectangular coordinates for Y_{ij}'s. Thus,

$$V_i = |V_i| \angle \theta_i \, ; \quad \theta_{ij} = \theta_i - \theta_j \, ; \; Y_{ij} = G_{ij} + jB_{ij}$$

Substituting these in Eq. (6.41) and Eq. (6.42), we get

$$P_i - |V_i| \sum_{j=1}^{n} \left(\left(G_{ij} \cos \theta_{ij} + B_{ij} \sin \theta_{ij} \right) |V_j| \right) = 0; \; i = 1, 2, \ldots, n \tag{6.43}$$

$$Q_i - |V_i| \sum_{j=1}^{n} \left(\left(G_{ij} \sin \theta_{ij} - B_{ij} \cos \theta_{ij} \right) |V_j| \right) = 0; \; i = 1, 2, \ldots, n \tag{6.44}$$

We now apply the various conditions at the buses (*P-Q*, *P-V*, and slack) from Table 6.1, to derive the necessary number of equations equal to the number of unknowns from Eq. (6.43) and Eq. (6.44).

The unknowns in the problem are θ_i at all except the slack bus, i.e., $n_1 + n_2$ in number, $|V_i|$ at all the *P-Q* buses, i.e., n_1 in number. Thus, we require $(2n_1 + n_2)$ equations. Let x be this vector of all unknowns as shown below

$$x = \begin{bmatrix} \theta_i \text{ each } P-Q \text{ bus} \\ \theta_i \text{ each } P-V \text{ bus} \\ |V_i| \text{ each } P-Q \text{ bus} \end{bmatrix} \begin{matrix} (n_1) \\ (n_2) \\ (n_1) \end{matrix} \tag{6.45a}$$

As can be seen, this x is a $(2n_1 + n_2)$ vector.

The known quantities are $|V_s|^{sp}$, θ_s at the slack bus, P_i^{sp} and Q_i^{sp} at the P-Q buses, and P_i^{sp} and $|V_i|^{sp}$ at the *P-V* buses. Let the vector of known quantities be denoted by y. Thus,

$$y = \begin{bmatrix} V_s & \text{at swing bus} \\ \theta_s & \text{at swing bus} \\ P_i^{sp} & \text{at } P-Q \text{ bus} \\ Q_i^{sp} & \text{at } P-Q \text{ bus} \\ P_i^{sp} & \text{at } P-V \text{ bus} \\ |V_i|^{sp} & \text{at } P-V \text{ bus} \end{bmatrix} \begin{matrix} 2 \\ \\ 2n_1 \\ \\ 2n_2 \\ \end{matrix} \tag{6.45b}$$

The set of $(2n_1 + n_2)$ equations is formed as follows

$$\begin{bmatrix} \text{Equation (6.43) with } P_i = P_i^{sp} \text{ for each } P-Q \text{ and } P-V \text{ bus} \\ \text{Equation (6.44) with } Q_i = Q_i^{sp} \text{ for each } P-Q \text{ bus} \end{bmatrix} \begin{matrix} (n_1 + n_2) \\ (n_1) \end{matrix} \tag{6.46}$$

The known variables assume the values indicated in Eq. (6.45b). Thus we have, from Eq. (6.46), a set of $(2n_1 + n_2)$ equations in the $(2n_1 + n_2)$ variables x in the required $F(x) = 0$ form.

These equations can be written in a notation which is closer to the industry notation as

$$\begin{matrix}(n_1+n_2)\\(n_1)\end{matrix}\begin{bmatrix}\Delta \boldsymbol{P}\\ \Delta \boldsymbol{Q}\end{bmatrix}=\begin{bmatrix}0\\0\end{bmatrix} \tag{6.47}$$

where the elements of $\Delta \boldsymbol{P}$ and $\Delta \boldsymbol{Q}$ are

$$\Delta P_i = P_i^{sp} - P_i = P_i^{sp} - |V_i|\left(\sum_{j=1}^{n}\left(G_{ij}\cos\theta_{ij} + B_{ij}\sin\theta_{ij}\right)|V_j|\right);\quad i=1,2,\ldots,n \quad i\neq s \tag{6.48}$$

$$\Delta Q_i = Q_i^{sp} - Q_i = Q_i^{sp} - |V_i|\left(\sum_{j=1}^{n}\left(G_{ij}\sin\theta_{ij} - B_{ij}\cos\theta_{ij}\right)|V_j|\right);\quad i=1,2,\ldots,n \quad i\neq s,\ i\neq P-V\text{ bus} \tag{6.49}$$

Now, the Newton iterates corresponding to Eq. (6.38) and Eq. (6.39) are arranged as shown below.

$$\begin{bmatrix}\Delta \boldsymbol{P}\\ \Delta \boldsymbol{Q}\end{bmatrix}_{(k-1)} = \begin{bmatrix}\overset{n_1+n_2}{\boldsymbol{H}} & \overset{n_1}{\boldsymbol{N}}\\ \boldsymbol{M} & \boldsymbol{L}\end{bmatrix}_{(k-1)} \begin{bmatrix}\Delta\theta\\ \dfrac{\Delta|\boldsymbol{V}|}{|\boldsymbol{V}|}\end{bmatrix}_{(k)} \tag{6.50}$$

$$\begin{bmatrix}\theta\\ |\boldsymbol{V}|\end{bmatrix}_{(k)} = \begin{bmatrix}\theta\\ |\boldsymbol{V}|\end{bmatrix}_{(k-1)} + \begin{bmatrix}\Delta\theta\\ \Delta|\boldsymbol{V}|\end{bmatrix}_{(k)} \tag{6.51}$$

The matrices $\boldsymbol{H}$, $\boldsymbol{N}$, $\boldsymbol{M}$, and $\boldsymbol{L}$ are respectively the partial derivatives $\dfrac{\partial \boldsymbol{P}}{\partial\theta}$, $\dfrac{\partial \boldsymbol{P}}{\partial|\boldsymbol{V}|}$, $\dfrac{\partial \boldsymbol{Q}}{\partial\theta}$, and $\dfrac{\partial \boldsymbol{Q}}{\partial|\boldsymbol{V}|}$ with the elements of $\boldsymbol{N}$ and $\boldsymbol{L}$ multiplied by the appropriate $|V_i|$. The columns and rows of **H** correspond to all the P-Q and P-V buses and do not contain the slack bus. Hence it is a square matrix of order $(n_1 + n_2)$. The columns and rows of **L** correspond to the P-Q buses and do not contain the P-V buses or the slack bus. Hence it is a square matrix of order n_1. Hence, the **N** and **L** matrices are rectangular matrices of order $(n_1 + n_2) \times n_1$ and $n_1 \times (n_1 + n_2)$, respectively. The division by $|V_i|$ of each of the components of $\Delta|\boldsymbol{V}|$ in Eq. (6.50) is done to facilitate a certain symmetry in some of the coefficients of the Jacobian matrix, as shown below. These are obtained from Eq. (6.48) and Eq. (6.49) as:

for $i \neq j$

$$H_{ij} = L_{ij} = |V_i|\,|V_j|\,(G_{ij}\sin\theta_{ij} - B_{ij}\cos\theta_{ij}) \tag{6.52}$$

$$N_{ij} = -M_{ij} = |V_i|\,|V_j|\,(G_{ij}\cos\theta_{ij} + B_{ij}\sin\theta_{ij}) \tag{6.53}$$

for $i = j$

$$H_{ii} = -Q_i - B_{ii}|V_i|^2 \tag{6.54}$$

$$L_{ii} = Q_i - B_{ii}|V_i|^2 \tag{6.55}$$

$$N_{ii} = P_i + G_{ii}|V_i|^2 \tag{6.56}$$

$$M_{ii} = P_i - G_{ii} |V_i|^2 \tag{6.57}$$

where

$$P_i = |V_i| \left(\sum_{j=1}^{n} (G_{ij} \cos \theta_{ij} + B_{ij} \sin \theta_{ij}) |V_j| \right) \tag{6.58}$$

$$Q_i = |V_i| \left(\sum_{j=1}^{n} (G_{ij} \sin \theta_{ij} - B_{ij} \cos \theta_{ij}) |V_j| \right) \tag{6.59}$$

Equation (6.50) is solved for $\Delta\boldsymbol{\theta}$ and $\dfrac{\Delta|V|}{|V|}$ and then $\boldsymbol{\theta}$ and $|V|$ are updated as in Eq. (6.51). Convergence is tested by the bus power mismatch criteria. This method generally converges in 2 to 5 iterations with a "flat" start. At the *P*-*V* buses, the voltage magnitudes are assumed to be the specified values.

At the end of each iteration, Q_i is computed at each bus and if the limits on Q_{gi} of the generators are violated, the bus is switched to a *P*-*Q* bus during the next iteration. When Q_{gi} is within limits, it is switched back to a *P*-*V* bus. This has been explained in Section 6.3.2.

6.4.2 Rectangular Version

If we use rectangular coordinates instead of the polar coordinates for the voltages, the Taylor series expansion of $\boldsymbol{F}(x)$ terminates with the second-ssdegree term. This property has been used to develop the so-called second-order load flow method. Although there is a lot of literature on this topic, it is not being used in the production grade programs. The interested reader can refer to the literature.

6.4.3 Computational Aspects

With a good initial guess, the Newton's method converges quadratically. In order to get a good initial guess, either one or two Gauss-Seidel iterations or the results of a DC load flow (to be explained next) can be used prior to switching to the Newton's method. The total computational time can be reduced by using either a constant Jacobian or changing the Jacobian every two iterations or so. The success of Newton's method depends on using sparse matrix techniques together with the LU decomposition method effectively. This is a topic of one of the subsequent sections. We shall next consider a series of approximate load flow models. These are: (i) decoupled load flow, (ii) Stott's fast decoupled load flow, and (iii) DC load flow.

6.5 SIMPLIFIED VERSIONS OF NEWTON'S METHOD FOR FAST COMPUTATION

6.5.1 Decoupled Load Flow

A unique and perhaps an important feature of the steady state operating conditions in a power system is the strong dependence between the injected real powers and the bus voltage angles, and between the injected reactive powers and the bus voltage magnitudes. In mathematical terms, this implies that there is a strong coupling between the P and θ variables and between the Q and $|V|$ variables. The coupling between the P and $|V|$ or between the Q and θ variables is weak. Hence in Eq. (6.50), we can set the matrices **N** and **M** = **0**. The resulting linear equations reduce to

$$\Delta \boldsymbol{P}_{(k-1)} = \boldsymbol{H}_{(k-1)} \Delta \boldsymbol{\theta}_{(k)} \tag{6.60}$$

$$\Delta \boldsymbol{Q}_{(k-1)} = \boldsymbol{L}_{(k-1)} \left[\frac{\Delta |\boldsymbol{V}|}{|\boldsymbol{V}|} \right]_{(k)} \tag{6.61}$$

The elements of **H** and **L** are defined by

$$H_{ij} = L_{ij} = |V_i||V_j|(G_{ij} \sin \theta_{ij} - B_{ij} \cos \theta_{ij}) \qquad i \neq j \tag{6.62}$$

$$H_{ii} = -Q_i - B_{ii} |V_i|^2 \tag{6.63}$$

$$L_{ii} = Q_i - B_{ii} |V_i|^2 \tag{6.64}$$

There are two ways of solving Eqs. (6.60) and (6.61):

1. Solve for $\Delta \boldsymbol{\theta}_{(k)}$ and $\left[\frac{\Delta |\boldsymbol{V}|}{|\boldsymbol{V}|} \right]_{(k)}$ simultaneously and then update both $\boldsymbol{\theta}$ and V.
2. Solve for $\Delta \boldsymbol{\theta}_{(k)}$ first. Then, update $\boldsymbol{\theta}$ as $\boldsymbol{\theta}_{(k)} = \boldsymbol{\theta}_{(k-1)} + \Delta \boldsymbol{\theta}_{(k)}$. Use this updated value of $\boldsymbol{\theta}$ in computing **L** and then solve Eq. (6.61) for $\left[\frac{\Delta |\boldsymbol{V}|}{|\boldsymbol{V}|} \right]_{(k)}$, and then update V.

The technique requires less memory space and the convergence is as good as in the Newton's method.

6.5.2 Fast Decoupled Load Flow

In decoupled load flow, in spite of the memory savings, the elements of the matrices **H** and **L** have to be calculated at each iteration. With some engineering judgment, we can do some approximations to make these matrices constant. These approximations are

$$\cos \boldsymbol{\theta}_{ij} \approx 1 \tag{6.65}$$

$$G_{ij} \sin \boldsymbol{\theta}_{ij} << B_{ij} \tag{6.66}$$

$$Q_i << B_{ii} |V_i|^2 \tag{6.67}$$

Introducing the approximations in Eqs (6.62) – (6.64), the resulting elements of the **H** and **L** matrices are

$$H_{ij} = L_{ij} = -|V_i||V_j| B_{ij} \qquad i \neq j \tag{6.68}$$

$$H_{ii} = L_{ii} = -|V_i|^2 B_{ii} \tag{6.69}$$

The linear decoupled Eqs (6.60) and (6.61) now become

$$\Delta \boldsymbol{P}_{(k-1)} = \boldsymbol{H}_{(k-1)} \Delta \boldsymbol{\theta}_{(k)} \tag{6.70}$$

$$\Delta \boldsymbol{Q}_{(k-1)} = \boldsymbol{L}_{(k-1)} \left[\frac{\Delta |V|}{|V|} \right]_{(k)} \tag{6.71}$$

where the elements of **H** and **L** are given in Eqs (6.68) and (6.69). In Eqs. (6.70) and (6.71), divide each equation by $|V_i|$ and in Eq. (6.72), set $|V_j| = 1$ on the right-hand side. The latter step of setting $|V_j| = 1$, is not needed in Eq. (6.71) because of the canceling effect due to the element in $\frac{\Delta |V|}{|V|}$. Now, define $B'_{ij} = -B_{ij}$. With the above modifications, the fast decoupled load flow (FDLF) equations become

$$\frac{\Delta P}{|V|} = B' \Delta \theta \tag{6.72}$$

$$\frac{\Delta Q}{|V|} = B'' \Delta V \tag{6.73}$$

where the elements of B' and B'' are formed by taking the negative of the appropriate elements of B. Both B' and B'' are real, sparse matrices and are arranged in the same manner as **H** and **L**, respectively. They are constant and can be solved using LU decomposition to be discussed in a later section. If there are no phase shifters, both B' and B'' are symmetric. The structure of Eq. (6.72) and Eq. (6.73) is such that very fast solutions for $\Delta\theta$ and ΔV can be obtained for each iteration using the LU factors of B' and B''. These solutions may be iterated with each other in some defined manner towards the exact solution, i.e., when $\frac{\Delta P}{|V|}$ and $\frac{\Delta Q}{|V|}$ are zero. The best strategy is to solve for $\Delta\boldsymbol{\theta}$ to update $\boldsymbol{\theta}$ for computing ΔQ. Then solve one solution of ΔV to update $|V|$. This is termed as the $(1\boldsymbol{\theta}, 1|V|)$ scheme, as explained in Fig. 6.4. The convergence criteria used for Eqs. (6.72) and (6.73) are

$$\underset{i \varepsilon \left(n_1 + n_2 \right)}{\text{Max}} |\Delta P_i| < C_p, \quad \underset{i \varepsilon\, n_1}{\text{Max}} |\Delta Q_i| < C_v$$

This method is also known as *Stott's decoupled load flow* in the literature.

6.5.3 DC Load Flow

In this method, we neglect the effect of Q altogether and assume all bus voltages as 1 p.u. Hence, Eq. (6.73) is neglected and $|V| = 1$ in Eq. (6.72). We also compute B' on the basis that all resistances are zero, i.e., $B'_{ij} = -B_{ij} = 1/x_{ij}$ where x_{ij} is the reactance of the line connecting the buses i and j. Thus

$$\Delta \boldsymbol{P} = \boldsymbol{B}' \Delta \boldsymbol{\theta} \tag{6.74}$$

This is the DC load flow model and we get information only about the bus angles. This model is widely used in the open-access market transactions.

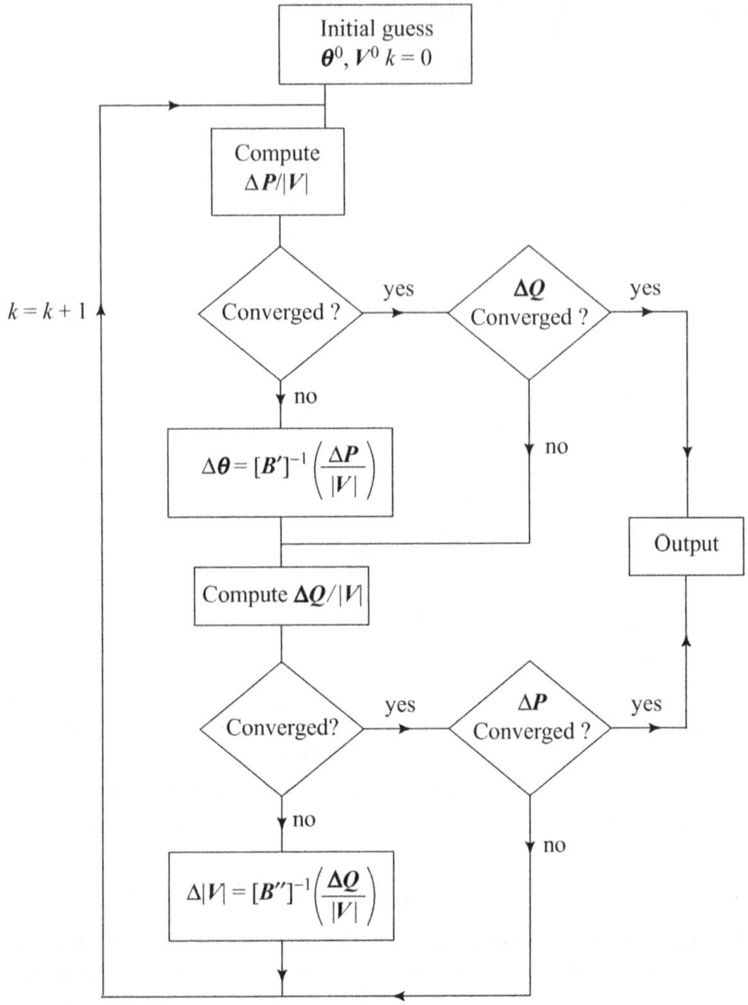

Fig. 6.4 *1θ, 1V iteration for fast decoupled power flow*

Example 6.1 For the network shown in Fig. 6.5, the line and bus data is given in Tables 6.2 and 6.3 respectively. Considering bus **1** as the slack bus, use the following methods to obtain one iteration of the load flow solution.
(a) Newton-Raphson with Y_{bus}, (b) Decoupled method, (c) Fast decoupled method, (d) DC load flow

Table 6.2 *Line data for the sample system (all are in p.u. quantities)*

Line number	*Between buses*	*Line impedance*	*Half line charging admittance*
1	1-2	$0 + j0.1$	0
2	2-3	$0 + j0.2$	0
3	1-3	$0 + j0.2$	0

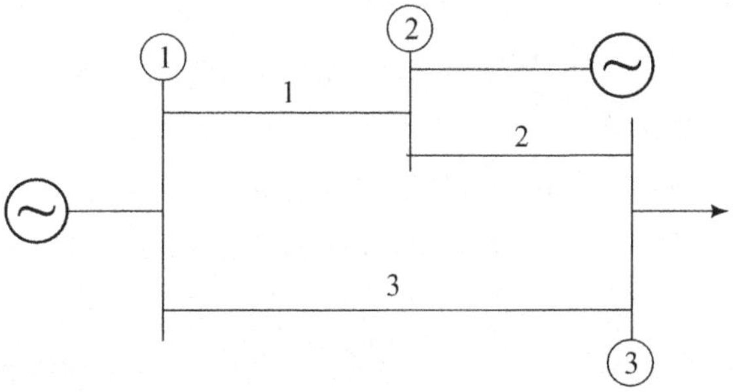

Fig. 6.5 *Sample system for Example 6.1*

Table 6.3 *Bus data (all are in p.u.)*

Bus number	Type	Generator		Load		Voltage magnitude	Reactive power limit	
		P	Q	P	Q	$\|V\|$	Q_{min}	Q_{max}
1	Slack	—	—	—	—	1.0	—	—
2	P-V	5.3217	—	—	—	1.0	0	5.3217
3	P-Q	—	—	3.6392	0.5339	—	—	—

The Y_{bus} matrix for the system written by inspection

$$Y_{bus} = \begin{array}{c} \\ 1 \\ 2 \\ 3 \end{array} \begin{array}{c} \begin{array}{ccc} 1 & 2 & 3 \end{array} \\ \begin{bmatrix} -j15 & j10 & j5 \\ j10 & -j15 & j5 \\ j5 & j5 & -j10 \end{bmatrix} \end{array}$$

Since bus **2** is a P-V bus and bus **3** is a P-Q bus, we need the expressions for the net real injected powers at buses 2 and 3, and the net injected reactive power at bus 3. These are given using Eqs (6.43) and (6.44) as (assuming $\theta_1 = 0$)

$$P_2 = |V_2|\,(10|V_1| \sin \theta_2 + 5\,|V_3| \sin (\theta_2 - \theta_3))$$

$$P_3 = |V_3|\,(5|V_1| \sin \theta_3 + 5\,|V_3| \sin (\theta_3 - \theta_2))$$

$$Q_3 = |V_3|\,(-5|V_1| \cos \theta_3 - 5\,|V_2| \cos (\theta_3 - \theta_2) + 10\,|V_3|$$

(a) Newton-Raphson Method The matrix equation for the solution of the load flow problem by the Newton–Raphson method is given by Eq. (6.52), as below:

$$\begin{bmatrix} \Delta \boldsymbol{P} \\ \Delta \boldsymbol{Q} \end{bmatrix}_{(k-1)} = \begin{bmatrix} \boldsymbol{H} & \boldsymbol{N} \\ \boldsymbol{M} & \boldsymbol{L} \end{bmatrix}_{(k-1)} \begin{bmatrix} \Delta \theta \\ \dfrac{\Delta |V|}{|V|} \end{bmatrix}_{(k)}$$

where

$$\Delta \boldsymbol{P} = \begin{bmatrix} \Delta P_2 \\ \Delta P_3 \end{bmatrix}, \quad \Delta \boldsymbol{Q} = \Delta \boldsymbol{Q}_3$$

$$\Delta\theta = \begin{bmatrix} \Delta\theta_2 \\ \Delta\theta_3 \end{bmatrix}, \quad \frac{\Delta|V|}{|V|} = \frac{\Delta|V_3|}{|V_3|}$$

since bus 2 is a *P*-*V* bus.

The change in the bus powers are obtained from the equations

$$\Delta P_i^{(k-1)} = P_i^{sp} - P_i^{(k-1)}$$

$$\Delta Q_i^{(k-1)} = Q_i^{sp} - Q_i^{(k-1)}$$

where P_i^{sp} and Q_i^{sp} are the net bus powers in per unit and are obtained from Table 6.3. The calculated bus powers are obtained from Eq. (6.43) and Eq. (6.44) at the *k*th iteration as follows ($k = 1, 2, ...$).

$$P_i^{(k-1)} = \left|V_i^{(k-1)}\right| \sum_{j=1}^{n} \left(\left(G_{ij} \cos\theta_{ij}^{(k-1)} + B_{ij} \sin\theta_{ij}^{(k-1)}\right)\left|V_j^{(k-1)}\right|\right), i = 2,\ 3$$

$$Q_i^{(k-1)} = \left|V_i^{(k-1)}\right| \sum_{j=1}^{n} \left(\left(G_{ij} \sin\theta_{ij} - B_{ij} \cos\theta_{ij}\right)\left|V_j\right|\right), i = 3$$

Using a flat start, the initial values of the voltages are

$$V_1^{(0)} = 1 + j0 = 1 \angle 0°,\ V_2^{(0)} = 1.1 + j0 = 1.1 \angle 0°,\ V_3^{(0)} = 1 + j0 = 1 \angle 0°$$

The change in the real and the reactive power for the buses at the first iteration ($k = 1$) are

$$\Delta P_2^{(0)} = 5.3217 - |V_2|\,(10\,|V_1| \sin\theta_2 + 5\,|V_3| \sin(\theta_2 - \theta_3)) = 5.3217$$

$$\Delta P_3^{(0)} = -3.6392 - |V_3|\,(5\,|V_1| \sin\theta_3 + 5\,|V_2| \sin(\theta_3 - \theta_2)) = -3.6392$$

$$\Delta Q_3^{(0)} = -0.5339 - |V_3|\,(-5\,|V_1| \cos\theta_3 - 5\,|V_2| \cos(\theta_3 - \theta_2) + 10|V_3|) = -0.0339$$

Elements of the matrices ***H***, ***M***. ***L***, and ***N*** are obtained from the expressions given in Eqs (6.52)–(6.57).

$$H_{22} = 11 \cos\theta_2 + 5.5 \cos(\theta_2 - \theta_3) = 16.5$$

$$H_{23} = -5.5 \cos(\theta_2 - \theta_3) = -5.5$$

$$N_{23} = 5.5 \sin(\theta_2 - \theta_3) = 0$$

$$N_{23} = 0 \qquad H_{32} = -5.5$$

$$M_{32} = 0 \qquad H_{33} = 10.5$$

$$M_{33} = 0$$

$$L_{33} = 9.5$$

For the first iteration, the equations to be solved are

$$\begin{array}{c} \\ 2 \\ 3 \\ 3 \end{array}\begin{bmatrix} 5.3217 \\ -3.6392 \\ -0.0339 \end{bmatrix}_{(0)} = \begin{array}{c} \\ 2 \\ 3 \\ 3 \end{array}\begin{array}{c} \begin{array}{ccc} 2 & 3 & 3 \end{array} \\ \left[\begin{array}{cc|c} 16.5 & -5.5 & 0 \\ -5.5 & 10.5 & 0 \\ \hline 0 & 0 & 9.5 \end{array}\right]_{(0)} \end{array} \begin{bmatrix} \Delta\theta_2 \\ \Delta\theta_3 \\ \dfrac{\Delta|V_3|}{|V_3|} \end{bmatrix}_{(1)}$$

Solving the above, we get

$$\Delta\theta_2^{(1)} = 0.25$$

$$\Delta\theta_3^{(1)} = -0.2152$$

$$\frac{\Delta|V_3^{(1)}|}{|V_3^{(1)}|} = -0.00357$$

Hence, we get

$$\theta_2^{(1)} = \theta_2^{(0)} + \Delta\theta_2^{(1)} = 0.25$$

$$\theta_3^{(1)} = \theta_3^{(0)} + \Delta\theta_3^{(1)} = -0.2152$$

and

$$|V_3^{(1)}| = |V_3^{(0)}| + \Delta|V_3^{(1)}| = 0.996$$

These values are used to compute the elements of the Jacobian and the bus power mismatches for the next iteration and an improved value of the voltage is obtained. The process is terminated when the changes in both the real and the reactive power at each bus is less than $C_p = 0.001$. In four iterations, the solution converges to the following values:

$$|V_3| = 0.90$$

$$\theta_2 = 0.2618 \text{ radians } (\cong 15°)$$

$$\theta_3 = -0.2618 \text{ radians } (\cong 15°)$$

(b) Decoupled Method In this example, since

$$[N] = 0, \qquad [M] - 0$$

the decoupled load flow solution will, therefore, be identical to the previous case.

(c) Fast Decoupled Method The matrix equation for the solution of load flow by the fast decoupled method is

$$\begin{bmatrix} \dfrac{\Delta P}{|V|} \\ \dfrac{\Delta Q}{|V|} \end{bmatrix}_{(k-1)} = \begin{bmatrix} B' & 0 \\ 0 & B'' \end{bmatrix} \begin{bmatrix} \Delta\theta \\ \Delta|V| \end{bmatrix}_{(k)}$$

Here $[B']$ and $[B'']$ are the corresponding elements of the $[-Y_{bus}]$ matrix. For this example, we obtain

$$\begin{bmatrix} \dfrac{\Delta P_2}{|V_2|} \\ \dfrac{\Delta P_3}{|V_3|} \end{bmatrix}_{(k-1)} = \begin{bmatrix} -B_{22} & -B_{23} \\ -B_{32} & -B_{33} \end{bmatrix} \begin{bmatrix} \Delta\theta_2 \\ \Delta\theta_3 \end{bmatrix}_{(k)}$$

and

$$\begin{bmatrix} \dfrac{\Delta Q_3}{|V_3|} \end{bmatrix}_{(k-1)} = [-B_{33}][\Delta|V_3|]_{(k)}$$

i.e.,

$$\begin{bmatrix} 4.838 \\ -3.6392 \end{bmatrix}_{(0)} = \begin{bmatrix} 15 & -5 \\ -5 & 10 \end{bmatrix} \begin{bmatrix} \Delta\theta_2 \\ \Delta\theta_3 \end{bmatrix}_{(1)}$$

$$[-0.0339] = [10]\left[\Delta|V_3^{(1)}|\right]$$

Solving the above, we get, $\Delta\theta_2^{(1)} = 0.2414$, $\Delta\theta_3^{(1)} = -0.2432$, $\Delta|V_3^{(1)}| = -0.00339$

Hence, $\theta_2^{(1)} = 0.2414$, $\theta_3^{(1)} = -0.2432$, $|V_3^{(1)}| = 0.996$

These values are used to compute the bus powers and currents for the next iteration. When the change in the ratio of both the real and the reactive power to the magnitude of voltage at each bus is less than C_p (= 0.001), the process is terminated.

(d) DC Power Flow The equations for the DC power flow are

$$\begin{bmatrix} \Delta P_2 \\ \Delta P_3 \end{bmatrix} = \begin{bmatrix} 15 & -5 \\ -5 & 10 \end{bmatrix} \begin{bmatrix} \Delta\theta_2 \\ \Delta\theta_3 \end{bmatrix}$$

Solving for $\Delta\theta_2$ and $\Delta\theta_3$, we get

$$\Delta\theta_2 = 0.2414 \text{ and } \Delta\theta_3 = -0.2432$$

$$\theta_2 = 0.2414 \text{ and } \theta_3 = -0.2432$$

Example 6.2 For the network shown in Fig. 6.6, with bus **1** as the slack bus, using the Gauss–Seidel method, obtain one iteration of the load flow solution. Also exhibit the structure of the equations for the Newton–Ralphson, decoupled, and the fast decoupled method. The line and bus data are given in Tables 6.4 and 6.5, respectively.

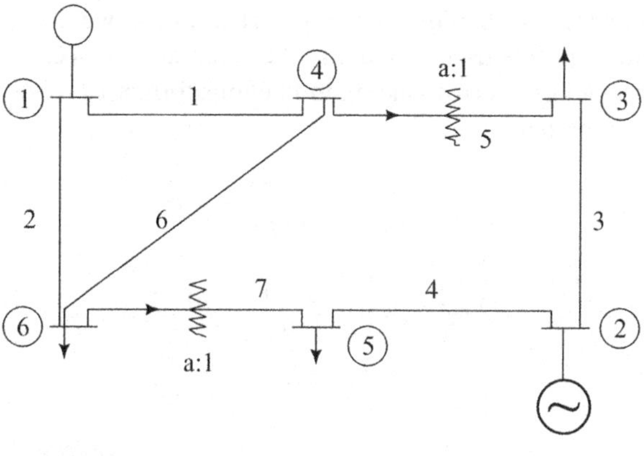

Fig. 6.6 *Sample system for Example 6.1*

Table 6.4 *Line data for the sample system (all values in p.u. on 50 MVA base)*

Line no.	*Between buses*		*Line Impedance (p.u.)*	*Half-charging admittance (p.u.)*	*Off-nominal turn ratio*
	From bus	*To bus*			
1	1	4	$0.08 + j\,0.37$	0.007	—
2	1	6	$0.123 + j\,0.518$	0.010	—
3	2	3	$0.723 + j\,1.05$	0	—
4	2	5	$0.282 + j\,0.64$	0	—
5	4	3	$0 + j\,0.133$	0	0.909
6	4	6	$0.097 + j\,0.407$	0.0076	—
7	6	5	$0 + j\,0.3$	0	0.976

Table 6.5 *Bus data*

Bus no	*Type*	*Generation*		*Load*		*Volt mag.*	*Reactive power limit*	
		P (MW)	*Q (MVAR)*	*P (MW)*	*Q (MVAR)*	$\|V\|$ *(p.u.)*	Q_{min}	Q_{max}
1	Slack	—	—	—	—	1.05	—	—
2	*P-V*	25	—	—	—	1.10	6.5	10.0
3	*P-Q*	—	—	27.5	6.5	—	—	—
4	*P-Q*	—	—	—	—	—	—	—
5	*P-Q*	—	—	15	9	—	—	—
6	*P-Q*	—	—	25	2.5	—	—	—

A static shunt capacitor at bus 4 is present with admittance $= j\,0.005$.

Using the data in Table 6.4, Y_{bus} is formulated. The transformers with off-nominal turns ratios in lines 4-3 and 6-5 are represented by their equivalent circuit. Various line and shunt admittances are

$$y_{14} = \frac{1}{0.08 + j0.37} = 0.56 - j2.57\,; \qquad y_{16} = \frac{1}{0.123 + j0.518} = 0.438 - j1.84\,;$$

$$y_{23} = \frac{1}{0.723 + j1.05} = 0.44 - j0.642; \qquad y_{25} = \frac{1}{0.282 + j0.64} = 0.58 - j1.31$$

For the π-equivalent of the off-nominal transformers, we have the following parameters. Subscript 0 is used to identify the shunt admittance. Thus, $y_i 0, j$ is the shunt admittance at bus i corresponding to line/transformer i-j. Figure 2.16(b) gives the equivalent circuit parameters.

$$y_{43} = \frac{y_t}{a} = -j\frac{7.55}{0.909} = -j8.3\,;$$

$$y_{40,3} = y_t\left(\frac{1}{a^2} - \frac{1}{a}\right) = -j0.83;$$

$$y_{30,4} = y_t\left(1 - \frac{1}{a}\right) = +j0.755\,;$$

$$y_{65} = \frac{y_t}{a} = -j\frac{3.333}{6.976} = -j3.42\,;$$

$$y_{60,5} = y_t\left(\frac{1}{a^2} - \frac{1}{a}\right) = -j0.01026\,;$$

$$y_{50,6} = y_t\left(1 - \frac{1}{a}\right) = j0.0999.$$

The line charging admittances at the various buses are given by

$$y_{10,4} = y_{40,1} = j0.007; \quad y_{10,6} = y_{60,1} = j0.01; \quad y_{40,6} = y_{60,4} = j0.0076.$$

$y_{40} = j0.005$ is the admittance due to the static capacitor at bus **4**. The $\boldsymbol{Y}_{\text{bus}}$ matrix with ground node as the reference, is obtained as shown below

$$\boldsymbol{Y}_{\text{bus}} = \begin{bmatrix} \begin{matrix}0.998\\-j4.413\end{matrix} & 0 & 0 & \begin{matrix}-0.56\\+j2.58\end{matrix} & 0 & \begin{matrix}-0.438\\+j1.84\end{matrix} \\ 0 & \begin{matrix}1.021\\-j1.955\end{matrix} & \begin{matrix}-0.445\\+j0.642\end{matrix} & 0 & \begin{matrix}-0.58\\+j1.31\end{matrix} & 0 \\ 0 & \begin{matrix}-0.445\\+j0.642\end{matrix} & \begin{matrix}0.44\\-j8.187\end{matrix} & j8.3 & 0 & 0 \\ \begin{matrix}-0.56\\+j2.58\end{matrix} & 0 & j8.3 & \begin{matrix}1.115\\-j14.025\end{matrix} & 0 & \begin{matrix}-0.555\\+j2.34\end{matrix} \\ 0 & \begin{matrix}-0.58\\+\text{j}1.31\end{matrix} & 0 & 0 & \begin{matrix}0.58\\-j4.63\end{matrix} & j3.42 \\ \begin{matrix}-0.438\\+j1.84\end{matrix} & 0 & 0 & \begin{matrix}-0.555\\+j2.34\end{matrix} & j.342 & \begin{matrix}0/993\\-j7.585\end{matrix} \end{bmatrix}$$

Assume a flat start, i.e., $V_3^{(0)} = V_4^{(0)} = V_5^{(0)} = V_6^{(0)} = 1 + j0$

and $$V_1 = 1.05 + j0, \quad V_2 = 1.1 + j0$$

Since bus **2** is a *P*-*V* bus, we shall first compute Q_2^{cal} from Eq. (6.22).

Noting $V_i = e_i + j\,fi$ in rectangular co-ordinates, Eq. (6.22) becomes

$$Q_2^{\text{cal}} = -e_2^2 B_{22} - f_2^2 B_{22} - e_2 \sum_{\substack{j=1\\ j\neq 2}}^{6} \left(B_{2j} e_j + f_j G_{2j}\right) + \sum_{\substack{j=1\\ j\neq 2}}^{6} f_2 \left(e_j G_{2j} - B_{2j} f_j\right)$$

Now, $\qquad e_i = 1, \quad f_i = 0 \qquad \text{for } i = 3, 4, 5, 6$

$$e_1 = 1, \quad f_1 = 0, \; e_2 = 1.1, \quad f_2 = 0$$

and hence, $\quad Q_2^{\text{cal}} = -1.21(-1.955) - 0 - 1.1(0.646 + 1.309) = 0.215$

$$Q_2^{\max} = \frac{10}{\text{Base MVA}} = 0.2 < Q_2^{\text{cal}}$$

Hence, we should treat bus **2** as the *P-Q* bus. Eq. (6.19) for the Gauss-Seidel iterative solution is

$$V_i = \frac{1}{Y_{ii}} \left[\frac{P_i^{sp} - jQ_i^{sp}}{V_i^*} - \sum_{\substack{i=1\\ i\neq j}}^{n} Y_{ij} V_j \right] \qquad \text{for } \begin{matrix} i=1, n \\ i \neq s \end{matrix}$$

Hence, for bus **2**, $\quad V_2^{(1)} = \dfrac{1}{Y_{22}} \left[\dfrac{P_2^{sp} - jQ_2^{sp}}{V_2^{(0)*}} - Y_{23} V_3^{(0)} - Y_{25} V_5^{(0)} \right]$

$$= \frac{1}{1.021 - j1.955} \left[\frac{0.5 - j0.2}{1.1} - (-0.445 + j0.642) - (-0.58 + j1.31) \right]$$

$$= \frac{1.47 \quad j2.13}{1.021 - j1.955} = 1.17 + j0.144.$$

Similarly, substituting for other buses,

$$V_3^{(1)} = \frac{1}{Y_{33}} \left[\frac{P_3^{sp} - jQ_3^{sp}}{V_3^{(0)*}} - Y_{32} V_2^{(1)} - Y_{34} V_4^{(0)} \right] = 1.14 + j0.0817$$

$$= \frac{1}{Y_{44}} \left[\frac{P_4 - jQ_4}{V_4^{(0)*}} - Y_{41} V_1^{(1)} - Y_{43} V_3^{(1)} - Y_{46} V_6^{(1)} \right] = 1.01 + j0.051$$

$$V_5^{(1)} = \frac{1}{Y_{55}} \left[\frac{P_5^{sp} - jQ_5^{sp}}{V_5^{(0)*}} - Y_{52} V_2^{(1)} - Y_{56} V_6^{(0)} \right] = 1.1 + j0.114,$$

and

$$V_6^{(1)} = \frac{1}{Y_{66}} \left[\frac{P_6^{sp} - jQ_6^{sp}}{V_6^{(0)*}} - Y_{61} V_1^{(1)} - Y_{64} V_4^{(1)} - Y_{65} V_5^{(1)} \right] = 1.06 + j0.12$$

These values of voltages are used to compute the new set of voltages in the next iteration.

The process is terminated when

$$\max_i \left| V_i^{(k)} - V_i^{(k-1)} \right| < C_v \quad \text{for } i = 2 \text{ to } 6$$

The value of C_v is usually taken as 0.0001. Using $C_v = 0.0001$ and the acceleration factor = 1.4, the solution converges after 17 iterations. The converged solution is shown in Table 6.7.

The above problem can also be solved using (a) the Newton–Raphson, (b) the decoupled, and (c) the fast decoupled method. We shall not carry out the solution but shall only indicate the structure of the submatrices of the linear matrix equation in these methods.

Newton–Raphson Method

$$\begin{bmatrix} \Delta P \\ \Delta Q \end{bmatrix}_{(k-1)} = \begin{bmatrix} H & N \\ M & L \end{bmatrix}_{(k-1)} \begin{bmatrix} \Delta\theta \\ \dfrac{\Delta|V|}{|V|} \end{bmatrix}_{(k)}$$

Table 6.6 *Converged solution for Example 6.2*

Bus no.	*Volts*	*Angle*	*Generation*		*Load*	
			MW	*MVAR*	*MW*	*MVAR*
1	1.05	0.0	48.2	23.8	0	0
2	1.02	–2.8656	25.0	11.0	0	0
3	0.8307	–13.3805	0	0	27.5	6.5
4	0.9310	–9.9184	0	0	0	0
5	0.8511	–13.0395	0	0	15.0	9.0
6	0.9058	–12.5114	0	0	25.0	2.5

Using Eqs (6.52) and (6.53), and the Y_{bus} matrix, we get for the first iteration,

$$H = \begin{array}{c} \\ 2 \\ 3 \\ 4 \\ 5 \\ 6 \end{array} \begin{array}{c} \begin{array}{ccccc} 2 & 3 & 4 & 5 & 6 \end{array} \\ \begin{bmatrix} 2.15 & -0.71 & 0.0 & -1.439 & 0.0 \\ -0.17 & 9.006 & -8.3 & 0.0 & 0.0 \\ 0.0 & -8.3 & 13.34 & 0.0 & -2.34 \\ -1.439 & 0.0 & 0.0 & 4.86 & -3.42 \\ 0.0 & 0.0 & -2.34 & -3.42 & 7.69 \end{bmatrix} \end{array}$$

$$N = \begin{array}{c} \\ 2 \\ 3 \\ 4 \\ 5 \\ 6 \end{array} \begin{array}{c} \begin{array}{cccc} 3 & 4 & 5 & 6 \end{array} \\ \begin{bmatrix} -0.489 & 0.0 & -0.634 & 0.0 \\ 0.40 & 0.0 & 0.0 & 0.0 \\ 0.0 & 1.112 & 0.0 & -0.554 \\ 0.0 & 0.0 & 0.52 & 0.0 \\ 0.0 & -0.554 & 0.0 & 0.988 \end{bmatrix} \end{array}$$

$$M = \begin{bmatrix} 0.489 & -0.489 & 0.0 & 0.0 & 0.0 \\ 0.0 & 0.0 & -1.112 & 0.0 & 0.554 \\ 0.634 & 0.0 & 0.0 & -0.634 & 0.0 \\ 0.0 & 0.0 & 0.554 & 0.0 & -0.988 \end{bmatrix} \begin{matrix} 3 \\ 4 \\ 5 \\ 6 \end{matrix}$$

$$L = \begin{bmatrix} 7.36 & -8.3 & 0.0 & 0.0 \\ -8.3 & 14.71 & 0.0 & -2.34 \\ 0.0 & 0.0 & 4.399 & -3.42 \\ 0.0 & -2.34 & -3.42 & 7.48 \end{bmatrix} \begin{matrix} 3 \\ 4 \\ 5 \\ 6 \end{matrix}$$

Knowing $\begin{bmatrix} \Delta P \\ \Delta Q \end{bmatrix}$ we can find $\begin{bmatrix} \Delta\theta \\ \dfrac{\Delta|V|}{|V|} \end{bmatrix}$ at the end of the first iteration, and the new values of $\theta^{(1)}$ and $|V^{(1)}|$ are found. The procedure is repeated until the required convergence is obtained.

(a) *Decoupled Load Flow* The equations to be solved are

$$[\Delta P] = [H][\Delta\theta]$$

$$[\Delta Q] = [L]\left[\frac{\Delta|V|}{|V|}\right]$$

For the first iteration, the values of H and L are given above. Hence, the new values of $\theta^{(1)}$ and $|V^{(1)}|$ are found.

(b) *Fast Decoupled Load Flow* The load flow equations using this method are

$$\left[\frac{\Delta P}{|V|}\right] = [B'][\Delta\theta]$$

$$\left[\frac{\Delta Q}{|V|}\right] = [B''][\Delta|V|]$$

$[B']$ and $[B'']$ are the elements of $[-B]$ matrix. Their values remain the same for all the iterations.

From Table 6.6, the $[-B]$ matrix is obtained by inspection as

	1	2	3	4	5	6
1	4.413	0.0	0.0	–2.58	0.0	–1.84
2	0.0	1.955	–0.642	0.0	–1.31	0.0
3	0.0	–0.642	8.187	–8.3	0.0	0.0
4	–2.58	0.0	–8.3	14.025	0.0	–2.34
5	0.0	–1.31	0.0	0.0	4.63	–3.42
6	–1.84	0.0	0.0	–2.34	–3.42	7.585

The matrices [B'] and [B''] are extracted easily form the above matrix as

$$[B'] = \begin{bmatrix} 1.955 & -0.642 & 0.0 & -1.31 & 0.0 \\ -0.642 & 8.187 & -8.3 & 0.0 & 0.0 \\ 0.0 & -8.3 & 14.025 & 0.0 & -2.34 \\ -1.31 & 0.0 & 0.0 & 4.63 & -3.42 \\ 0.0 & 0.0 & -2.34 & -3.42 & 7.585 \end{bmatrix}$$

and

$$[B''] = \begin{bmatrix} 8.187 & -8.3 & 0.0 & 0.0 \\ -8.3 & 14.025 & 0.0 & -2.34 \\ 0.0 & 0.0 & 4.63 & -3.42 \\ 0.0 & -2.34 & -3.42 & 7.585 \end{bmatrix}$$

Using the values of $\left[\frac{\Delta P}{|V|}\right]$ and $\left[\frac{\Delta Q}{|V|}\right]$, the above equations are solved alternately using the most recent voltage values, until the solution converges within the specified limits.

(c) *DC Load Flow* Here, we simply have the equation

$$\Delta P = B' \, \Delta\theta$$

where the elements of B' are obtained by first constructing Y_{bus}, neglecting the line resistances, and then obtaining B' the usual way. It is explained below in detail.

The matrix B' corresponding to Eq. (6.74) for DC load flow is constructed first by constructing Y_{bus}, neglecting the line resistances in Table 6.4, and for the tap-changing transformer, we use the susceptance from the equivalent circuit. Thus Y_{bus} is

	1	2	3	4	5	6
1	−4.63	0	0	2.7	0	1.93
2	0	−2.51	0.95	0	1.56	0
3	0	0.95	−9.25	8.3	0	0
4	2.7	0	8.3	−13.46	0	2.46
5	0	1.56	0	0	−4.95	3.42
6	1.93	0	0	2.46	3.42	−7.81

The B' matrix is

	2	3	4	5	6
2	2.51	−0.95	0	−1.56	0
3	−0.95	9.25	−8.3	0	0
4	0	−8.3	13.46	0	−2.46
5	−1.56	0	0	4.95	−3.42
6	0	0	−2.46	−3.42	7.81

The reason for a big variation in some of the entries between this matrix and the B' matrix in the fast decoupled load flow, is due to the fact that line resistances are high in the lines 2, 3, and 4.

6.6 LOAD FLOW SOLUTION IN LARGE-SCALE SYSTEMS

Newton's method with its fast decoupled version remains as the industry technique for solving the power flow problems. Since the matrices $[B']$ and $[B'']$ are constant and sparse, the techniques employed are the so-called LU decomposition method combined with sparsity programming. The main aspect in Newton's method is solving the set of linear equations of the form

$$\boldsymbol{Ax} = \boldsymbol{b} \tag{6.75}$$

where $\boldsymbol{A}$ is of large dimension, and sparse. Given $\boldsymbol{b}$, we have to solve for $\boldsymbol{x}$. Solving Eq. (6.75) repeatedly for $\boldsymbol{x}$ for different sets of $\boldsymbol{b}$ vectors is called repeat solution. Sometimes $\boldsymbol{A}$ may get modified slightly from iteration to iteration, e.g., when the P-V bus gets switched to P-Q and vice versa, or in the case of a line switching. Two aspects are very important for performing all these tasks in an efficient manner, namely

(i) Solution of Eq. (6.75) without directly inverting $\boldsymbol{A}$, as the direct inversion of matrix $\boldsymbol{A}$ for a system with large number of buses is extremely difficult and time taking. The inversion is carried out by Gaussian elimination or ***LU*** decomposition.

(ii) Efficient storage of the elements of the $\boldsymbol{Y}_{\text{bus}}$ and the Jacobian matrices.

6.6.1 Gaussian Elimination

The method is illustrated for three equations first.

$$a_{11}x_1 + a_{12}x_2 + a_{13}x_3 = b_1 \tag{6.76}$$

$$a_{12}x_1 + a_{22}x_2 + a_{23}x_3 = b_2 \tag{6.77}$$

$$a_{31}x_1 + a_{32}x_2 + a_{33}x_3 = b_3 \tag{6.78}$$

Step 1—Elimination of x_1 from Eqs (6.76) and (6.77).

(i) Divide Eq. (6.76) by a_{11} to obtain

$$x_1 + \left(\frac{a_{12}}{a_{11}}\right)x_2 + \left(\frac{a_{13}}{a_{11}}\right)x_3 = \frac{b_1}{a_{11}}$$

i.e.,

$$x_1 + a_{12}^{(1)}x_2 + a_{13}^{(1)}x_3 = b_1^{(1)} \tag{6.79}$$

(ii) Multiply Eq. (6.79) by a_{21} and subtract from Eq. (6.77) to obtain

$$\left(a_{22} - a_{21}a_{12}^{(1)}\right)x_2 + \left(a_{23} - a_{21}a_{13}^{(1)}\right)x_3 = \left(b_2 - a_{21}b_1^{(1)}\right)$$

$$a_{22}^{(1)}x_2 + a_{23}^{(1)}x_3 = b_2^{(1)} \tag{6.80}$$

(iii) Multiply Eq. (6.79) by a_{31} and subtract form Eq. (6.78) to get

$$\left(a_{32} - a_{31}a_{12}^{(1)}\right)x_2 + \left(a_{33} - a_{31}a_{12}^{(1)}\right)x_3 = \left(b_3 - a_{31}b_1^{(1)}\right)$$

$$a_{32}^{(1)}x_2 + a_{33}^{(1)}x_3 = b_3^{(1)} \tag{6.81}$$

Step 2—Elimination of x_2 from Eq. (6.81)

(i) Divide Eq. (6.80) by a_{22}, i.e., $x_2 + \dfrac{a_{23}^{(1)}}{a_{22}^{(1)}}x_3 = \dfrac{b_2^{(1)}}{a_{22}^{(1)}}$

$$x_2 + a_{23}^{(2)}x_3 = b_2^{(2)} \tag{6.82}$$

(ii) Multiply Eq. (6.82) by $a_{32}^{(1)}$ and subtract from Eq. (6.81)

$$\left(a_{33}^{(1)} - a_{32}^{(1)}a_{23}^{(2)}\right)x_3 = \left(b_3^{(1)} - a_{32}^{(1)}b_2^{(2)}\right)$$

$$a_{33}^{(2)}x_3 = b_3^{(2)} \tag{6.83}$$

We now have Eq. (6.79), Eq. (6.82), and Eq. (6.83) as

$$x_1 + a_{12}^{(1)}x_2 + a_{13}^{(1)}x_3 = b_1^{(1)} \tag{6.84}$$

$$x_2 + a_{23}^{(2)}x_3 = b_2^{(2)} \tag{6.85}$$

$$a_{33}^{(2)}x_3 = b_3^{(2)} \tag{6.86}$$

Step 3—This is a trivial step of dividing Eq. (6.86) by $a_{33}^{(2)}$, to get x_3.

$$x_3 = \frac{b_3^{(2)}}{a_{33}^{(2)}} = b_3^{(3)} \tag{6.87}$$

Finally, we have Eqs (6.84), (6.85), and (6.87) as

$$x_1 + a_{12}^{(1)}x_2 + a_{13}^{(1)}x_3 = b_1^{(1)} \tag{6.88}$$

$$x_2 + a_{23}^{(2)}x_3 = b_2^{(2)} \tag{6.89}$$

$$x_3 = b_3^{(3)} \tag{6.90}$$

We have thus reduced the original equations to the *upper triangular form.* The process of obtaining Eqs. (6.88)–(6.90) is known as forward substitution. These can be successively solved for x_3, x_2, and x_1 in that order as

$$x_3 = b_3^{(3)} \tag{6.91}$$

$$x_2 = b_2^{(2)} - a_{33}^{(2)}x_3 \tag{6.92}$$

$$x_1 = b_1^{(1)} - a_{13}^{(1)}x_3 - a_{12}^{(1)}x_2 \tag{6.93}$$

This is called *back substitution.* The procedure can be extended to any number of equations.

6.6.2 Use of *LU* and *L′DU* Factorization

If we have to solve $\boldsymbol{x}$ for a different set of vectors $\boldsymbol{b}$, then it will be convenient to systematize the procedure in the preceding section. One such method is to factorize the matrix A as

$$A = LU \tag{6.94}$$

The matrix $\boldsymbol{L}$ is a lower triangular matrix with non-zero diagonal element, and the matrix $\boldsymbol{U}$ is an upper triangular matrix with diagonal elements equal to 1. The details of the factorization method are provided in Appendix A.5. After factorization, Eq. (6.75) becomes

$$\boldsymbol{LUx} = \boldsymbol{b} \tag{6.95}$$

i.e.,
$$\boldsymbol{Ly} = \boldsymbol{b} \tag{6.96}$$

where
$$\boldsymbol{Ux} = \boldsymbol{y} \tag{6.97}$$

Since $\boldsymbol{L}$ is a lower triangular matrix, we can solve for $y_1,\ldots, y_n$ successively from

$$\begin{bmatrix} \ell_{11} & 0 & 0 & 0 & 0 \\ \ell_{21} & \ell_{22} & 0 & \cdots & 0 \\ \vdots & \vdots & \vdots & \vdots & \vdots \\ \ell_{n1} & \ell_{n2} & \cdots & \cdots & \ell_{nn} \end{bmatrix} \begin{bmatrix} y_1 \\ \vdots \\ y_n \end{bmatrix} = \begin{bmatrix} b_1 \\ \vdots \\ b_n \end{bmatrix} \tag{6.98}$$

Knowing y_i's, we can solve for x_i' s from

$$\begin{bmatrix} 1 & u_{12} & \cdots & \cdots & u_{1n} \\ 0 & 1 & u_{23} & \cdots & u_{2n} \\ \vdots & \vdots & \vdots & \vdots & \vdots \\ 0 & \cdots & \cdots & \cdots & 1 \end{bmatrix} \begin{bmatrix} x_1 \\ \vdots \\ x_n \end{bmatrix} = \begin{bmatrix} y_1 \\ \vdots \\ y_n \end{bmatrix} \tag{6.99}$$

An alternative version is to express $A = \boldsymbol{L'DU}$ where $\boldsymbol{L'}$ is a lower triangular matrix with diagonal entries equal to 1, $\boldsymbol{D}$ is a diagonal matrix, and $\boldsymbol{U}$ is an upper triangular matrix as before. The advantage of this factorization is that if A is symmetric, then both $\boldsymbol{L'}$ and $\boldsymbol{U}$ are transposes of each other. Thus in the computer, we need to store only $\boldsymbol{L'}$ or $\boldsymbol{U}$. Recall that in the fast decoupled power flow method, both $\boldsymbol{B'}$ and $\boldsymbol{B''}$ are both symmetric if there are no phase shifters.

Example 6.3

Given
$$A = \begin{bmatrix} 2 & 1 & 3 \\ 2 & 3 & 4 \\ 3 & 4 & 7 \end{bmatrix} \quad \text{and} \quad \begin{bmatrix} 12 \\ 22 \\ 33 \end{bmatrix}$$

Solve for $\boldsymbol{x}$ using Gaussian elimination.

Solution: Following the procedure above, we get

$$x_1 + (1/2)\,x_2 + (3/2)\,x_3 = 6$$
$$x_2 + (1/2)\,x_3 = 5$$
$$x_3 = 2$$

Back substitution yields $x_3 = 2$; $x_2 = 4$; and $x_1 = 1$. The details are left to the reader to work out.

6.6.3 Efficient Storage of the Elements of the Y_{bus} and the Jacobian Matrices [5]

If we consider the 6 bus system shown in Example 6.2, the Y_{bus} has a dimension of 6×6, where 16 of the 36 elements are zeros. The elements of the Jacobian matrix depends on the elements of the Y_{bus} matrix, hence a large number of elements of the Jacobian matrix will also be zeros. While carrying out the above process of solving for x, one has to handle the Jacobian matrix. So if the matrix is stored as it is, then the procedure will involve handling and storing of a lot of zero elements which is a wastage of time and memory. A measure of this is given by the sparsity, S, of the Y_{bus} matrix which is defined as

$$S = \frac{z}{n^2} \tag{6.100}$$

where z is the total number of zero elements in the Y_{bus} matrix, and n is the total number of buses in the system. For a real system with hundreds or even thousands of buses, each bus is usually connected to only a few other buses. So the sparsity for these cases is more than 90%. Therefore, to reduce the memory requirement and time, only the non-zero elements of the matrix are stored by adopting some special methodology/convention which is described here. In this method, the non-zero elements, along with their positions in the Y_{bus} matrix, are stored in only 4 vectors as described below.

(i) **DIAG:** This contains the diagonal elements of the Y_{bus} in the order of the rows, i.e., the ith element of the 'DIAG' is the (i, i) element of Y_{bus}. The length of the DIAG is equal to n.

(ii) **OFFDIAG:** This contains the non-zero off-diagonal elements of the Y_{bus} matrix. The position of the off-diagonal elements in the Y_{bus} matrix can be obtained using the information stored in the next two vectors — Y_{ROW} and Y_{COL}. The length of the OFFDIAG is equal to $(n^2 - n - z)$.

(iii) **Y_{ROW}:** Each element of this vector indicates the element number of the OFFDIAG at which the first (non-zero off-diagonal element) of a new row of Y_{bus} is stored. To elaborate, if the jth element of Y_{ROW} is 'k', then it means that the kth element of OFFDIAG is the first non-zero off-diagonal element of the jth row of Y_{bus}. It can be understood that the first element of Y_{ROW} is always 1 and the length of Y_{ROW} is equal to n.

(iv) **Y_{COL}:** This stores the column number (in Y_{bus}) of each element of the OFFDIAG. That means, if the pth element of Y_{COL} is 'q', then the pth element of OFFDIAG is present in the qth column of the Y_{bus} matrix. The length of Y_{COL} is equal to that of OFFDIAG.

Following this convention, the Y_{bus} matrix of Example 6.2 will be stored and shown as follows.

$DIAG = [0.998 - j4.413 \;\; 1.021 - j1.955 \;\; 0.44 - j8.187 \;\; 1.115 - j14.025 \;\; 0.58 - j4.63$ $0/993 - j7.585]$

$$OFFDIAG = \begin{bmatrix} -0.56 + j2.58 \\ -0.438 + j1.84 \\ -0.445 + j0.642 \\ -0.58 + j1.31 \\ -0.445 + j0.642 \\ j8.3 \\ -0.56 + j2.58 \\ j8.3 \\ -0.555 + j2.34 \\ -0.58 + j1.31 \\ j3.42 \\ -0.438 + j1.84 \\ -0.555 + j2.34 \\ j3.42 \end{bmatrix}^T$$

$$Y_{\text{ROW}} = [1 \quad 3 \quad 5 \quad 7 \quad 10 \quad 12]$$

$$Y_{\text{COL}} = [4 \quad 6 \quad 3 \quad 5 \quad 2 \quad 4 \quad 1 \quad 3 \quad 6 \quad 2 \quad 6 \quad 1 \quad 4 \quad 5]$$

The total number of elements to be stored is the sum of the lengths of these four vectors, i.e., $2n + 2(n^2 - n - z) = 2(n^2 - z) = 2n^2(1 - S)$. For a system of 100 buses having a sparsity $S = 0.9$, the number of elements to be stored is 2000. This is much less as compared to the total number of elements of the Y_{bus} matrix which is 10000. The advantage becomes even more prominent in the case of real life systems with thousands of buses and sparsity more than 90%.

6.7 INCLUSION OF FACTS DEVICES [25, 26]

With the increasing number of installations of FACTS controllers and HVDC transmission lines, it is becoming necessary to include the effect of these in the load flow solutions. With simplified load flow methods like the fast decoupled load flow or the DC load flow, there is convergence problem when the power system includes these power electronics based devices. Therefore, generally, the Newton–Raphson load flow (NRLF) method is used. The modification in the NRLF method for including the FACTS devices are discussed here.

6.7.1 Thyristor-Based Shunt FACTS Controller—Static VAR Compensator (SVC)

The different available configurations of the shunt connected thyristor-based variable impedance type FACTS devices are discussed in Sec. 2.7.1. Effectively, all of them can be represented as a variable reactance whose value depends on the firing angle α of the thyristors. For SVC, this reactance is connected in shunt to an existing power system bus or a newly created bus (say bus 'j'), as shown in Fig. 6.7. Usually the SVC includes a controller which tries to maintain the voltage of bus 'j' at some pre-specified value by varying the firing angles. However, there is a limit to the value of the reactance that the SVC can attain, as explained in Sec. 2.7.1 and Fig. 2.22. Once that limit is reached, the SVC stays at that limit and the voltage can not be maintained any further. Beyond this point, the SVC behaves like a fixed reactance, with the value of the reactance being the limiting reactance of the SVC. Hence, in the load flow, the SVC can be included by considering the bus 'j' (where the SVC is connected) as a P-V bus with a limit on the reactance. Since the SVC does not exchange any active power, the specified active power P^{sp} for the bus remains the same as the specified power for the system without the SVC (P^{sp0}).

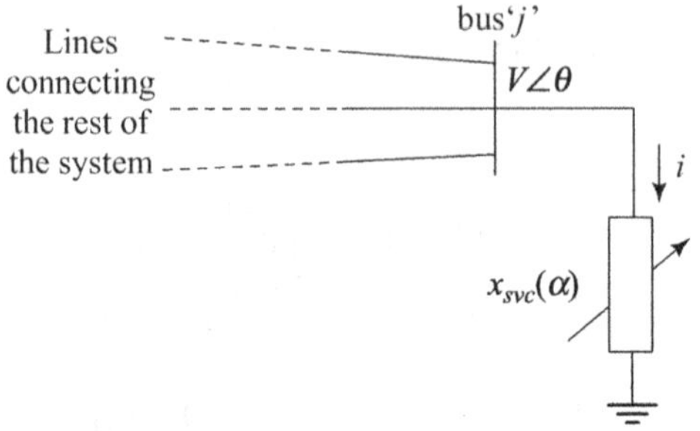

Fig. 6.7 *Representation of SVC connected to a power system bus*

While acting as a P-V bus, the magnitude of the bus voltage ($|V|$) is specified at some required value. NRLF is carried out to obtain the angle of the bus voltage (θ) and the reactive power injection at the bus (Q). The reactive power injection by the SVC, Q_{SVC}, can be obtained by subtracting the specified reactive power at the bus for the system without the SVC (Q^{sp0}).

$$Q_{\text{SVC}} = Q - Q^{sp0} \tag{6.101}$$

The corresponding reactance of the SVC can be obtained using the relation

$$x_{\text{SVC}} = \frac{|V_j|^2}{Q_{\text{SVC}}} \tag{6.102}$$

Once the limit is reached, the SVC is represented by a fixed reactance to ground (of value equal to the maximum possible reactance of the SVC) connected to it. This reactance is included in the diagonal element of the bus admittance matrix corresponding to the bus where the SVC is connected (i.e., bus 'j' in this case) as shown below.

$$Y_{\text{bus}}(j,j) = Y_{\text{bus}}(j,j) + 1/jx_{\text{SVC}}; \tag{6.103}$$

The bus 'j' is considered as P-Q bus, with specified active and reactive power equal to P^{sp0} and Q^{sp0} respectively.

6.7.2 Thyristor-Based Series FACTS Controller—Thyristor Controlled Series Capacitor (TCSC)

The detailed configuration of TCSC is discussed in Sec. 2.7.1. The equivalent representation of the TCSC connected in the line between the buses 'i' and 'j' is shown in Fig. 6.8. Effectively, it is a variable reactance, x_{tcsc}, in series with the impedance of the line i-j ($z_{\text{line}} = r_{\text{line}} + jx_{\text{line}}$). One additional bus is considered between the line impedance and the TCSC reactance as shown in Fig. 6.8. Thus the total number of buses in the system becomes $(n+1)$; and let the additional bus be denoted as the $(n+1)^{\text{th}}$ bus. This $(n + 1)^{\text{th}}$ bus is considered as the P-Q bus with the specified active and reactive power injection equal to zero, i.e. $P^{sp}_{(n+1)} = 0;\quad Q^{sp}_{(n+1)} = 0;$

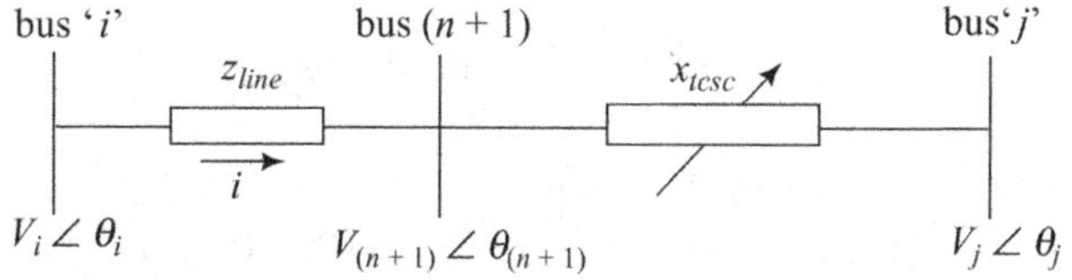

Fig. 6.8 *Representation of SVC connected to a power system bus*

The value of the TCSC reactance, x_{tcsc} is regulated by changing the firing angle of the thyristors in order to control the active power flow $P_{(n+1)j}$ through this line $(n+1) - j$. Therefore, the active power flow through this line, $P_{(n+1)j}$, is also specified at a particular value $P^{sp}_{(n+1)j}$ and one additional equation is obtained as

$$\Delta P_{(n+1)j} = P^{sp}_{(n+1)j} - P_{(n+1)j} = P^{sp}_{(n+1)j} - \frac{|V_{(n+1)}||V_j|}{x_{\text{tcsc}}}\sin(\theta_{(n+1)} - \theta_j) \tag{6.104}$$

Two more equations are obtained corresponding to the power injection at bus $(n+1)$

$$\Delta P_{(n+1)} = P^{sp}_{(n+1)} - P_{(n+1)}$$

$$= 0 - |V_{(n+1)}|\left(\sum_{\substack{k=1\\k\neq j}}^{n+1}\left(G_{(n+1)k}\cos\theta_{(n+1)k} + B_{(n+1)k}\sin\theta_{(n+1)k}\right)|V_k|\right) - P_{(n+1)j} \tag{6.105}$$

$$\Delta Q_{(n+1)} = Q^{sp}_{(n+1)} - Q_{(n+1)}$$

$$= 0 - |V_{(n+1)}|\left(\sum_{\substack{k=1\\k\neq j}}^{n+1}\left(G_{(n+1)k}\sin\theta_{(n+1)k} - B_{(n+1)k}\cos\theta_{(n+1)k}\right)|V_k|\right) - Q_{(n+1)j} \tag{6.106}$$

The equations for power injection at bus j also get modified as

$$\Delta P_j = P^{sp}_j - P_j = 0 - |V_j|\left(\sum_{k=1}^{n}\left(G_{jk}\cos\theta_{jk} + B_{jk}\sin\theta_{jk}\right)|V_k|\right) - P_{j(n+1)} \tag{6.107}$$

$$\Delta Q_j = Q^{sp}_j - Q_j = 0 - |V_j|\left(\sum_{k=1}^{n}\left(G_{jk}\sin\theta_{jk} - B_{jk}\cos\theta_{jk}\right)|V_k|\right) - Q_{j(n+1)} \tag{6.108}$$

Here, G_{jk} and B_{jk} are the elements of the modified Y_{bus} obtained by including the new bus $(n+1)$. Though, it should be kept in mind that the line reactance of the line connecting bus $(n+1)$ and bus j is the TCSC reactance which is a variable. So it can not be entered in the Y_{bus} and the corresponding element is kept as zero. The effect of the TCSC reactance is, however, taken care of by the inclusion of the power flow terms $P_{(n+1)j}$, $Q_{(n+1)j}$, $P_{j(n+1)}$ and $Q_{j(n+1)}$ in Eqs. (6.105) – (6.108). The expression for

$P_{(n+1)j}$ can be obtained from Eq. (6.104). The expressions for $Q_{(n+1)j}$, $P_{j(n+1)}$ and $Q_{j(n+1)}$ are as shown below.

$$Q_{(n+1)j} = \frac{|V_{(n+1)}|^2}{x_{\text{tcsc}}} - \frac{|V_{(n+1)}||V_j|}{x_{\text{tcsc}}} \cos(\theta_{(n+1)} - \theta_j) \quad (6.109)$$

$$P_{j(n+1)} = \frac{|V_j||V_{(n+1)}|}{x_{\text{tcsc}}} \sin(\theta_j - \theta_{(n+1)}) \quad (6.110)$$

$$Q_{j(n+1)} = \frac{|V_j|^2}{x_{\text{tcsc}}} - \frac{|V_j||V_{(n+1)}|}{x_{\text{tcsc}}} \cos(\theta_j - \theta_{(n+1)}) \quad (6.111)$$

The equations for the remaining $(n-1)$ buses can be obtained using Eqs (6.48)–(6.49) with the number of buses modified to $(n+1)$.
These equations can be solved along with Eqs. (6.104)–(6.108) to obtain

(i) The TCSC reactance, x_{tcsc}
(ii) The magnitude of the voltages of the P-Q buses, including bus $(n+1)$
(iii) The voltage angles of the P-V buses and the P-Q buses (including bus $(n+1)$)

If the solved value of x_{tcsc} is negative, it means that the TCSC is in the capacitive mode whereas a positive x_{tcsc} means the TCSC is in the inductive mode.

As mentioned in Sec. 2.7.1, there is a limit to the magnitude of the value of x_{tcsc}. If the solved value exceeds that, then the power flow $P^{sp}_{(n+1)j}$ cannot be maintained at the specified value. In that case, x_{tcsc} is considered equal to the limiting value. Since it is no more a variable, Eq. (6.104) is not to be solved. The limiting value of the TCSC reactance is taken as the constant reactance of the line connecting bus $(n+1)$ and bus j, and the Y_{bus} is modified to include the additional bus and the constant TCSC reactance. The power injection equations for all the $(n+1)$ buses can be obtained using Eqs. (6.48)–(6.49). These equations are solved to get the bus voltage magnitudes and their angles.

Here the TCSC is considered to be placed near one of the terminal buses (bus j). If the TCSC is considered at the middle of the line, then two new additional P-Q buses are to be included and the equations are to be modified accordingly.

6.7.3 Switched Converter-Based Shunt FACTS Controller—STATCOM

As described in Sec. 2.7.2, static synchronous compensator (STATCOM) consists of a voltage sourced converter (VSC) which is connected to a system bus through a coupling transformer. Depending on the compensation requirement, the output voltage of the VSC is controlled till the current limit of the converter is reached. Therefore the STATCOM is represented as a variable voltage source, as shown in Fig. 6.9(a). Once the current limit is reached, the STATCOM behaves like a constant current source, as shown in Fig. 6.9(b). Usually the STATCOM is controlled to maintain the voltage at the bus where the STATCOM is connected (bus m in Fig. 6.9) at some specified value. This is done by controlling the output voltage of the converter ($|V_{sh}|\angle\theta_{sh}$) and thus the reactive power injection by the STATCOM.

Hence, for including the STATCOM in load flow analysis, bus m is considered as a P-V bus similar to the case of the SVC (Sec 6.7.1). If the losses in the power electronic switches are ignored, the STATCOM does not exchange any active power with the power system. So, the specified active power for the bus m remains the same as the specified power for the system without the STATCOM $\left(P_m^{sp0}\right)$. The magnitude of the bus voltage ($|V_m|$) is specified at some required value.

NRLF is carried out to obtain the angle of the bus voltage (θ_m) and the reactive power injection at the bus (Q_m). The reactive power injection by the STATCOM, Q_{SC}, can be obtained as

$$Q_{SC} = Q_m - Q_m^{sp0} \tag{6.112}$$

where Q_m^{sp0} is the specified reactive power at bus m for the system without the STATCOM.

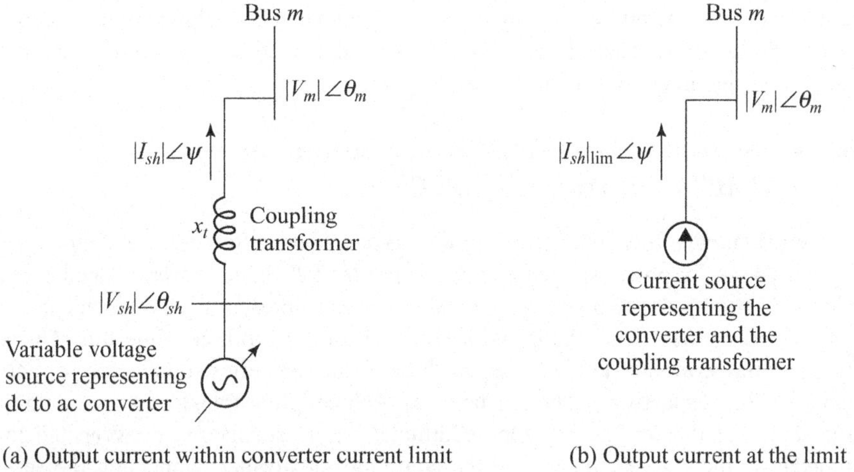

Fig. 6.9 *Single line representation of STATCOM*

The STATCOM current can be obtained as

$$|I_{sh}|\angle\psi = \frac{-jQ_{SC}}{(V_m\angle\theta_m)^*} \tag{6.113}$$

Hence the voltage of the variable voltage source (i.e., the output of the VSC) can be obtained as

$$|V_{sh}|\angle\theta_{sh} = |V_m|\angle\theta_m + j * x_t\,(|I_{sh}|\angle\psi) \tag{6.114}$$

When the value of $|I_{sh}|$ exceeds the converter current limit, $|I_{sh}|_{\lim}$, the STATCOM fails to maintain the voltage of bus m and operates like a constant current source of value $|I_{sh}|_{\lim}$. So the bus m is now considered as a P-Q bus with specified active and reactive power equal to P_m^{sp0} and Q_m^{sp0} respectively. Since the STATCOM does not exchange any active power with the system, the STATCOM current is in quadrature with the voltage of bus m, i.e., $\psi = (\theta_m \pm 90°)$. Hence,

$$I_{sh} = |I_{sh}|_{\text{limit}}\angle(\theta_m \pm 90°) = \pm j|I_{sh}|_{\text{limit}}\angle\theta_m \tag{6.115}$$

So the reactive power exchanged by the STATCOM under this condition, is given by

$$(Q_{SC})_{\text{limit}} = \pm |V_m||I_{sh}|_{\text{limit}} \tag{6.116}$$

The positive and the negative sign indicate reactive power injection and reactive power absorption by the STATCOM respectively. So this is included in Eq. (6.49) as

$$\begin{aligned} \Delta Q_i &= Q_i^{sp} - |V_i|\left(\sum_{j=1}^{n}\left(G_{ij}\sin\theta_{ij} - B_{ij}\cos\theta_{ij}\right)|V_j|\right) + (Q_{SC})_{\text{limit}} \\ &= Q_i^{sp} - |V_i|\left(\sum_{j=1}^{n}\left(G_{ij}\sin\theta_{ij} - B_{ij}\cos\theta_{ij}\right)|V_j|\right) \pm |V_m||I_{sh}|_{\text{limit}} \end{aligned} \tag{6.117}$$

where, $i = 1, 2, \ldots, n$ and $i \neq s$, $i \neq P\text{-}V$ bus

Since the STATCOM current is constant, the term corresponding to $(Q_{SC})_{\text{limit}}$ is a function of only the bus voltage magnitude $|V_m|$, and can be included in the Jacobian matrix. So Eq. (6.117) and Eq. (6.48) can be solved using the NRLF method to obtain the bus voltage magnitudes and angles.

6.7.4 Switched Converter-Based Shunt-series FACTS Controller—UPFC

The unified power flow controller (UPFC) was described in Sec. 2.7.2. The equivalent representation for power system was shown in Fig. 2.27(b) and is reproduced here in Fig. 6.10. It includes two variable voltage sources (representing the voltage source converters) connected in shunt and series with the transmission line through two coupling transformers. The voltages of these shunt and series sources are $|V_{sh}|\angle\theta_{sh}$ and $|V_{se}|\angle\theta_{se}$ respectively. The reactance of each coupling transformer is x_t and the impedance of the line is z_{line}. One additional bus is considered between the line impedance and the reactance x_t of the coupling transformer connecting the series voltage source with the system, as shown in Fig. 6.10. Thus the total number of buses in the system becomes $(n+1)$; and let the additional bus be denoted as the $(n+1)^{\text{th}}$ bus. This $(n+1)^{\text{th}}$ bus is considered as the P-Q bus with specified active and reactive power injection equal to zero, i.e., $P_{(n+1)}^{sp} = 0;\ \ Q_{(n+1)}^{sp} = 0;$

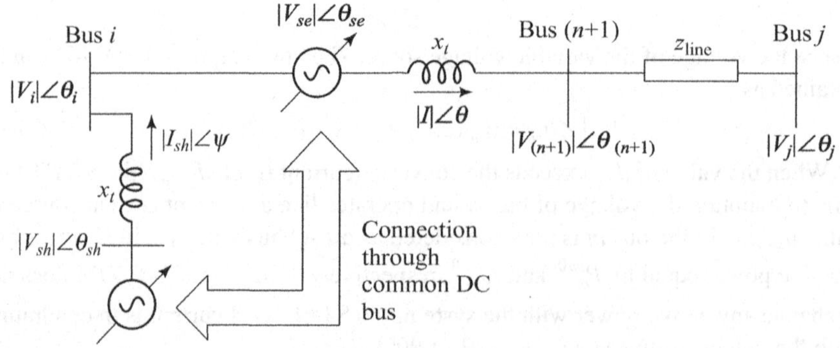

Fig. 6.10 *Equivalent representation of UPFC*

The sources have coupling among themselves through a common DC bus, thus allowing only active power transfer. Usually the variables controlled by the converter controllers are the voltage of the bus where the shunt arm of the UPFC is connected ($|V_i|$), and the active and reactive power flow through the transmission line where the series arm of the UPFC is connected ($P_{i(n+1)}$ and $Q_{i(n+1)}$ respectively). It should be noted that the power flow through the line between bus (n+1) and bus j is also $P_{i(n+1)}$ and $Q_{i(n+1)}$ respectively. Let the specified values for these variables be $|V_i|^{sp}$, $P_{i(n+1)}^{sp}$ and $Q_{i(n+1)}^{sp}$ respectively. The expressions of the active and reactive power flow through the line entering bus (n+1) can be obtained using Eqs (2.95)–(2.96). Thus we get two additional equations as

$$\Delta P_{i(n+1)} = P_{i(n+1)}^{sp} - P_{i(n+1)}$$

$$= P_{i(n+1)}^{sp} - \left[\frac{|V_i||V_{n+1}|}{x_t}\sin\theta_{i(n+1)} + \frac{|V_{se}||V_{n+1}|}{x_t}\sin\left(\theta_{se} - \theta_{(n+1)}\right)\right] \quad (6.118)$$

$$\Delta Q_{i(n+1)} = Q_{i(n+1)}^{sp} - Q_{i(n+1)}$$

$$= Q_{i(n+1)}^{sp} - \left[\frac{|V_i||V_{(n+1)}|}{x_t}\cos\theta_{i(n+1)} - \frac{|V_{(n+1)}|^2}{x_t} + \frac{|V_{se}||V_{(n+1)}|}{x_t}\cos\left(\theta_{se} - \theta_{(n+1)}\right)\right] \quad (6.119)$$

The voltage magnitude of bus i being specified, it is considered as the P-V bus. The specified active power at bus i is the same as that for the system without the UPFC $\left(P_i^{sp0}\right)$, but the active power mismatch equation is modified to include the active power flowing through the line $P_{i(n+1)}$, and the active power flowing towards the shunt branch converter of the UPFC (P_{sh}). Neglecting the losses in the converters, the algebraic sum of the active power exchanged by the series branch converter (P_{se}) and the shunt branch converter (P_{sh}) of the UPFC with the system must be equal to zero.

$$P_{sh} + P_{se} = 0 \quad \Rightarrow \quad P_{sh} = -P_{se} \quad (6.120)$$

The active power exchanged by the series branch converter can be obtained from Eq. (2.92) as

$$P_{se} = \frac{|V_{se}|}{x_t}\left[|V_i|\sin(\theta_i - \theta_{se}) - |V_{n+1}|\sin(\theta_{n+1} - \theta_{se})\right] \quad (6.121)$$

Therefore, the active power mismatch equation for bus i becomes

$$\Delta P_i = P_i^{sp} - P_i = 0 - \left[|V_i|\left(\sum_{k=1}^{n}(G_{ik}\cos\theta_{ik} + B_{ik}\sin\theta_{ik})|V_k|\right) + P_{i(n+1)} - P_{se}\right] \quad (6.122)$$

The active and reactive power mismatch equations for bus (n+1) are given as

$$\Delta P_{(n+1)} = P^{sp}_{(n+1)} - P_{(n+1)}$$

$$= 0 - |V_{(n+1)}| \left(\sum_{\substack{k=1 \\ k \neq i}}^{n+1} \left(G_{(n+1)k} \cos \theta_{(n+1)k} + B_{(n+1)k} \sin \theta_{(n+1)k} \right) |V_k| \right) - P_{(n+1)i} \qquad (6.123)$$

$$\Delta Q_{(n+1)} = Q^{sp}_{(n+1)} - Q_{(n+1)}$$

$$= 0 - |V_{(n+1)}| \left(\sum_{\substack{k=1 \\ k \neq i}}^{n+1} \left(G_{(n+1)k} \sin \theta_{(n+1)k} - B_{(n+1)k} \cos \theta_{(n+1)k} \right) |V_k| \right) - Q_{(n+1)i} \qquad (6.124)$$

Here, G_{ik} and B_{ik} are the elements of the modified Y_{bus} obtained by including the new bus (n+1). Though it should be kept in mind that the line reactance of the line connecting bus j and bus (n+1), (i.e. the reactance of the transformer coupling the series converter with the system) is not entered in the Y_{bus} and the corresponding element is kept as zero. The effect of the reactance and the series branch converter is, however, taken care of by the inclusion of the power flow terms $P_{i(n+1)}$, $P_{(n+1)i}$ and $Q_{(n+1)i}$ in Eqs (6.122)–(6.124). The expressions for $P_{i(n+1)}$ can be obtained from Eq. (6.118) whereas the expressions for $P_{(n+1)i}$ and $Q_{(n+1)i}$ are as shown below.

$$P_{(n+1)i} = \frac{|V_{n+1}||V_i|}{x_t} \sin \theta_{(n+1)i} + \frac{|V_{n+1}||V_{se}|}{x_t} \sin\left(\theta_{(n+1)} - \theta_{se}\right) \qquad (6.125)$$

$$Q_{(n+1)j} = \frac{|V_{(n+1)}|^2}{x_t} - \frac{|V_{(n+1)}||V_i|}{x_t} \cos \theta_{(n+1)i} - \frac{|V_{(n+1)}||V_{se}|}{x_t} \cos\left(\theta_{(n+1)} - \theta_{se}\right) \qquad (6.126)$$

The power mismatch equations for the remaining (n–1) buses can be obtained using Eqs (6.48)–(6.49) with the number of buses modified to (n+1). Those equations are to be solved along with Eqs (6.122)–(6.124) and Eqs (6.118)–(6.119) to obtain

(i) The UPFC series branch source voltage magnitude and angle, $|V_{se}|\angle\theta_{se}$
(ii) The magnitude of the voltages of the P-Q buses, including bus (n+1)
(iii) The voltage angles of the P-V buses including bus i and the P-Q buses including bus (n+1).

Using the solved values of bus voltages and their angles, the net reactive power injection at all the P-V buses (including bus i) are obtained. The reactive power injection by the shunt connected converter of the UPFC (Q_{sh}) can be obtained as

$$Q_{sh} = Q_i - Q_i^{sp0} \qquad (6.127)$$

where Q_i is the calculated net reactive power injection of bus i and Q_i^{sp0} is the specified reactive power of bus i before the inclusion of the UPFC.

The current injected at bus i from the shunt connected converter of the UPFC can be found as

$$|I_{sh}|\angle\psi = \frac{P_{sh} - jQ_{sh}}{(V_i\angle\theta_i)^*} = \frac{-P_{se} - jQ_{sh}}{(V_i\angle\theta_i)^*} \tag{6.128}$$

Hence, the UPFC shunt branch source voltage magnitude and angle, $|V_{sh}|\angle\theta_{sh}$ can be obtained as

$$|V_{sh}|\angle\theta_{sh} = |V_i|\angle\theta_i + j * x_t\,(|I_{sh}|\angle\psi) \tag{6.129}$$

Example 6.4 For the power system network shown in Example 2 (Fig. 6.6), find the power flow through line 2-3 using the converged solution given in Table 6.7. Now, a TCSC is placed at the middle of line 2-3 to maintain the power flow at 10.5 MW. The reactance of the fixed capacitor of the TCSC is equal to half of the line 23 reactance and the TCSC inductor has a reactance one-third that of the fixed capacitor. Using Newton–Raphson load flow method, solve the network with TCSC to find the value of the TCSC reactance required.

Solution: From the converged solution of Table 6.7, the current through line 2-3

$$I_{23} = \frac{1.02\angle-2.8656 - 0.8307\angle-13.05}{0.723 + j*1.05} = 0.1814 - j0.0746$$

The active power flowing out of bus 2 towards bus 3

$$P_{23} = \text{Re}[(1.02\angle-2.8656).(0.1814 - j0.0746)^*] = 0.1886 \text{ p.u.} = 9.43 \text{ MW}$$

Now, the power flow is to be maintained at 10.5 MW or 0.21 p.u. Hence, the TCSC has to be operated in the capacitive mode.

Since the TCSC is placed at the middle of the line, two additional buses, say bus 7 and bus 8 are considered. Both of them are P-Q buses with

$$P_7^{sp} = 0;\quad Q_7^{sp} = 0;\quad P_8^{sp} = 0;\quad Q_8^{sp} = 0;$$

Therefore, the new line connecting bus 2 with bus 7, and the line connecting bus 8 with bus 3 both are considered to have half the reactances (and half of the line charging susceptance) of the original line 2-3.

The mismatch equation for the power flow through the TCSC (line 7-8) can be written as

$$\Delta P_{78} = 0.21 - P_{78} = P_{78}^{sp} - \frac{|V_7||V_8|}{x_{\text{tcsc}}}\sin(\theta_7 - \theta_8)$$

The power mismatch equations for bus 7 and 8 are written following Eqs (6.105) and (6.106). Similarly, the power mismatch equations for bus 2 and bus 3 are modified as shown in Eq. (6.107) and Eq. (6.108). The Y_{bus} matrix can be calculated as shown below. As the matrix is symmetrical, only the upper triangular part is shown.

$$\begin{bmatrix}
0.9922-j4.3925 & 0 & 0 & -0.5583+j2.5820 & 0 & -0.4339+j1.8275 & 0 & 0 \\
 & 1.4663-j2.6006 & 0 & 0 & -0.5765+j1.3085 & 0 & -0.8897+j1.2921 & 0 \\
 & & 0.8897-j8.8109 & j8.2715 & 0 & 0 & 0 & -0.8897+j1.2921 \\
 & & & 1.1124-j13.9919 & 0 & -0.5541+j2.3249 & 0 & 0 \\
 & & & & 0.5765-j4.6418 & j3.4153 & 0 & 0 \\
 & & & & & 0.9880-j7.6341 & 0 & 0 \\
 & & & & & & 0.8897-j1.2921 & 0 \\
 & & & & & & & 0.8897-j1.2921
\end{bmatrix}$$

With the modified and additional equations and the Y_{bus}, NRLF solution can be found. The solved value of TCSC reactance is –0.9742 p.u.

6.8 INCLUSION OF HVDC TRANSMISSION [25, 26]

The HVDC transmission system was described in Sec. 2.8. A two terminal HVDC system consists of two converter stations at the two terminals connected by a DC line. One of these converters acts as the rectifier and the other acts as the inverter. From the equivalent circuit of Fig. 2.31, the expression of the current flowing through the line is given as

$$I_d = \frac{V_{d0r}\cos\alpha - V_{d0i}\cos\beta}{R_{cr} + R_l + R_{ci}} \tag{6.130}$$

Here, the subscript 'r' refers to the rectifier quantities and the subscript 'i' refers to the inverter quantities. The expression can also be written in terms of the extinction angle of the inverter, γ as

$$I_d = \frac{V_{d0r}\cos\alpha - V_{d0i}\cos\gamma}{R_{cr} + R_l - R_{ci}} \tag{6.131}$$

where $\gamma = \pi - \delta$ (6.132)

Usually the load flow analysis in the AC system is carried out in per unit. So, for including the HVDC link in the AC system, it is necessary to define the per unit system for DC and its relation with the per unit AC system.

$$(V_{\text{dc}})_{\text{base}} = \frac{3\sqrt{2}}{\pi}(V_{\text{ac}})_{\text{base}} = V_{d0}; \quad (P_{\text{ac}})_{\text{base}} = (P_{\text{dc}})_{\text{base}};$$
$$(I_{\text{dc}})_{\text{base}} = \frac{(P_{\text{dc}})_{\text{base}}}{(V_{\text{dc}})_{\text{base}}}; \quad (Z_{\text{dc}})_{\text{base}} = \frac{(V_{\text{dc}})_{\text{base}}}{(I_{\text{dc}})_{\text{base}}}; \tag{6.133}$$

When expressed in per unit, the following relations can be established.

$$x_c = 2R_c; \quad I_{\text{ac}} = \frac{3\sqrt{2}}{\pi} I_d; \tag{6.134}$$

Now, if the HVDC link is connected between bus '*j*' and bus '*k*' of the AC system, then Eq. (6.131) can be written as

$$I_d = \left(KV_j \cos\alpha - KV_k \cos\gamma\right)/R_{\text{tot}} \tag{6.135}$$

$$\text{where } K = \frac{3\sqrt{2}}{\pi}; \quad R_{\text{tot}} = R_{cr} + R_l - R_{ci} = \frac{x_{cr}}{2} + R_l - \frac{x_{ci}}{2}; \tag{6.136}$$

In a two terminal HVDC system, usually the rectifier operates on current control and the inverter operates on constant extinction angle control. Hence, I_d is specified at a value I_d^{sp} and γ is kept constant. So there is only one variable, the firing angle α of the rectifier and there is one equation as shown below.

$$\Delta I_d = I_d^{sp} - \left(KV_j \cos\alpha - KV_k \cos\gamma\right)/R_{\text{tot}} \tag{6.137}$$

However, the active and reactive power mismatch equations at the buses '*j*' and '*k*' also get modified by the inclusion of the power going into/coming out of the DC system. The active and reactive power from bus '*j*' going into the HVDC system can be obtained as

$$(P_j)_{\text{dc}} = V_j I_j \cos\phi_j; \quad (Q_j)_{\text{dc}} = V_j I_j \sin\phi_j; \tag{6.138}$$

Here, ϕ is the angle of the voltage at the bus. Using Eq. (6.134), the expression of the active power becomes

$$\begin{aligned}(P_j)_{dc} &= V_j I_j \cos\phi_j = KV_j I_d \cos\phi_j = \left(KV_j \cos\phi_j\right)\left(\frac{KV_j \cos\alpha - KV_k \cos\gamma}{R_{\text{tot}}}\right) \\ &= \frac{1}{R_{\text{tot}}}\left[K^2V_j^2 \cos\phi_j \cos\alpha - K^2V_jV_k \cos\phi_j \cos\gamma\right]\end{aligned} \tag{6.139}$$

Similarly, the expression for the reactive power becomes

$$\begin{aligned}(Q_j)_{dc} &= V_j I_j \sin\phi_j = KV_j I_d \sin\phi_j = \left(KV_j \sin\phi_j\right)\left(\frac{KV_j \cos\alpha - KV_k \cos\gamma}{R_{\text{tot}}}\right) \\ &= \frac{1}{R_{\text{tot}}}\left[K^2V_j^2 \sin\phi_j \cos\alpha - K^2V_jV_k \sin\phi_j \cos\gamma\right]\end{aligned} \tag{6.140}$$

The specified power at bus 'j' is kept unchanged and the power mismatch equations take the form

$$\Delta P_j = P_j^{sp} - P_j = P_j^{sp0} - |V_j| \left(\sum_{\substack{m=1 \\ m \neq k}}^{n} \left(G_{jm} \cos \theta_{jm} + B_{jm} \sin \theta_{jm} \right) |V_m| \right) - \left(P_j \right)_{\text{dc}} \quad (6.141)$$

$$\Delta Q_j = Q_j^{sp} - Q_j = Q_j^{sp0} - |V_j| \left(\sum_{\substack{m=1 \\ m \neq k}}^{n} \left(G_{jm} \sin \theta_{jm} - B_{jm} \cos \theta_{jm} \right) |V_m| \right) - \left(Q_j \right)_{\text{dc}} \quad (6.142)$$

The power mismatch equations at bus 'k' are also modified in a similar manner. These modified equations, power mismatch equations at other buses, and Eq. (6.137) is solved by Newton's method to obtain the bus voltage magnitude, phase angle of the voltages, and the rectifier firing angle α.

6.9 CONCLUSION

This chapter discussed the analytical aspects of solving the load flow problem. The advantages of Newton's method in relation to the Gauss–Seidel method have been pointed out. The current trend in the industry seems to use the fast decoupled load flow, since in terms of the speed of the solution, it is also suitable for on-line applications. For large systems, a combination of optimally ordered elimination techniques for solving the linear equations together with sparsity programming is used to make Newton's method a practical tool. The modifications required to include FACTS devices and HVDC transmission system into an existing power system is also discussed.

Problems

6.1 Show that when there is no mutual coupling, the diagonal and off-diagonal elements in the Y_{bus} matrix can be computed from the formula

$$Y_{ii} = \sum_{j=1}^{n} y_{ij}$$

$$Y_{ij} = -y_{ij}$$

where Y_{ij} represents the admittance between the buses i and j, and Y_{ii} is the admittance between bus i and the ground bus.

6.2 Consider the three-bus system in Fig. P.6.1. The line impedances are given in p.u. and the line charging is neglected.
Treat bus 1 as the slack bus with voltage = 1.05 $\angle$ 0°.
At bus 2, P_{G2} = 25 MW, Q_{G2} = 15 MVAR, P_{L2} = 50 MW, and Q_{L2} = 25 MVAR.
At bus 3, P_{L3} = 60 MW and Q_{L3} = 30 MVAR.
Assume 100 MVA as the base MVA. Carry out 3 iterations of load flow using the Gauss–Seidel method with 1.4 as the acceleration factor.

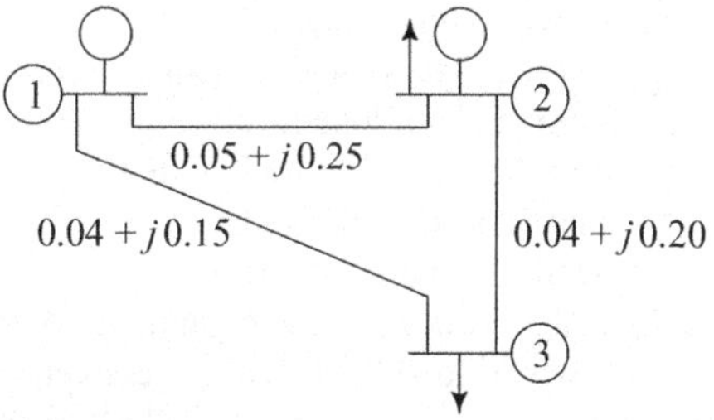

Fig. P.6.1 *Three-bus system*

Load/Gen. Data				
Bus	***Gen.***		***Load***	
	MW	***MVAR***	***MW***	***MVAR***
1	0	0	0	0
2	100	50	50	25
3	0	0	75	30
4	0	0	100	50

6.3 In Problem 6.2, treat bus **2** as the voltage regulated (P-V) bus with voltage held at 1.02 and reactive limits between 30 and –10 MVAR. Carry out 2 or 3 iterations using the Gauss–Seidel method.

6.4 Assume total line charging admittance of j 0.10 for the line 1-2 in Problem 6.2. Construct the new Y_{bus} and using the Gauss–Seidel method, carry out 2 or 3 iterations of load flow. Repeat if bus **2** is a P-V bus as specified in Problem 6.3. Acceleration factor may be taken as 1.4. Use the method explained in Section 6.3.5.

6.5 The four-bus system in Fig. P.6.2 has the line and load/gen. data shown below. All line charging is neglected. Take base MVA = 100.

Line Data (p.u.)	
Bus code	***Impedance***
1-3	j 0.05
1-4	j 0.05
2-3	j 0.05
3-4	j 0.10

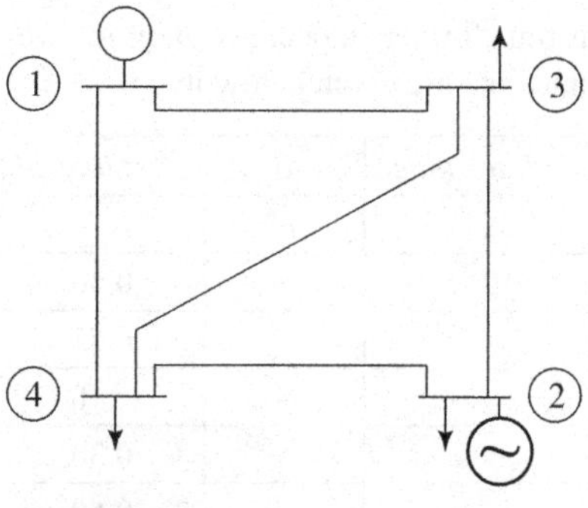

Fig. P.6.2 *The four-bus system*

Bus 1 is the slack bus with voltage equal to 1.02 ∠ 0°.

1. Using the Y_{bus} and the Gauss–Seidel method with an acceleration factor of 1.6, perform 2-3 iterations of load flow.
2. Using the Newton–Raphson method with polar coordinates, exhibit the structure of the matrices ***H***, ***N***, ***M***, and ***L***.

6.6 In Problem 6.5, show the $\boldsymbol{B}'$ and $\boldsymbol{B}''$ matrices.

6.7 For Problem 6.2, using polar coordinates and the Newton–Raphson method, perform 2 or 3 iterations of load flow. Test the convergence with the bus power mismatch of 0.01 p.u.

6.8 Consider the system in Fig. P.6.3 popularly known as the Ward–Hale system.

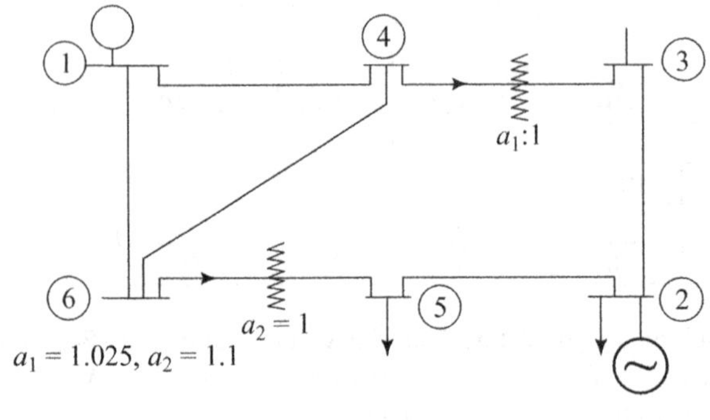

Fig. P.6.3 *The Ward and Hale system*

The line data are given below:

Bus code	*Impedance* z_{ij}	*Admittance* y_{ij}	*Half-line charging admittance*
1-6	$0.123 + j\,0.518$	$0.434 - j\,1.827$	$j\,0.02$
1-4	$0.08 + j\,0.37$	$0.56 - j\,2.58$	$j\,0.014$
2-5	$0.282 + j\,0.64$	$0.575 - j\,1.308$	0
4-6	$0.097 + j\,0.407$	$0.554 - j\,2.325$	$j\,0.015$
5-6	$j\,0.286$	$-j\,3.50$	0
2-3	$0.723 + j\,1.05$	$0.445 - j\,0.646$	0
3-4	$j\,0.11$	$-j\,9.1$	0

The off-nominal transformer impedances/admittances are referred to the unity side. The load/gen. data are given below in p.u.

Bus number	*Voltage mag.*	δ	*Net injected P*	*Net injected Q*
1	1.05	0°	—	—
2	1.10	—	0.50	—
3	—	—	– 0.55	– 0.13
4	—	—	0	0
5	—	—	– 0.30	– 0.18
6	—	—	– 0.50	– 0.50

Bus 2 is a voltage regulated (P-V) bus;

1. construct the Y_{bus} matrix for the system,
2. perform two iterations using the Gauss–Seidel method, and
3. exhibit the structure of fast decoupled load flow equations.

6.9 Do the load flow study for Problem 6.2 using fast decoupled load flow equations.

6.10 Exhibit the structure with numerical values for the decoupled load flow and fast decoupled load flow equations in Problem 6.5. Show the DC load flow equations also.

7 OPTIMUM POWER FLOW

7.1 INTRODUCTION

In the Optimal Power Flow (OPF) problem, the goal is to find the optimal settings of a given power system that optimizes the objective function such that the power flow and all constraints are satisfied. The objective function could be the minimization of total generation cost, or the total system losses, or the shift of generation and other controls from an optimal operating point. In this section, the total generation cost is considered. We consider the minimization of total generation cost.

7.2 CLASSICAL METHOD OF ECONOMIC LOAD DISPATCH WITHOUT CONSIDERING LOSSES

7.2.1 Cost Curves of Thermal Power Station

In a thermal power plant, coal is burned to convert its stored chemical energy to heat. This heat is then further converted to mechanical and finally electrical form of energy by the use of the turbine and generator. Hence, one way of assessing the efficiency of such a plant is to calculate the amount of heat energy (of coal) required to generate one kilowatt-hour of electrical energy. This is denoted as *heat rate* of the plant. A typical plot of heat rate of a thermal power plant is shown in Fig. 7.1(a). The plot is between the output in MW power and the input expressed in Kcal/h or MBtu/h. $(P_g)_{\text{minimum}}$ and $(P_g)_{\text{maximum}}$ are the minimum and maximum limits of output of the plant. The minimum limit is because of the constraint that a minimum amount of water must be flown through the boiler tube (for its safety). The incremental heat rate, $\dfrac{\partial H}{\partial P_{g_i}}$ is considered linear as shown in Fig. 7.1(b). For the ith plant, it can be expressed as

$$\text{Incremental heat rate, } \frac{\partial H}{\partial P_{g_i}} = m_i P_{g_i} + n \tag{7.1}$$

Hence, the cost function, then, can be obtained by integrating the incremental heat rate and multiplying that by the cost of the fuel (coal here).

The form of the equation will be

$$\text{cost function,} \quad f_i = \frac{1}{2} a_i P_{g_i}^2 + b_i P_{g_i} + c_i \tag{7.2}$$

The coefficients a_i, b_i, c_i are usually obtained empirically.

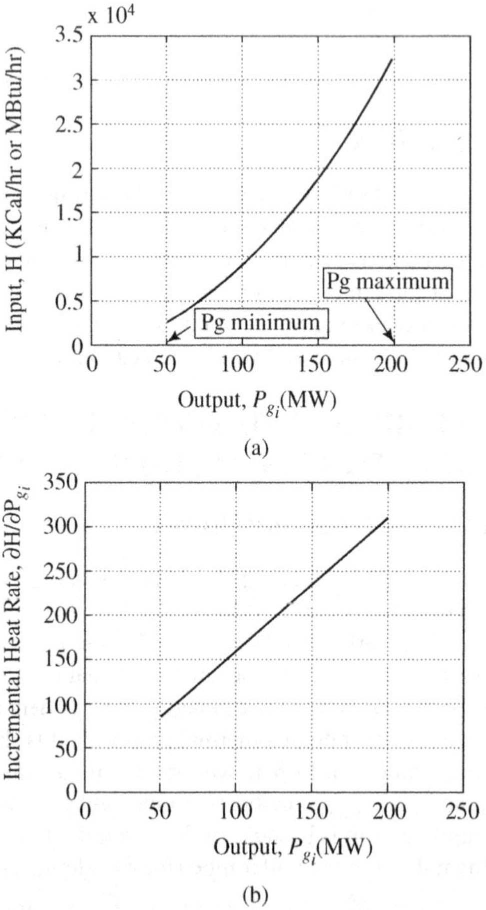

Fig. 7.1 *Variation of (a) the heat rate, and (b) the incremental heat rate with power output of a thermal power plant*

7.2.2 Problem Statement of Economic Load Dispatch

Total cost, $F_T = \sum_{i=1}^{N_g} f_i\left(P_{g_i}\right)$, N_g is the total number ot the rmal power plants.

The objective is to minimize F_T. Now, the total generation is equal to the total load plus losses.

$$\sum_{i=1}^{N_g} P_{g_i} = P_{\text{load}} + P_{\text{loss}} \tag{7.3}$$

Also there are some lower and upper limits on the generation of each plant as shown below:

$$\left(P_{g_i}\right)_{\text{minimum}} \le P_{g_i} \le \left(P_{g_i}\right)_{\text{maximum}} \tag{7.4}$$

At first, let us ignore (i) the loss, and (ii) the limits on the generation.
Hence, the constraint becomes

$$h = \sum_{i=1}^{N_g} P_{g_i} - P_{\text{load}} = 0 \tag{7.5}$$

The Lagrange function becomes

$$L = F_T - \lambda h \tag{7.6}$$

So, conditions for minimum cost operation are

$$\frac{\partial L}{\partial P_{g_i}} = 0 \;\Rightarrow \frac{\partial f_i}{\partial P_{g_i}} - \lambda = 0 \;\Rightarrow a_i P_{g_i} + b_i - \lambda = 0 \quad i = 1, 2, \ldots, N_g \tag{7.7}$$

$$\frac{\partial L}{\partial \lambda} = 0 \;\Rightarrow \sum_{i=1}^{N_g} P_{g_i} - P_{\text{load}} = 0 \tag{7.8}$$

Equation (7.6) can also be expanded as

$$(IC)_1 = (IC)_2 = \cdots\cdots = (IC)_{N_g} = \lambda \tag{7.9}$$

where $(IC)_i$ denotes the incremental cost of the ith machine.

Inclusion of Inequality Constraints for the Generation Limits As shown in Fig. 7.1(a), the power plants have some upper and lower limits on their output active power. Similarly, the incremental heat rate (and hence, incremental cost) is also limited by these limits on the generation. As the total load P_{load} increases, the solution of (7.8)–(7.9) may give some required λ and generation values which may be beyond the limits of that plant. To explain this, let us take the example of a power system with three power plants—Plant 1, Plant 2 and Plant 3 having generations P_{g_1}, P_{g_2} and P_{g_3} respectively. The cost functions of these plants are as given below.

$$\begin{aligned}
\text{Plant 1:}\quad & f_1 = \frac{1.5}{2} P_{g_1}^2 + 10 P_{g_1} + 120 \\
\text{Plant 2:}\quad & f_2 = \frac{1.2}{2} P_{g_2}^2 + 15 P_{g_2} + 150 \\
\text{Plant 3:}\quad & f_3 = \frac{1.0}{2} P_{g_3}^2 + 20 P_{g_3} + 180
\end{aligned} \tag{7.10}$$

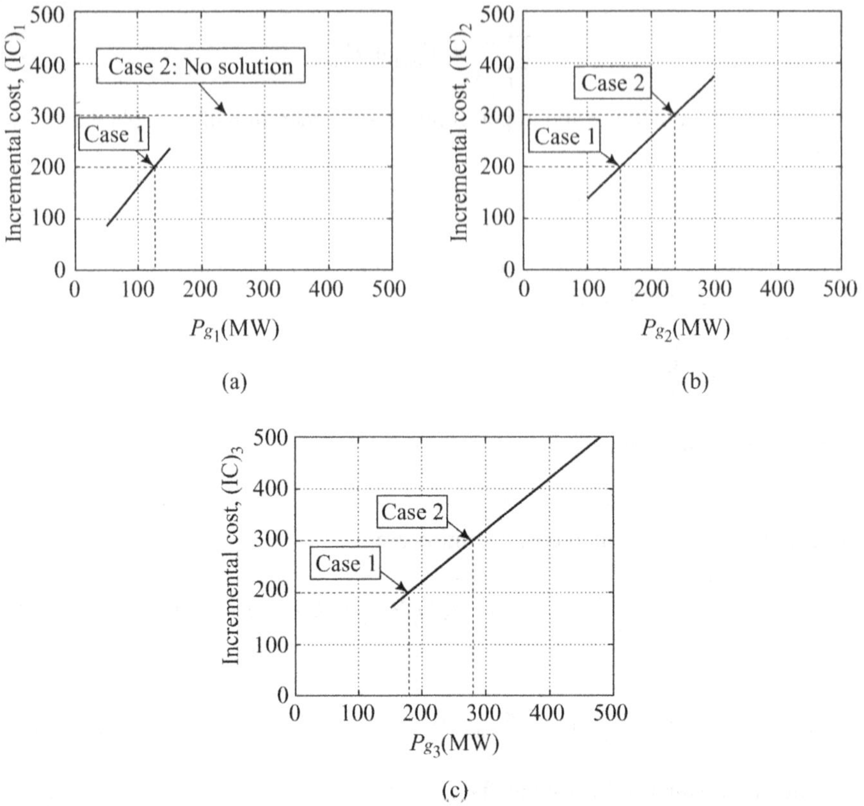

Fig. 7.2 *Variation of the incremental cost with power output of (a) Power plant 1, (b) Power plant 2, and (c) Power plant 3*

The plots of the incremental costs $(IC)_1$, $(IC)_2$ and $(IC)_3$ are shown in Fig. 7.2 (a), (b) and (c). The limits on the generation are as given below.

$$\begin{aligned} &\text{Plant 1:} \quad 50\text{ MW} \le P_{g_1} \le 150\text{ MW} \\ &\text{Plant 2:} \quad 100\text{ MW} \le P_{g_2} \le 300\text{ MW} \\ &\text{Plant 3:} \quad 150\text{ MW} \le P_{g_3} \le 500\text{ MW} \end{aligned} \tag{7.11}$$

Case 1: Suppose the total load in the power system (P_{load}) is 460 MW. The loss is neglected. The solution of (7.8) and (7.9) gives the following results:

$$\lambda = 200;\ Pg_1 = 126.4\text{ MW};\ Pg_2 = 153.9\text{ MW};\ Pg_3 = 179.7\text{ MW}$$

Case 2: Now, if the load increases to 710 MW, the solution becomes

$$\lambda = 300;\ Pg_1 = 193\text{ MW};\ Pg_2 = 237\text{ MW};\ Pg_3 = 280\text{ MW}$$

However, it can be seen from Fig. 7.2(a) that this is not a feasible solution as $(IC)_1$ cannot be 300. Also, the required generation exceeds the maximum limit for plant 1. In such cases, the generation of the plant (for which the limit is getting violated) is fixed at the limiting value and solution is carried out for the remaining plants. Here, we consider P_{g_1}=150 MW. Therefore, Eq. (7.8) gets modified to

$$P_{g_2} + P_{g_3} = P_{\text{load}} - P_{g_1} = 710 - 150 = 560 \tag{7.12}$$

Equation (7.9) is now applicable for only Plant 2, and Plant 3 and takes the following form:

$$1.2P_{g_2} + 15 = P_{g_3} + 20 = \lambda \tag{7.13}$$

The solution of (7.1) and (7.12) gives

$$\lambda = 323.2;\ P_{g_2} = 256.8 \text{ MW};\ P_{g_3} = 303.2 \text{ MW}$$

It can be found that for Plant 1, the value of $\lambda = 235$. Notice that the values of the Lagrangiean multiplier λ for plants with outputs inside their limits are the same, but the Lagrangiean multiplier for the generator(s) with output at its limit (e.g., Plant 1 here) is different from one another and also different from the Lagrangiean multiplier for generators with outputs inside their limits.

Thus, the procedure to solve the economic dispatch problem can be summarized as follows:

Step 1: Solve equations (7.8) and (7.9) to obtain the initial solution.
Step 2: Check output limits for each generating plant according to (7.4). If $P_{g_i} \geq (P_{g_i})_{\text{maximum}}$, then set $P_{g_i} = (P_{g_i})_{\text{maximum}}$. If $P_{g_i} \leq (P_{g_i})_{\text{maximum}}$, then set $P_{g_i} = (P_{g_i})_{\text{maximum}}$.
Step 3: After fixing the output of the plants with limit violations at their limits, remove components corresponding to these plants from (7.3). For simplicity, let us consider that the generation of plants 1 to N_k are violating limits and their corresponding limiting values are $P_{g_1}^{\text{lim}}, P_{g_2}^{\text{lim}}, \cdots, P_{g_{N_k}}^{\text{lim}}$. Then Eq. (7.4) can be written as

$$h - \sum_{i=N_k+1}^{N_G} P_{g_i} - \left(P_L - \left[P_{g_1}^{\text{lim}} + P_{g_2}^{\text{lim}} + \cdots + P_{g_{N_k}}^{\text{lim}}\right]\right) = 0 \tag{7.14}$$

Step 4: Go back to Step 1 until outputs of all generators satisfy Eq. (7.4).

7.3 ECONOMIC LOAD DISPATCH CONSIDERING LOSSES

Now the loss in the power system network is also considered. Hence, the equality constraint (7.5) becomes

$$h = \sum_{i=1}^{N_g} P_{g_i} - P_{\text{loss}} - P_{\text{load}} \tag{7.15}$$

This P_{loss} is a function of the generations of the power plants, i.e. P_{g_i}. So Eq. (7.7) takes the form

$$\frac{\partial L}{\partial P_{g_i}} = (IC)_i - \lambda\left(1 - \frac{\partial P_{\text{loss}}}{\partial P_{g_i}}\right) = 0 \quad \text{or,} \quad \lambda = (IC)_i \cdot \left(\frac{1}{1 - \dfrac{\partial P_{\text{loss}}}{\partial P_{g_i}}}\right)$$

$$\Rightarrow (IC)_i \cdot \left[\frac{1}{1-(ITL)_i}\right] = (IC)_i \cdot (PF)_i \tag{7.16}$$

where, $(ITL)_i$ = incremental transmission loss = $\dfrac{\partial P_{\text{loss}}}{\partial P_{g_i}}$

and (PF) = penalty factor = $\dfrac{1}{1-ITL}$

Loss Formula For solving Eq. (7.16), the relation between the network loss and the generation of the power plants is required. This can be obtained using the load–flow equations, and in that case, the method is called optimal power flow, described in the next section. However, an alternative way was proposed by Kirchmayer and Stagg in [28] where an approximate relation is given as

$$P_{\text{loss}} = \sum_{i=1}^{N_g}\sum_{j=1}^{N_g} P_{g_i} B_{ij} P_{gj} + \sum_{i=1}^{N_g} B_i P_{g_i} + B_0 \tag{7.17}$$

The coefficients B_{ij}, B_i and B_0 are called *loss coefficients* or *B-coefficients* and can be obtained from the bus impedance matrix as shown in [17].

It is observed that the first term of Eq. (7.17) is more dominant than the remaining two terms. So neglecting the second and third terms, we can get

$$\frac{\partial P_{\text{loss}}}{\partial P_{g_i}} = 2\sum_{j=1}^{N_g} B_{ij} P_{gj} \tag{7.18}$$

Using this in Eq. (7.16) we get

$$a_i P_{g_i} + b_i - \lambda\left(1 - 2\sum_{j=1}^{N_g} B_{ij} P_{gj}\right) = 0 \quad \Rightarrow (a_i + 2\lambda B_{ii}) P_{g_i} = \lambda - b_i - 2\lambda \sum_{\substack{j=1 \\ \neq i}}^{N_g} B_{ij} P_{gj}$$

$$\Rightarrow P_{g_i} = \frac{\lambda - b_i - 2\lambda \sum_{\substack{j=1 \\ \neq i}}^{N_g} B_{ij} P_{gj}}{(a_i + 2\lambda B_{ii})} \tag{7.19}$$

Equation (7.19) is used in Step 2 below while calculating the value of generation.

Solution Steps Since Eq. (7.16) is non-linear, the solution requires some iterative method. One frequently used method is the *Lambda Iterative Technique* [27], the steps of which are described below:

Step 1: Choose initial values of P_{g_i} and λ. One possible choice might be the solution of (7.8) and (7.9) neglecting losses. Let these be the kth iteration value.

Step 2: Calculate the values of generation for the $(k+1)^{\text{th}}$ iteration, $P_{g_i}^{k+1}$ from (7.19) using the kth iteration values.

Step 3: Check if $\left|P_{g_i}^{k+1} - P_{g_i}^{k}\right| \le \varepsilon_1$, where ε_1 is the predefined tolerance. If the condition is not satisfied, then set $P_{g_i}^{k} = P_{g_i}^{k+1}$ and go back to Step 2. If the condition is satisfied, then go to Step 4.

Step 4: Calculate P_{loss} using the solved values of P_{g_i}. Find h using Eq. (7.15). Check if $|h| \le \varepsilon$. If the condition is satisfied, then that is the optimal solution. Otherwise go to Step 5.

Step 5: Update the value of λ.

If h = positive, $\lambda^{\text{new}} = \lambda^{\text{old}} - \Delta\lambda$

If h = negative, $\lambda^{\text{new}} = \lambda^{\text{old}} + \Delta\lambda$

$\Delta\lambda$ = 1–5% of original λ

Then go back to Step 2.

7.4 OPTIMAL POWER FLOW

For more accurate results, power-flow equations are used instead of the loss formula. The standard form of the optimization becomes

$$\text{Minimize } C(P_g) = \sum_{i=1}^{N_G} C_i(P_{g_i}) \tag{7.20}$$

subject to the equality constraints

$$h_P = P_g - P_L - P(V,\theta) = 0 \tag{7.21}$$

$$h_Q = Q_g - Q_L - Q(V,\theta) = 0 \tag{7.22}$$

And the inequality constraints

$$g_S = S(V,\theta) - S^M \le 0 \tag{7.23}$$

$$g_{VM} = V - V^M \le 0 \tag{7.24}$$

$$g_{VM} = V^m - V \le 0 \tag{7.25}$$

$$g_{PM} = P_g - P_g^M \le 0 \tag{7.26}$$

$$g_{PM} = P_g^m - P_g \le 0 \tag{7.27}$$

$$g_{QM} = Q_g - Q_g^M \le 0 \tag{7.28}$$

$$g_{QM} = Q_g^m - Q_g \le 0 \tag{7.29}$$

where

$$P_i(V,\theta) = V_i \sum_{j=1}^{N} V_j (G_{ij} \cos\theta_{ij} + B_{ij} \sin\theta_{ij})$$

$$Q_i(V,\theta) = V_i \sum_{j=1}^{N} V_j (G_{ij} \sin\theta_{ij} - B_{ij} \cos\theta_{ij})$$

and N is the number of buses.

Let the admittance of line connecting the bus i and the bus k be $y_{ik} = g_{ik} + jb_{ik}$. When neglecting the line charging, the apparent power flowing on the line $i - k$ is

$$S_{ik}(V,\theta) = \left| \begin{matrix} V_i^2 g_{ik} - V_i V_k (g_{ik} \cos\theta_{ik} + b_{ik} \sin\theta_{ik}) \\ +j\left[V_i^2 b_{ik} - V_i V_k (b_{ik} \cos\theta_{ik} - g_{ik} \sin\theta_{ik})\right] \end{matrix} \right|$$

The new objective function is constructed by adding constraints into the original objective function with penalty factors as

$$\begin{aligned} F(X) = C(X) &+ \sum_{i=1}^{N} f_{Pi} h_{Pi}^2(X) + \sum_{i=1}^{N} f_{Qi} h_{Qi}^2(X) + \sum_{i=1}^{N} f_{VMi} g_{VMi}^2(X) + \sum_{i=1}^{N} f_{Vmi} g_{Vmi}^2(X) + \\ &\sum_{i=1}^{N} f_{PMi} g_{PMi}^2(X) + \sum_{i=1}^{N} f_{Pmi} g_{Pmi}^2(X) + \sum_{i=1}^{N} f_{QMi} g_{QMi}^2(X) + \sum_{i=1}^{N} f_{Qmi} g_{Qmi}^2(X) + \\ &\sum_{i=1}^{N_L} f_{Si} g_{Si}^2(X) \end{aligned} \tag{7.30}$$

where

X: state vector consisting of P_g, Q_g, V, and θ

f_{Pi}, f_{Qi}, f_{Vmi}, f_{Vmi}, f_{Pmi}, f_{Pmi}, f_{Qmi}, f_{Qmi}, f_{Si}: penalty factors associated with the constraints (7.21)–(7.29). If a constraint is not violated, its associated penalty factor is 0.

N_L: number of lines considered in Eq. (7.23)

This is an unconstrained optimization that can be solved using Newton's method. The search direction of the Newton's method is

$$\Delta X^k = -\left(\nabla^2 F(X^k)\right)^{-1} \nabla F(X^k) \tag{7.31}$$

where

$\nabla^2 F(X^k)$: Hessian matrix of the augmented objective function

The iteration based on the search direction is

$$X^{k+1} = X^k + \alpha^k \Delta X^k \tag{7.32}$$

where α^k is a scalar step length obtained by a line search method such as quadratic interpolation.

The iteration (7.31) will stop when a stopping criterion is met, i.e.

$$\left\| X^{k+1} - X^k \right\| \le \varepsilon_1 \tag{7.33}$$

or

$$\frac{\left|F(X^{k+1}) - F(X^k)\right|}{\left|F(X^k)\right|} \leq \varepsilon_2 \tag{7.34}$$

where ε_1, ε_2 are tolerances.

Because the inversion of the Hessian matrix in Eq. (7.31) can be memory constrained and computationally expensive, it can be rewritten in the form of Eq. (7.35) and then solved as a linear equation by the Gauss elimination process.

$$\nabla^2 F(X^k)\Delta X^k = -\nabla F(X^k) \tag{7.35}$$

It is also noted that the Hessian matrix of the augmented objective function is very sparse. Therefore, the sparsity computational method can be exploited [reference].

For a given stopping criterion, the procedure to solve the OPF (7.20)–(7.29) using the Newton method is as follows:

Step 1: Choose a set of initial values for the penalty factors.
Step 2: Form the augmented objective function (7.30).
Step 3: Solve the initial power flow to obtain the initial state variables X^0, set $k = 0$.
Step 4: Calculate the augmented objective function, its gradient, and it's Hessian at X^k.
Step 5: Calculate the search direction ΔX^k using (7.31).
Step 6: Calculate the step length α^k using a line search method such as quadratic interpolation.
Step 7: Update state variables using (7.32), then update the augmented objective function (7.30), its gradient, and its Hessian using the updated state variables.
Step 8: Check the stopping criterion. If the problem converges, go to Step 9. If not, set $k = k + 1$, and go back to Step 5.
Step 9: Check if all constraints are satisfied. If yes, done. If not, scale the penalty factors for the violated constraints by a factor of 2.
Step 10: Reset $k = 0$, and go back to Step 4.

Example 7.1 The three-machine, nine-bus system shown in Fig. 7.3 is used to set up the OPF. The system parameters are given in Table 7.1.

For all buses, the lower and upper voltage limits are

$V^m = 0.9$ pu
$V^M = 1.1$ pu

Loads at buses 5, 6, and 8 are

$P_5 + jQ_5 = 125 + j50$ MVA;
$P_6 + jQ_6 = 90 + j\,30$ MVA;
$P_8 + jQ_8 = 100 + j\,35$ MVA;

No reactive limit on generators is considered. The generator costs and their real output limits are given in Table 7.2.

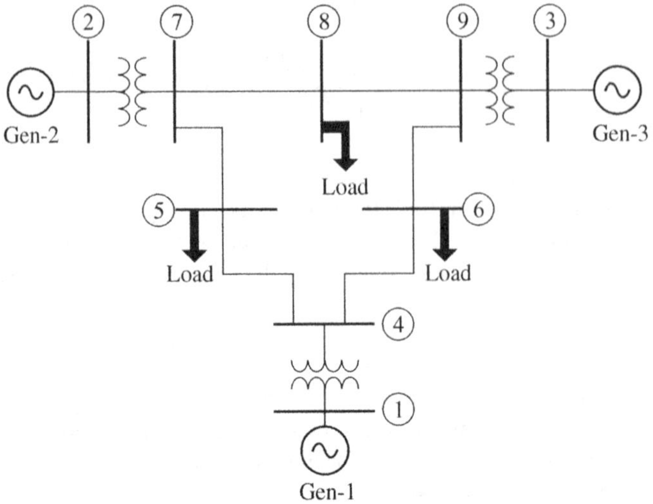

Fig. 7.3 *The three-machine, nine-bus system*

Table 7.1 *System parameters*

From	*To*	*R (pu)*	*X (pu)*	*B (pu)*	S^M *(MVA)*
1	4	0.0000	0.0576	0.0000	250
4	6	0.0170	0.0920	0.1580	250
6	9	0.0390	0.1700	0.3580	150
3	9	0.0000	0.0586	0.0000	300
6	8	0.0119	0.1008	0.2090	150
7	8	0.0085	0.0720	0.1490	250
7	2	0.0000	0.0625	0.0000	250
7	5	0.0320	0.1610	0.3060	250
5	4	0.0100	0.0850	0.1760	250

Table 7.2 *Generator data*

Generator	P_g^M*(MW)*	P_g^m*(MW)*	*Cost Function (\$/h):* $a_2 P^2 + a_1 P + a_0$		
			a_2	a_1	a_0
1	20	200	0.0060	2.0	140
2	20	150	0.0075	1.5	120
3	10	100	0.0070	1.8	80

The MATPOWER[1] software package is used to solve the OPF. The results are given in Table 7.3.

[1] R. Zimmerman, C. Murillo-Sanchez, and D. Gan, MATPOWER: A MATLAB Power System Simulation Package, http://www.pserc.cornell.edu/matpower/

Table 7.3 *OPF results*

Bus #	*Voltage magnitude (pu)*	*Voltage angle (degree)*	*P (MW)*	*Q (MVAr)*	*Total cost ($/h)*
1	1.100	0	105.53	12.18	1131.14
2	1.097	2.441	113.07	–1.53	
3	1.086	2.382	99.30	–22.74	
4	1.095	–2.893			
5	1.100	–0.409			
6	1.073	–5.516			
7	1.100	–0.915			
8	1.089	–2.68			
9	1.085	–4.613			

7.5 CONCLUSION

Determination of the most economic combination of generation is becoming very important considering the fact that the load demand is increasing day by day and saving even a small fraction of that means the saving of a lot of money. In the era of a restructured and deregulated power system, this gives an edge to the operation of a utility. Inclusion of network losses makes the equations non-linear and hence, accurate computation requires the use of optimal power-flow methods.

8 STABILITY STUDIES

8.1 INTRODUCTION

In this Chapter, we shall discuss the transient behavior of a power system resulting from major disturbances, such as a fault followed by switching operations, sudden rejection of load or generation. A major disturbance upsets the balance between the mechanical input and the electrical output of a generator, with the result that some generators may accelerate while others may decelerate. The rotor angles will undergo wide variations and in this process the 'synchronism' of the system gets affected. The 'synchronism' must be maintained by timely operation of the circuit breakers; otherwise both voltages and frequency may deviate so widely from the normal values that cascading outages and system separation may occur. Transient stability simulation studies are carried out to study these phenomena and the results enable the engineer to plan and coordinate the protection scheme effectively. Critical clearing times of the circuit breakers can be computed and the protection zones of distance relays during transient swings can be adjusted. Compared to short circuit and load flow studies, transient stability studies are more complex since they involve the electromechanical dynamics of the synchronous machines and its associated controls such as the excitation system, and the governor. There is good commercial and educational software for doing transient stability studies. Here, stress is given on problem formulation and the mathematical modeling of the process.

8.2 CLASSIFICATION OF POWER SYSTEM STABILITY

As per the IEEE/CIGRE Joint Task Force on Stability Terms and Definitions (P. Kundur et al), "*Power system stability is the ability of an electric power system, for a given initial operating condition, to regain a state of operating equilibrium after being subjected to a physical disturbance, with most system variables bounded so that practically the entire system remains intact*". The ability of the synchronous machines of an interconnected power system to remain in synchronism after being subjected to a disturbance is termed as rotor angle stability. It depends on the capacity of the system to maintain/restore equilibrium between the electromagnetic torque and the mechanical torque of each synchronous machine. It is customary to classify the rotor angle stability into two types depending on the magnitude and nature of the disturbance, namely *steady-state* (*or small signal*) *stability*, and *transient stability*. The disturbances altering the operation of a power system can be small and 'gradual' (e.g. controlled changes of power input and output) or large and 'sudden' (e.g. uncontrolled loss of large amounts of generation or load, or the occurrence of transmission line faults). The terms 'gradual' and 'sudden' should be interpreted here in relation to the major time constants of the system or part of the system under study.

8.2.1 Small-signal Stability

Let us assume that the system is operating in a certain steady-state condition. If, for small changes in the state of the system, the system returns to the steady state as $t \rightarrow \infty$, we call the particular operating condition as steady-state or small-signal stable. The state of a power system can be characterized by several variables, depending on the desired complexity of the model. It is obvious that steady-state stability depends on the operating condition of the system. The range of operating conditions for which the system is steady-state stable, constitutes a measure of the steady-state limit. If this limit is specified in terms of the power transfer from one machine to another, then it precisely constitutes what is known as the steady-state power limit. The mathematical tool for studying steady-state stability is through a linear model of the power system followed by eigenvalue analysis. This topic is not discussed in the book.

8.2.2 Transient Stability

Transient stability refers to the ability of the system to remain in synchronism when subjected to large disturbances like faults in the transmission lines. Disturbances like these result in large excursions of generator rotor angles that are governed by the non-linear power-angle relationship. Usually, the instability is manifested in the form of angular separation between the generators of the system. This angular separation may be of a single machine from the rest of the machines, or of a group of machines from another group. When the focus is on the stability of a particular machine, then the rest of the power system can be represented by an ***infinite bus***, which is equivalent to an ideal voltage source having constant voltage and frequency irrespective of the power exchanged by it. This assumption is reasonable as the capacity of the single machine is very small as compared to that of the rest of the system and hence, the change in the voltage and frequency caused by the power exchanged by the machine is negligible. Thus the whole system is equivalently represented by a single machine (the machine of interest) connected to the infinite bus (*SMIB system*) through some transmission line, as shown in Fig. 8.1. While operating under steady state, the machine is subjected to a large disturbance resulting in rotor speed deviations. The machine is called *transiently stable* if the rotor ultimately reaches a new stable equilibrium position.

However, the SMIB representation is not sufficient when the focus of the study is on the stability of the complete multi-machine power system, i.e., if all the synchronous machines of the system retain synchronism. In that case, the study should include the models of all the machines and the connecting network. While all the machines are operating in steady-state equilibrium, a large disturbance is imposed on the system. The system is called *transiently stable* if the relative rotor angles of all the machines reach steady-state values. The relative rotor angle is given by the difference of the individual rotor angles from a reference. While the rotor angle of any one of the machines can be chosen as the reference, generally the machine with the largest inertia, namely, the machine connected to the swing bus, is taken as reference for computing the relative rotor angles.

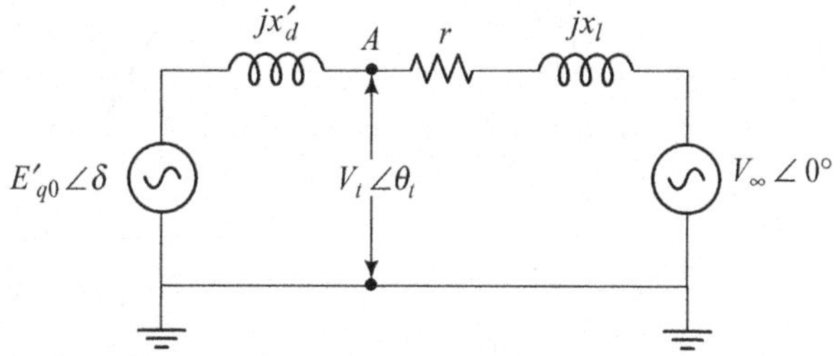

Fig. 8.1 *Schematic diagram of the SMIB system*

The disturbances usually considered in transient stability studies are the most severe ones such as a short circuit on the transmission lines, buses, or at transformer terminals. Also, these faults may be symmetrical, three-phase to ground, single-phase to ground, or two-phase to ground. When a fault occurs, it is necessary to determine when to clear the fault without any of the rotating machines losing their synchronism. The maximum time through which the fault can be left on the system and yet have the system stable after its clearance, is called the *critical clearing time.* Determination of the critical clearing time constitutes an important facet of transient stability study since such information enables the setting of relays to clear the fault, etc. Other disturbances considered are the sudden disconnection of generating units or loads.

8.3 EQUAL AREA CRITERIA FOR THE DETERMINATION OF STABILITY CONDITION OF SMIB SYSTEM [3, 17]

In the single machine infinite bus of Fig. 8.1, the synchronous machine is represented by the classical model (described in Sec. 2.3), i.e., a voltage of constant magnitude behind transient reactance with the angle of the voltage being the rotor angular position, δ. The equations representing the generator, Eqs (2.21)–(2.23) can be combined and written as

$$\frac{d^2\delta}{dt^2}=\frac{\omega_s}{2H}[P_m-P_e-k_D\Delta\omega_r]=\frac{\omega_s}{2H}\left[P_m-\frac{E'_{q0}V_t}{x'_d}\sin(\delta-\theta_t)-k_D\Delta\omega_r\right] \quad (8.1)$$

The mechanical power, P_m can be considered as constant as its time constant is much higher as compared to that of the electrical power, P_e. Considering a $P_m = 1.0$ p.u. and neglecting the damping co-efficient k_D, the plot of P_e versus δ is shown in Fig. 8.2. P_m is also shown in the figure.

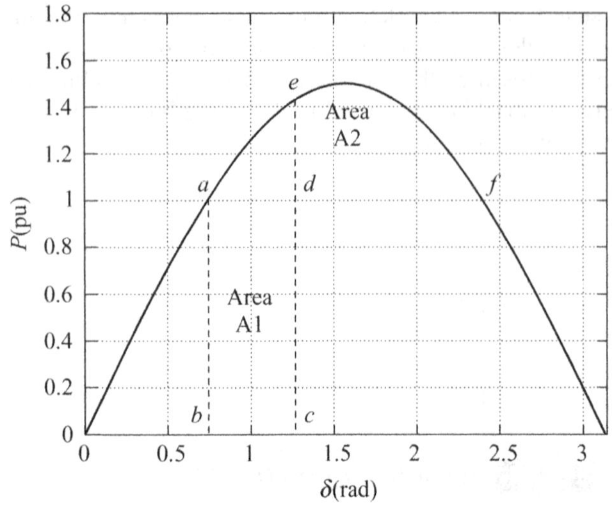

Fig. 8.2 *Plots of electrical power (P_e) and mechanical power (P_m) of alternator with rotor angle, δ*

There can be two possible steady-state operating points (where $P_m = P_e$), shown as point '*a*' and point '*f*'. However, the operation at point '*f*' is not stable as any disturbance takes the rotor angle beyond point '*f*' and the system cannot come back to stable operation. Hence, the machine is considered to operate at point '*a*'. Now if there is a short-circuit fault at the machine terminal making V_t and thus P_e zero, the rotor starts accelerating following Fig. (8.1). The movement is along the path $a \rightarrow b \rightarrow c$ in Fig. 8.2. Suppose that the fault is cleared at time t_{cl} (corresponding to the point '*c*') and the rotor angle has increased to a value δ_{cl} by this time. As δ has increased, the electrical power P_e is equal to the value shown by point '*e*' in Fig. 8.2. As P_e is more than P_m, the machine starts decelerating; however, the speed is still positive and hence, δ continues increasing. However, the decelerating force will bring the speed to zero and then make it negative provided P_e is still higher than P_m, i.e., the rotor angle has not gone beyond point '*f*'. The rotor angle will start reducing and will go beyond point '*a*' when again P_m becomes higher than P_e causing acceleration. Thus the rotor angle oscillates about point '*a*' if there is no damping, and in the presence of damping, the rotor angle finally stabilizes at point '*a*'. However, as mentioned before, this return to a stable operating point takes place only if the fault is cleared within a time such that the rotor angle does not go beyond point '*f*'. It can be shown that this happens only if the areas A1 (accelerating area) and A2 (decelerating area) shown in Fig. 8.2 are related as

Area A1 $\geq$ Area A2.

This is called the *Equal Area Criterion*. The limiting value of the fault clearing time for which this relation holds is the *critical clearing time* (t_{cr}) for this fault. The corresponding rotor angle is denoted by δ_{cr}. An expression of δ_{cr} in terms of the initial rotor angle δ_0, the mechanical power P_m, and the machine constants can be obtained by equating A1 and A2 as shown below.

$$\text{Area A1} = \int_{\delta_0}^{\delta_{cr}} P_m d\delta = P_m [\delta_{cr} - \delta_0]$$

$$= \text{Area A2} = \int_{\delta_{cr}}^{\pi-\delta_0} (P_{\max} \sin\delta - P_m) d\delta = P_{\max} [\cos\delta]_{\pi-\delta_0}^{\delta_{cr}} - P_m [\pi - \delta_0 - \delta_{cr}]$$

$$\Rightarrow P_m [\pi - 2\delta_0] = P_{\max} [\cos\delta_{cr} + \cos\delta_0]$$

$$\Rightarrow \delta_{cr} = \cos^{-1}\left[\frac{P_m}{P_{\max}}(\pi - 2\delta_0) - \cos\delta_0\right] = \cos^{-1}[(\pi - 2\delta_0)\sin\delta_0 - \cos\delta_0] \quad (8.2)$$

In the accelerating area A1,

$$\frac{d^2\delta}{dt^2} = \frac{\omega_s}{2H} P_m \Rightarrow \frac{d\delta}{dt} = \int_0^t \frac{\omega_s}{2H} P_m dt = \frac{\omega_s}{2H} P_m t$$

$$\Rightarrow \delta = \frac{\omega_s P_m}{4H} t^2 + \delta_0 \quad (8.3)$$

Using Eq. (8.2) in Eq. (8.3), the critical clearing time, t_{cr} can be found as

$$t_{cr} = \sqrt{\frac{4H(\delta_{cr} - \delta_0)}{\omega_s P_m}} \quad (8.4)$$

8.4 MATHEMATICAL REPRESENTATION FOR USE IN TRANSIENT STABILITY SIMULATION

The equal area criterion is not sufficient when the stability of a multi-machine system with interconnected networks and more detailed model of the machine is to be studied. In that case, models of the machines and the network are to be integrated with different relevant assumptions and solved using numerical methods. Those methods will be discussed in the next section. However, before that, the mathematical description of stability, a generalized model of the system, and methods of interface of the machine and network equations are discussed here.

8.4.1 Mathematical Definitions of Stability

Consider a dynamical system whose description is given by a set of first-order nonlinear differential equations of the form

$$\dot{x}_i = f_i(x_1, x_2, \ldots, x_n) \quad i = 1, 2, \ldots, n \quad (8.5)$$

or equivalently in vector form as

$$\dot{X} = f(x) \quad (8.6)$$

where x and f are n dimensional vectors. x is termed as the state of the system, since knowing x at time $t = t_0$, Eq. (8.5) or Eq. (8.6) uniquely define the solution of $x(t)$ for $t > t_0$.

The equilibrium states (i.e., steady state condition) of the system are the solutions of the equation

$$\mathbf{0} = f(x) \tag{8.7}$$

Eq. (8.7) may have several solutions since it constitutes a set of nonlinear algebraic equations. Let the solutions be designated by $x_e^1, x_e^2, \ldots$. In the absence of any disturbing forces, let the system be at some equilibrium state say x_e^i so that $\mathbf{0} = f(x_e^i)$. For small disturbances around x_e^i, Eq. (8.6) can be linearized by expanding it around x_e^i in a Taylor series and neglecting the second and higher order terms. Let the deviations around x_e^i be denoted by $\Delta x = x - x_e^i$. Then, the linearization will yield

$$\Delta \dot{x} = A \Delta x \tag{8.8}$$

where

$$A = \left.\frac{\partial f}{\partial x}\right|_{x=x_e^i} = \left.\begin{bmatrix} \dfrac{\partial f_1}{\partial x_1} & \dfrac{\partial f_1}{\partial x_2} & \cdots & \dfrac{\partial f_1}{\partial x_n} \\ \vdots & \vdots & & \vdots \\ \dfrac{\partial f_n}{\partial x_1} & \dfrac{\partial f_n}{\partial x_2} & \cdots & \dfrac{\partial f_n}{\partial x_n} \end{bmatrix}\right|_{x=x_e^i} \tag{8.9}$$

For a given x_e^i, A is an $n \times n$ matrix with real coefficients. If A has all eigenvalues with negative real parts, we say that x_e^i is stable for small disturbances, or equivalently, has small-signal stability. In a power system, if x_e^i represents a particular (steady-state) operating condition, we say the system is *steady-state stable* for that particular operating condition if A has all eigenvalues with negative real parts.

In contrast to steady-state stability, transient stability deals with large disturbances in the system. The linearized model is, therefore, no longer valid. Consequently, it will be necessary to solve the system of nonlinear differential equations of the nature of Eq. (8.6) corresponding to the faulted and the post-fault state. The configuration of the transmission system in the faulted and the post-fault states are different, since either the fault would have cleared itself, or by means of switching out the transmission lines, the fault would have been isolated. Let the corresponding mathematical description on the right-hand side of Eq. (8.6) be $f_1(x)$ and $f_2(x)$ for the faulted and the post-faulted state, respectively, and let the assumed fault clearing time be t_{cl}. Occurrence of a fault will result in an abrupt change of a system variable, such as voltage at a bus or mechanical power. However, the system state will be continuous at $t = 0$, i.e., $x(t_0^-) = x(t_0^+) = x(t_0)$. Mathematically, we have therefore to solve

$$\dot{x} = f_1(x), \quad 0^+ < t \le t_{cl}, \quad x(t_0) \tag{8.10}$$

and

$$\dot{x} = f_2(x) \quad \text{for } t > t_{cl} \tag{8.11}$$

Eqs (8.10) and (8.11) are solved numerically to find whether the state x will settle down to a steady-state value. For different assumed clearing times t_{cl} or sequence of switching operations, the swing curves are computed. From these swing curves it is ascertained whether the system is stable or not. Repetitive computations of the swing curves are necessary to find both the critical clearing time and the switching operation sequence.

8.4.2 Mathematical Models for Transient Stability Simulation

The equations of the form of Eq. (8.6) representing the dynamics of the synchronous machines and the associated exciter were discussed in Sec. 2.3. Two reduced order models for the alternator, namely the flux-decay model and the classical model were presented. In the flux-decay model given by Eqs (2.14) – (2.16) and Eq. (2.20), the state variables are: the generator rotor angle (δ), the rotor speed deviation ($\Delta\omega_r$), the q-axis component of the voltage behind the transient reactance (E'_q), and the exciter e.m.f. (E_{fd}). E'_q is actually a representation of the rotor (field) flux. The generator is represented by a voltage source behind a transient reactance, as shown in Fig. 2.1. The magnitude and the angle of the source voltage are functions of the state variables and the generator terminal bus voltage as shown in Eq. (2.7). Further reduction of this model with certain assumptions gives the classical model as described in Eqs (2.21) – (2.22). In this model, the state variables are the generator rotor angle and the rotor speed deviation. The generator is represented by a voltage of constant magnitude behind the transient reactance and the angle of the voltage is nothing but the rotor angle (δ). A derivation of the classical model from the first principle is shown below. The assumptions on which the classical model can be used to represent the alternators are also mentioned with justifications.

When the steady-state equilibrium of a power system gets disturbed, there is momentarily a difference between the mechanical power input and the electrical power output in each synchronous machine. This tends to accelerate or decelerate the rotor depending on whether this difference is positive or negative. The mechanical power input gets adjusted according to the governor action. The synchronous machine dynamics together with that of the exciter dynamics and the voltage at the buses modifies the electrical output of the generator. For the ith machine, we can write

$$\frac{d}{dt}\left(W_{\text{K.E},i}\right)+P_{di}=P_{mi}-P_{ei} \tag{8.12}$$

where, $W_{\text{K.E},i}$ = kinetic energy in MW sec; P_{di} = damping power in MW

P_{mi} = mechanical input power in MW; P_{ei} = electrical output power in MW

The synchronous frequency is denoted by f_0. Since the kinetic energy is proportional to the square of the frequency, we have

$$W^0_{\text{K.E},i}=W^0_{\text{K.E},i}\left(\frac{f_i}{f_0}\right)^2 \tag{8.13}$$

where $W_{\text{K.E},i}$ = kinetic energy at synchronous frequency.

Since $f_i = f_0 + \Delta f_i$, we have

$$W_{\text{K.E},i} = W^0_{\text{K.E},i}\frac{(f_0+\Delta f_i)^2}{f_0^2} \cong W^0_{\text{K.E},i}\left(1+\frac{2\Delta f_i}{f_0}\right) \tag{8.14}$$

since Δf_i is small during a transient swing. Differentiating the expression for $W_{\text{KiE},i}$ we have

$$\frac{d}{dt}\left(W_{\text{K.E},i}\right) = 2\frac{W^0_{\text{K.E},i}}{f_0}\frac{d}{dt}\Delta f_i \tag{8.15}$$

Let $\delta_i = \theta_i - \omega_0 t$ and $\dfrac{d\delta_i}{dt} = \dfrac{d\theta_i}{dt} - \omega_0 = \omega_i - \omega_0 = 2\pi(f_i - f_0) = 2\pi\Delta f_i$

where θ_i is the absolute rotor angle of machine i with respect to a fixed reference, and δ_i is the rotor angle of machine i measured with respect to a synchronously rotating reference frame. Both θ_i and δ_i are measured in radians. The change in frequency Δf_i is given by the expression $\omega_i - \omega_0 = 2\pi\Delta f_i$. We shall take the damping power to be proportional to the frequency deviation from the synchronous frequency. With these assumptions, substitution of Eq. (8.15) in Eq. (8.12) and division by base MVA yields

$$\frac{H_i}{\pi f_0}\frac{d^2\delta_i}{dt^2} + D_i\frac{d\delta_i}{dt} = P_{mi} - P_{ei}, \quad i = 1, 2, \ldots, m \tag{8.16}$$

where $H_i = \dfrac{W^0_{\text{K.E},i}}{\text{Base MVA}}$ = inertia constant of machine i in p.u.;

D_i = damping constant in p.u.; P_{mi}, P_{ei} are the mechanical and eletrical power in p.u. for machine i; m = number of machines ;

The mechanical shaft input power to the generator is a function of the governor turbine dynamics. The electrical power can be expressed using Park's equations in the d–q variables, which in turn are influenced by the network equations, and the voltage regulator and the excitation system. However, so far as the first swing duration is concerned (of the order of 1–3 s), we can make a number of physically justifiable assumptions that reduces considerably the mathematical complexity of the model. These will be explained in terms of the block diagram shown in Fig. 8.3 for just one machine. If there are several machines in the system, each machine is influenced by, and in turn each machine influences other machines through the transmission network equations.

1. The network is modeled as a lumped circuit in the phasor variables. The implication here is that the nominal AC frequency is much larger compared to the frequencies of the machine oscillations under transient condition. This results in an algebraic description of the network in terms of the phasor variables. Strictly, the network must also be modeled in terms of differential equations. But this becomes prohibitively inefficient and not justified in terms of the normally desired accuracy.

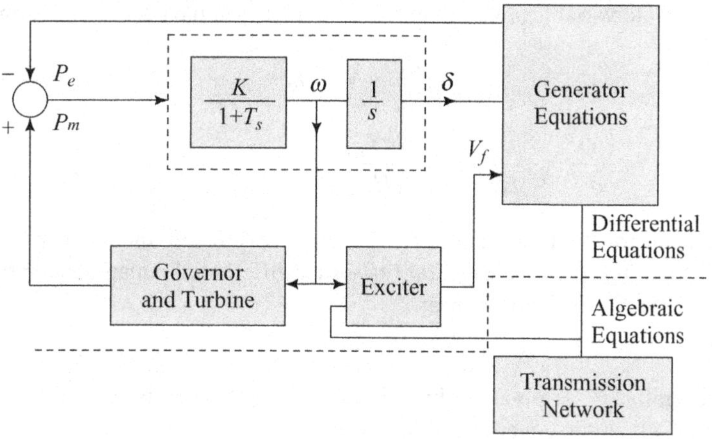

Fig. 8.3 *Block diagram showing the inter-relationships of variables with one machine connected to network K-1/D, T = H/πf_0D*

2. The generator is represented by a voltage source of constant magnitude (determined from prefault load flow) in series with a constant reactance called the *direct-axis transient reactance* x'_{di}. This is a highly simplified model of the synchronous machine.
3. The voltage regulator action is not fast enough to be effective during the first swing. Hence, V_f remains constant.
4. Power output from the governor and the turbine P_m remains essentially constant during the first swing (equal to the prefault value).
5. The mechanical angle δ_i of each machine rotor coincides with the phase angle of the voltage behind the transient reactance. This again stems from an approximation of the synchronous machine mode.
6. The loads are converted to an equivalent admittance based on voltages at the buses calculated from the pre-fault load flow study.

The above assumptions simplify the treatment of the transient stability problem. We now have, for the *m* machine system, the following equations:

$$\frac{H_i}{\pi f_0}\frac{d^2\delta_i}{dt^2} + D_i\frac{d\delta_i}{dt} = P_{mi} - P_{ei}, \quad i = 1, 2, \ldots, m \tag{8.17}$$

P_{mi} = Constant and is determined from pre-fault load flow. P_{ei} is determined from the solution of the algebraic network equations . We now introduce the state variables.

$$x_i - \delta_i \tag{8.18}$$

$$x_{i+m} = \frac{d\theta_i}{dt} = \frac{d\delta_i}{dt} + \omega_0 = \omega_i \quad i = 1, \ldots, m \tag{8.19}$$

x_i represents the rotor angle with respect to a synchronously rotating reference frame, and x_{i+m}, the frequency of the *i*th machine.

With these state variables, we can express Eq. (8.17) in the state space form as

$$\frac{d\delta_i}{dt} = \omega_i - \omega_0 \tag{8.20}$$

$$\frac{d\omega_i}{dt} = -\frac{D_i}{H_i}\pi f_0 x_i + \frac{\pi f_0}{H_i}(P_{mi} - P_{ei}) \quad i = 1, 2, \ldots, m \tag{8.21}$$

If we assume for the time being that P_{ei} can be expressed analytically in terms of δ_i $(i = 1, 2, ..., m)$, then we have $2m$ first-order differential equations corresponding to Eqs (8.20) and (8.21) of the form

$$\dot{x} = f(x) \tag{8.22}$$

More generally, P_{ei} is obtained by solving at each time instant, a set of nonlinear differential algebraic equations of the form

$$\dot{x} = F(x, y) \tag{8.23}$$

$$0 = G(x, y) \tag{8.24}$$

If we assume loads as constant impedances, then it is possible to eliminate the algebraic variables y and have a mathematical description equivalent to Eq. (8.22) only. This will be discussed in the following sub-section.

8.4.3 Interface of Generators, Transmission Networks and Load

Consider a transmission system having n buses (excluding the ground bus) with m generator buses $(m < n)$ having a multiport representation as shown in Fig. 8.4. Let the bus admittance matrix be denoted by Y_{bus}. Without loss of generality, the first m buses are designated as generator buses.

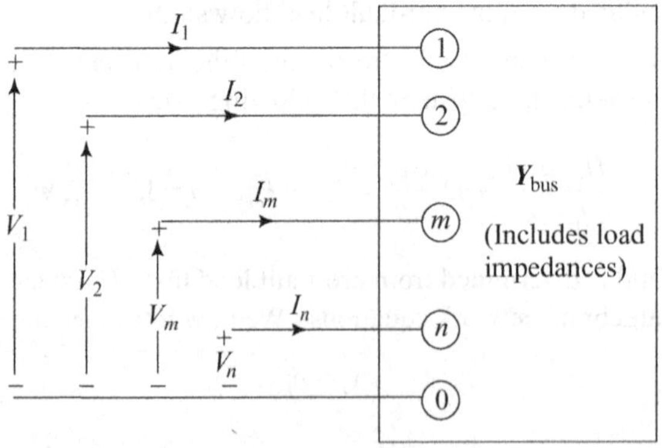

Fig. 8.4 *Terminal representation of a transmission network after absorbing load impedances*

1. Load flow study for conditions existing prior to the occurrence of fault, i.e., at $t = 0^-$ will yield the following information:

(a) $P_{Li}, Q_{Li}, P_{Gi}, Q_{Gi}$ $\quad i = 1, 2, ..., m$
(b) $P_{Li}, Q_{Li},$ $\quad i = m + 1, ..., n$
(c) $|V_i| \angle\psi_i = V_i$ $\quad i = 1, 2, ..., n$

P and Q are the real and reactive powers respectively, at the buses and the subscripts L and G refer to the load and generation, respectively. V_i is the phasor voltage at the ith bus.

2. Loads are converted into equivalent admittances y_{Li} between the ith bus and the ground bus as

$$y_{Li} = (P_{Li} - jQ_{Li})/|V_i|^2 \qquad i = 1, 2, ..., n \tag{8.25}$$

If there is no load at a particular bus i, then, $y_{Li} = 0$.

3. With the addition of elements corresponding to the load admittances, multiport representation gets modified as

$$\boldsymbol{Y}_{\text{bus}} = \boldsymbol{Y}_{\text{bus(old)}} + \text{diag}(y_{Li}) \tag{8.26}$$

Note that only the diagonal elements of the bus admittance matrix get modified. Fig. 8.4 shows the n-port representation with the modified Y_{bus}.

4. At the generator buses, the complex power $P_{Gi} + jQ_{Gi}$ is converted into an equivalent voltage source in series with the direct axis transient reactance of the machine as follows

$$E_i = V_i + jx'_{di} I_i = V_i + jx'_{di} \frac{(P_{Gi} - jQ_{Gi})}{V_i^*} = |E_i| \angle\delta_i \tag{8.27}$$

$$i = 1, 2, \dots, m$$

where x'_{di} is the direct-axis transient reactance of the ith machine and E_i is the voltage behind the transient reactance of the ith machine. The complete representation is shown in Fig. 8.5.

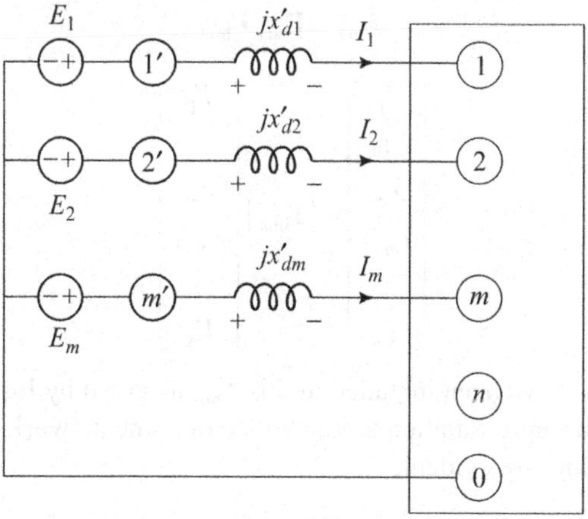

Fig. 8.5 *Network representation of Fig. 8.4 augmented by synchronous machine representation*

8.5 SIMULATION METHODS

Basically, two methods are used in the power industry. The first method calls for alternate computation cycles of the differential and algebraic equations, i.e., Eq. (8.23) and Eq. (8.24). It is called the *partitioned explicit method* (PE). The second method called the *simultaneous solution with implicit integration* (SI) involves making Eq. (8.23) algebraic by using an implicit integration method and solving these along with Eq. (8.24), simultaneously at each time step as a set of nonlinear algebraic equations. The PE approach is used in the industry in the production grade programs because of its programming simplicity and robust nature of the algorithm. Because of the alternate solution cycles, it is prone to numerical instability. The SI is better when the system contains both slow and fast dynamics. However, if the algebraic variables (y) are eliminated from Eqs (8.23) and (8.24) and the resulting description is Eq. (8.22), then we can have straightforward numerical integration. This is the case when the loads are considered as constant impedance type.

8.5.1 Partitioned Explicit Method

In this method, the network equations and the differential equations are solved alternately. The network equations are obtained as follows:

Network Equations Let the generator reactances be denoted by the steady-state v–i relations

$$v_k = jx'_{dk}\, i_k \qquad k = 1, 2, \ldots, m \tag{8.28}$$

The voltage sources are specified as

$$E_k = \text{specified} = |E_k| \angle \delta_k \qquad k = 1, 2, \ldots, m \tag{8.29}$$

The network representation is

$$\boldsymbol{I}_{\text{bus}} = \boldsymbol{Y}_{\text{bus}}\, \boldsymbol{V}_{\text{bus}}$$

i.e.,

$$\begin{bmatrix} I_1 \\ \vdots \\ I_m \\ I_{m+1} \\ \vdots \\ I_n \end{bmatrix} = [\boldsymbol{Y}_{\text{bus}}] \begin{bmatrix} V_1 \\ \vdots \\ V_m \\ V_{m+1} \\ \vdots \\ V_n \end{bmatrix} \tag{8.30}$$

Note that the loads have been included in this $\boldsymbol{Y}_{\text{bus}}$ as given by Eq. (8.26).

When the generator representation is augmented onto the network representation, the following constraints are evident:

$$\begin{aligned} i_k &= I_k && k = 1, 2, \ldots, m \\ v_k &= E_k - V_k && k = 1, 2, \ldots, m \\ I_k &= 0 && k = m+1, \ldots, n \end{aligned} \tag{8.31}$$

Inverting Eq. (8.28), we get $i_k = y_k v_k$ with $y_k = -j/x'_{dk}$.
Using the constraint equations Eqs (8.31) and substituting them in Eq. (8.30), we get

$$
\begin{aligned}
&(Y_{11} + y_1)\, V_1 + Y_{12} V_2 + \cdots + Y_{1m} V_m \qquad + \cdots + Y_{1n} V_n = y_1 E_2 \\
&Y_{21} + V_1 + (Y_{22} + Y_2) V_2 + \cdots + Y_{2m} V_m \qquad + \cdots + Y_{2n} V_n = y_2 E_2 \\
&\vdots \qquad\qquad\qquad\qquad\qquad\qquad\qquad\qquad \vdots \\
&Y_{m1} V_1 + Y_{m2} V_2 + \cdots + \cdots + (Y_{mm} + y_m) V_m + \cdots + Y_{mn} V_n = y_m E_m \\
&Y_{m+1,\, m+1} V_1 + \cdots \qquad + \cdots \qquad + \cdots + Y_{m+1,\, n}\, V_n = 0 \\
&\vdots \qquad\qquad\qquad\qquad\qquad\qquad\qquad\qquad \vdots \\
&Y_{n1} V_1 + \cdots \qquad + \cdots \qquad \cdots + Y_{nn} V_n = 0
\end{aligned}
\tag{8.32}
$$

These are the network equations, which can be solved for V_1, ..., V_n knowing E_1, ... E_m. The electrical power output of each machine is given by

$$P_{ei} = \text{Re}\left[E_i I_i^*\right] \tag{8.33}$$

where

$$I_i = (E_i - V_i) y_i \qquad \text{i} = 1, 2, ..., m$$

Eq. (8.33) together with the network equations Eqs (8.32), collectively are represented symbolically as

$$\mathbf{0} = \boldsymbol{G(x, y)}$$

where $\boldsymbol{y}$ is the set of algebraic variables $V_1, ..., V_m$.

Solution Scheme For numerical integration, the assumed fault clearing time t_{cl} is divided into intervals of Δt such that $k\Delta t = t_{cl}$ where k is an integer. Assume that the voltages behind the transient reactances have been computed as per Eq. (8.27) based on pre-fault load flow. The solution proceeds as follows:

1. For $t = 0^+$, the fault conditions are incorporated in Eq. (8.32). For example, if a symmetrical three-phase to ground fault occurs at the pth bus, set V_p = 0. Then Eq. (8.32) is solved for the voltages V_i. The generator currents I_k (k = 1, 2, ..., m) are computed and hence, the power output P_{ei} of each generator using Eq. (8.33).
2. Equations (8.20) and (8.21) are then integrated for each machine with the initial conditions $\delta_i(0) = \delta_i$ from Eq. (8.27) and $\omega_i(0) = \omega_0$, to obtain δ_i and ω_i at $t = \Delta t$. The value of δ_i at $t = \Delta t$ is denoted by $\delta_i = (\Delta t)$.
3. At $t = \Delta t$, the phasor voltages behind the transient reactances, or equivalently the voltage sources E_i (i = 1, 2, ..., m) are given by

$$E_i(\Delta t) = |E_i| \angle \delta_i(\Delta t) \tag{8.34}$$

Note that by virtue of assumption 2 in Sec. 8.3, the magnitude of this voltage is held constant at its value at $t = 0^-$, but its phase angle is different. With $E_i(\Delta t)$ given by Eq. (8.34), Eq. (8.32) is again solved for voltages V_i and hence, I_i(i = 1, 2, ..., m) at $t = \Delta t$. The power outputs of all the generators at $t = \Delta t$ are then computed using Eq. (8.33).

4. Equations (8.20) and (8.21) are then integrated for each machine using the values δ_i and ω_i at $t = \Delta t$ obtained in step 2 as the initial conditions. Thus values of δ_i and ω_i at $t = 2\,\Delta t$ are obtained for each machine.
5. Steps 3 and 4 are repeated for obtaining δ_i at $t = 3\Delta t$ and so on until $t = t_{cl}$, the fault clearing time.
6. At $t = t_{cl}$, + we have a new transmission configuration as a result of the clearing of the fault. The network equations Eqs (8.32) are solved with this new configuration and the values of δ_i and ω_i are computed for $t_{cl} + \Delta t$.
7. The procedure is continued for about 5 s or so. The variation of the rotor angles δ_i as a function of time constitutes the swing curves. The nature of the swing curves will indicate whether some or all machines are stable or not.
8. The study is repeated for different clearing times. The maximum value of clearing time for which all machines indicate stability is called *the critical clearing time*, denoted by t_{cr}. In other words, if the fault is cleared earlier than this, the system is transiently stable and if cleared later than this, one or more machines will become unstable. Computation of t_{cr}, therefore, involves repetitive simulations.

8.5.2 Simultaneous Solution with Implicit Integration Method

The trapezoidal rule for integration of a differential equation

$$\dot{x} = f(x), \qquad x(0) = x_0 \tag{8.35}$$

is given by

$$x_{n+1} = x_n + \frac{\Delta t}{2}[f(x_n) + f(x_{n+1})] \tag{8.36}$$

Using this idea, we convert Eqs (8.23) and (8.24) as

$$x_{n+1} = x_n + \frac{\Delta t}{2}[f(x_n, y_n) + f(x_{n+1}, y_{n+1})] \tag{8.37}$$

$$0 = G(x_{n+1}, y_{n+1}) \qquad n = 0, 1, 2, \ldots \tag{8.38}$$

In the above equations, the subscripts n and $n + 1$, refer to the time instants t_n and t_{n+1}, respectively. Equations (8.37) and (8.38) are a set of nonlinear algebraic equations of the form

$$\hat{F}(x_{n+1}, y_{n+1}) = 0 \tag{8.39}$$

$$G(x_{n+1}, y_{n+1}) = 0 \tag{8.40}$$

Applying the Newton–Raphson method, we get at $(k + 1)$ iteration (at $t = t_{n+1}$)

$$x_{n+1}^{(k+1)} = x_{n+1}^{(k)} + \Delta x_{n+1}^{(k)} \tag{8.41}$$

$$y_{n+1}^{(k+1)} = y_{n+1}^{(k)} + \Delta y_{n+1}^{(k)} \tag{8.42}$$

where $\Delta x_{n+1}^{(k)}$ and $\Delta y_{n+1}^{(k)}$ are obtained from solving the linear equations

$$\begin{bmatrix} -\widehat{F}\left(x_{n+1}^{(k)}, y_{n+1}^{(k)}\right) \\ -\widehat{G}\left(x_{n+1}^{(k)}, y_{n+1}^{(k)}\right) \end{bmatrix} = \left.\begin{bmatrix} \dfrac{\partial \widehat{F}}{\partial x} & \dfrac{\partial \widehat{F}}{\partial y} \\ \dfrac{\partial G}{\partial x} & \dfrac{\partial G}{\partial y} \end{bmatrix}\right|_{(k)} \begin{bmatrix} \Delta x_{n+1}^{(k)} \\ \Delta y_{n+1}^{(k)} \end{bmatrix} \tag{8.43}$$

Computation is carried out $n = 1, 2, ...,$ knowing the values of x, y at $n = 0$.

8.5.3 Direct Numerical Integration of Swing Equations after Eliminating Algebraic Equations

In the previous method, the solution of the network equations is required for the faulted and the post-fault systems in order to obtain the electrical power output at each step of integration. If we are not interested in bus voltages as a function of time, these buses can be eliminated, and a multiport representation for both the faulted and the post-fault networks can be obtained at the internal buses. This is possible if all the loads are represented as constant impedances. In this method, we obtain a multiport representation at the internal nodes of the generators by eliminating all the other buses. Using the parameters of this multiport representation, an analytical expression for $P_{ei} = (i = 1, 2, ..., m)$ is obtained involving only the δ_i's $(i = 1, 2, ..., m)$. Each time the transmission configuration gets changed as a result of switching operations, the multiport representation also gets changed. The procedure is as follows:

1. Load impedances are absorbed in the $\boldsymbol{Y}_{\text{bus}}$ of the transmission network as discussed in Section 8.4.3.
2. This modified $\boldsymbol{Y}_{\text{bus}}$ is augmented by the impedances of the synchronous machines corresponding to the direct axis transient reactances x'_{di}. Let the voltages of the new buses be denoted by $E_1, ..., E_m$ and admittances of the machines be $y_1, ..., y_m$. Thus, m new internal buses are created and the overall bus admittance matrix is obtained as follows (for the purpose of clarity, the internal buses are denoted by $1', 2', ... m'$).

   Let $\boldsymbol{Y}_{\text{bus}}$ (after inclusion of load impedances) be partitioned as

$$\boldsymbol{Y}_{\text{bus}} = \begin{array}{c} \\ m \\ n-m \end{array}\begin{array}{c} \begin{array}{cc} m & n-m \end{array} \\ \begin{bmatrix} Y_1 & Y_2 \\ Y_3 & Y_4 \end{bmatrix} \end{array} \tag{8.44}$$

where $\boldsymbol{Y}_1$ of dimension $m \times m$ corresponds to the buses where generators are connected. $\boldsymbol{Y}_2$, $\boldsymbol{Y}_3$, and $\boldsymbol{Y}_4$ are the other sub matrices. Let the machine admittance matrix be denoted by $y = \text{diag}\,(y_1, ..., y_m)$. Then the new augmented $\boldsymbol{Y}_{\text{bus}}$ of dimension $(n + m) \times (n + m)$ is symbolically represented as

$$\boldsymbol{Y}_{\text{bus}}^{\text{aug}} = \begin{array}{c} \\ m \\ m \\ n-m \end{array}\begin{array}{c} \begin{array}{ccc} m & m & n-m \end{array} \\ \begin{bmatrix} y & -y & 0 \\ -y & Y_1 + y & Y_2 \\ 0 & Y_3 & Y_4 \end{bmatrix} \end{array} \tag{8.45}$$

The bus admittance at the internal nodes is obtained by applying the network reduction formula to the above matrix. Before applying the network reduction formula, the fault condition must be appropriately reflected in the matrix. Thus, if a symmetrical three-phase to ground fault at bus p is to be simulated, the row and column corresponding to the pth bus is deleted, before doing the network reduction. Let the elements of the bus admittance matrix at the internal buses be denoted by $Y_{ij}(i, j = 1, 2, ..., m)$. Since $m' = m$, we retain the notation m for the sake of convenience.

We are now left with a multiport representation shown in Fig. 8.6.

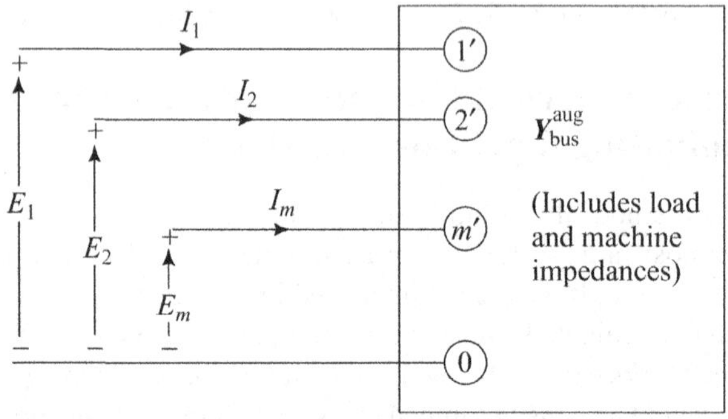

Fig. 8.6 *Terminal representation at internal nodes of generators*

At any bus i, the current into the system is

$$I_i = \sum_{j-1}^{m} Y_{ij} E_j \tag{8.46}$$

The active power into bus i (which is the active power output of generator i) is given by

$$P_{ei} = \text{Re}\left[E_i I_i^*\right] = |E_i|^2 Y_{ii} \cos \theta_{ii} + \sum_{\substack{j=1 \\ j \neq i}}^{m} |E_i||E_j||Y_{ij}| \cos\left(\theta_{ij} - \delta_{ij}\right) \tag{8.47}$$

where $\qquad Y_{ij} = |Y_{ij}| \angle \theta_{ij} = G_{ij} + jB_{ij}$ and $E_i = |E_i| \angle \delta_i$

The generator dynamics is given by Eq. (8.17) or Eqs (8.20)–(8.21). Since P_{ei} is expressed in terms of δ_i s, the above equations can be integrated straightaway, because the initial conditions are known.

$$\delta_i\big|_{t=0^+} = \delta_i \text{ from (8.27) and } \left.\frac{d\delta_i}{dt}\right|_{t=0^+} = 0, \quad i = 1, \ldots, m \tag{8.48}$$

In the expression for the P_{ei} in Eq. (8.47), $|E_i|(i = 1, 2, ..., m)$ are held constant at the values calculated form the load flow as given by Eq. (8.27). If the transfer conductances are neglected, then $G_{ij} = 0$ and $\theta_{ij} = 90°$. The expression for P_{ei} then reduces to

$$P_{ei} = |E_i|^2 G_{ii} + \sum_{\substack{j=1 \\ j \neq i}}^{m} |E_i||E_j||B_{ij}| \sin\left(\delta_i - \delta_j\right) \tag{8.49}$$

since $$G_{ij} = Y_{ii} \cos \theta_{ij} \text{ and } |Y_{ij}| = B_{ij}$$

Coming to the final step involved in this method, the general expression for P_{ei} under the faulted and the post-fault stages can be computed by substituting appropriate values of $|Y_{ij}|$ and θ_{ij} in Eq. (8.47). For the sake of brevity, let the two expressions be P^f_{ei} and P^{pf}_{ei} respectively. It now remains to integrate the following two systems of equations.

Faulted State

$$\frac{H_i}{\pi f_0}\frac{d^2\delta_i}{dt^2} + D_i\frac{d\delta_i}{dt} = P_{mi} - P^f_{ei}, \quad 0 \le t \le t_{cl} \tag{8.50}$$

Post-fault State

$$\frac{H_i}{\pi f_0}\frac{d^2\delta_i}{dt^2} + D_i\frac{d\delta_i}{dt} = P_{mi} - P^{pf}_{ei}, \quad t \ge t_{cl} \tag{8.51}$$

The initial values of the state variables, for solving Eq. (8.50) is obtained as shown below.

$$\delta_i\big|_{t=0^+} = \delta_i \text{ from (8.27) and } \left.\frac{d\delta_i}{dt}\right|_{t=0^+} = 0, \quad i = 1, 2, \ldots, m$$

The solution obtained at $t = t_{cl}$ is the initial values of the state variables for Eq. (8.51). Well-known techniques in numerical analysis such as modified Euler, R-K method, can be used to integrate these equations to obtain the swing curves. The swing curves can be calculated for different values of t_{cl} and the critical clearing time can be computed as discussed earlier.

Conceptually, both the methods, i.e., the PE method and the direct integration method, highlight some interesting aspects of the problem. In the PE method involving network solution, computation proceeds in alternate cycles and the retention of the identity of all the nodes, including load buses, permits the representation of loads. Although we represented the loads by constant impedance, nonlinear representation of the load is possible. Solution of the network equations poses no problem. Monitoring any of the bus voltages as a function of time is possible and this may be desirable in studies involving voltage fluctuations, relay coordination during transient swings, etc.

In the direct simulation method, we have to integrate a set of nonlinear differential equations. These can be easily converted into $2m$ first-order differential equations as indicated in Eq. (8.20) and Eq. (8.21) which can then be integrated easily. The direct method also leads naturally to a proper understanding of the qualitative methods, such as the energy function methods, for transient stability study. If only the first swing excursions are to be observed, the direct method offers a quick way. This is especially true if interest is focused on the critical clearing time, and the bus voltages during the transient period are not required.

Example 8.1 The synchronous machine shown in Fig. 8.7 is generating 100 MW and 75 MVAR. The voltage of the infinite bus **q** is $1 + j0$ and the tie line reactance is 0.08 per unit on a 100 MVA base. The machine transient reactance is 0.20 p.u. and the inertia constant is 4.0 p.u. on a 100 MVA base. Calculate the changes in phase angle and the speed of the generator for a three-phase to ground fault at bus **p** for a duration of 0.10 s. Use (a) modified Euler method, and (b) Runge-Kutta method with a time increment of 0.02 s. The time for simulation can be taken as 0.2 s. The frequency of supply is 50 Hz.

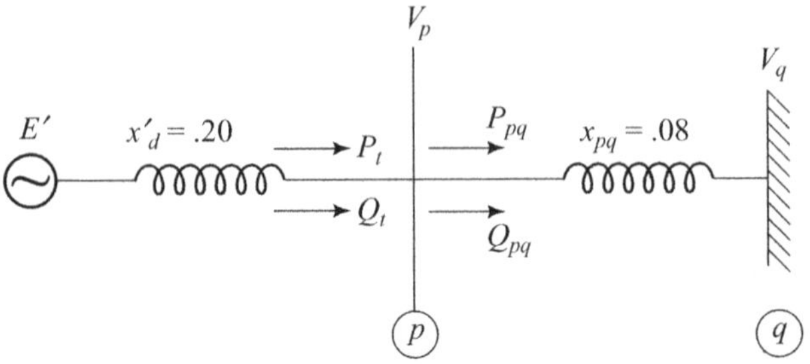

Fig. 8.7 *Single machine infinite bus system*

Solution: ***Base Load Flow*** Line current I_{pq} is given by

$$I_{pq} = \frac{V_p - V_q}{jx_{pq}}$$

Power flow in the line is

$$V_p I^*_{pq} = P_{pq} + jQ_{pq}$$

Substituting the values of I_{pq}

$$V_p \frac{(V^*_p - V^*_q)}{-jx_{pq}} = P_{pq} + jQ_{pq}$$

$$V_p \left(\frac{V^*_p - (1 + j0)}{-j0.08} \right) = 1.0 + j0.75$$

since

$$P_{pq} = P_t, \text{ and } Q_{pq} = Q_t.$$

If we put $V_p = e_q + jf_p$, the above equation simplifies to

$$(e_p + jf_p)\,[(e_p - jf_p) - (1 + j0)] = 0.06 - j0.8$$

Separating the real and imaginary parts, we get

$$f_p = 0.08;\; e_p^2 + (0.08)^2 - e_p = 0.06$$

Solving the quadratic equation and taking only the real value of e_p, we get $e_p = 1.051$. Hence,

$$V_p = 1.051 + j0.08$$

Calculation of Voltage Behind Transient Reactance The machine current is given by

$$I_t = \frac{P_t - jQ_t}{V_p^*} = \frac{1 - j0.75}{1.051 - j0.08} = 1.186\angle -32.51^\circ = 1 - j0.638$$

The voltage behind the transient reactance E' is given by

$$E' = V_p + j x_d' I_t$$

$$= 1.051 + j0.08 + j0.2(1 - j0.638)$$

$$= 1.1785 + j0.28$$

$$= 1.21131 \angle 13.365^\circ$$

Simulation The fault at bus **p** is simulated by setting the voltage at bus **p** equal to zero. A new load flow is obtained at the instant of the fault to calculate P_e. As the fault is on the terminal of the generator, $P_e(0) = 0$.

The differential equations to be solved are given by Eq. (8.20) and Eq. (8.21). Since δ is already defined to be $\theta - \omega_0 t$

$$\frac{d\delta}{dt} = \omega - \omega_0$$

$$\frac{d\omega}{dt} = \frac{\pi f_0}{H}(P_m - P_e)$$

since damping $D \equiv 0$.

$P_m = 1.0$ p.u. throughout the calculations since the governor action is not taken into consideration. The initial conditions are $\delta(0) = 13.365^\circ$ and $\omega(0) = \omega_0 = 2\pi f_0 = 314.1593$ rad/s.

(a) *Solution by Modified Euler's Method*

For $t = 0.02$ s

Step 1 (predictor step)

$$\left.\frac{d\delta}{dt}\right|_0 = \omega(0) - 2\pi f_0 = 0$$

$$\left.\frac{d\omega}{dt}\right|_0 = \frac{\pi f_0}{H}(P_m - P_e(0))$$

$$= \frac{\pi \times 50}{H}(1 - 0) = 39.27$$

The first estimates of δ, ω at $t = 0.02$ s are obtained as

$$\delta_{(0.02)}^{(1)} = \delta(0) + \Delta t \left.\frac{d\delta}{dt}\right|_0$$

$$= 13.365 + 0$$

$$= 13.365^\circ$$

$$\omega_{(0.02)}^{(1)} = \omega(0) + \Delta t \left.\frac{d\omega}{dt}\right|_0$$

$$= 314.15926 + (0.02)\,(39.27)$$

$$= 314.94466 \text{ rad/s}$$

The first estimate of the phase angle is used in the load flow calculation to determine a first estimate of the generator power. The generator power is zero and it will remain zero until the fault exists. Hence, $P_{e(0.02)}^{(1)} = 0$.

Step 2 (corrector step) Using the first estimates of δ, ω, and P_e, the derivatives at the end of the step are calculated. Then the average of this and the derivative in step 1 are used to compute improved estimates.

$$\left.\frac{d\delta}{dt}\right|_{(0.02)}^{(1)} = \omega_{(0.02)}^{(1)} - 2\pi f_0 = 314.95 - 314$$

$$= 0.95$$

$$\left.\frac{d\omega}{dt}\right|_{(0.02)}^{(1)} = \frac{\pi f_0}{H}\left(P_m - P_{e(0.02)}^{(1)}\right)$$

$$= 39.27$$

Hence the improved estimates for δ and ω at $t = 0.02$ are obtained as

$$\delta_{(0.02)}^{(2)} = \delta(0) + \frac{\Delta t}{2}\left(\left.\frac{d\delta}{dt}\right|_0 + \left.\frac{d\delta}{dt}\right|_{(0.02)}^{(1)}\right)$$

$$= 13.365 + \left(\frac{180}{\pi} \times \frac{0.02}{2}\right)(0 + 0.95) = 13.81511°$$

$$\omega_{(0.02)}^{(2)} = \omega(0) + \frac{\Delta t}{2}\left(\left.\frac{d\omega}{dt}\right|_0 + \left.\frac{d\omega}{dt}\right|_{(0.02)}^{(1)}\right) = 314.94466$$

A load flow is obtained using the second estimate of the phase angle $\delta_{(0.02)}^{(2)}$ in order to obtain the voltage and generator power at $t = 0.02$. We can improve the estimates by repeating the corrector step to get a better average.

The calculation is next repeated for $t = 0.04$ s by incrementing the time by $t = 0.02$ s. The procedure is continued until $t = 0.10$ s, at which point the fault clears.

Computation at Fault Clearing Instant After the clearance of fault at $t = 0.10$ s, the post-fault condition at $t = 0.10^+$ must be evaluated, which is then used for further calculation. For example, at $t = 0.10^+$, since the fault is removed,

$$I_t = \frac{E' - (1 + j0)}{j(0.2 + 0.08)}$$

where $E' = 1.21131 \angle 24.615°$ is obtained from calculation at $t = 0.10^-$ s. I_t is calculated as $1.8019 - j0.36154$.

Hence,

$$V_p = E' - jx'_d I_t$$

$$= 1.02892 + j0.14415$$

Therefore,

$$P_{e(0.10^+)} = \text{Re}\left[V_p I_p^*\right]$$

$$= 1.80188$$

From this point onwards, the calculations for $t = 0.12$ s are carried out, by repeating steps 1 and 2 (predictor and corrector steps) using P_e as above at $t = 0.10^+$. Computations are carried out until $t = 0.2$ s and the values are tabulated as in Table 8.1.

Table 8.1 *Solution of Example 8.1 by modified Euler's method*

Time t s	Speed rad/s	Generator values		Real power P_e p.u. MW	Bus p	
		Voltage magnitude |E'|p.u. volts	Phase angle δ degrees		Voltage magnitude |Vp| p.u. V	Phase angle δ_p degrees
0	314.1593	1.2113	13.3657	0	0	0
0.0200	314.9447	1.2113	13.8157	0	0	0
0.0400	315.7301	1.2113	15.1657	0	0	0
0.0600	316.5155	1.2113	17.4157	0	0	0
0.0800	317.3009	1.2113	20.5657	0	0	0
0.1000	318.0863	1.2113	24.6157	0	0	0
0.1200	317.3375	1.2113	28.7548	2.811	1.0312	–9.2908
0.1400	316.3954	1.2113	31.9103	2.2867	1.0245	–10.2855
0.1600	315.3213	1.2113	33.8938	2.4124	1.0200	–10.9068
0.1800	314.1795	1.2113	34.5898	2.4558	1.0183	–11.1240
0.2000	313.0355	1.2113	33.9579	2.4164	1.0198	–10.9268

(b) Solution by Runge-Kutta Method

The fourth order Runge-Kutta method (explained in Appendix B) is employed. Specifically we use scheme I given by (B.27) with $h = 1$. The Runge–Kutta method requires the calculation of four estimates for changes in the phase angles (k_1, k_2, k_3, k_4) and in the speed $(\ell_1, \ell_2, \ell_3, \ell_4)$ to obtain values at the next time increment. For $t = 0$, the initial conditions are known. For $t = 0.02$,

Step 1

First Estimate: $$k_1 = \left.\frac{d\delta}{dt}\right|_{(0)} \Delta t = (\omega(0) - 2\pi f_0)\Delta t = 0$$

$$\ell_1 = \left.\frac{d\omega}{dt}\right|_{(0)} \Delta t = \frac{\pi f_0}{H}(P_m - P_e(0))\Delta t = \frac{\pi \times 50}{4}(1-0)\times 0.02 = 0.785$$

Step 2

Second Estimates: $k_2 = \left(\left(\omega(0) + \frac{\ell_1}{2}\right) - 2\pi f_0\right)\Delta t$

$$= \left(\left(314.16 + \frac{0.786}{2}\right) - 100\pi\right) - 0.02 = 0.00785$$

$$\ell_2 = \frac{\pi f_0}{H}\left(P_m - P_{e(0)}^{(1)}\right)\Delta t$$

where $P_{e(0)}^{(1)}$ is the machine power obtained from a load flow solution using the phase angle $\delta(0) + \frac{k_1}{2}$. Thus, before ℓ_2 can be calculated, a load flow calculation must be made. Since the fault is on the terminal of the generator, $P_{e(0)}^{(1)} = 0$. Therefore, after substitution

$$\ell_2 = 0.785$$

Step 3
Third Estimates:

$$k_3 = \left(\left(\omega(0) + \frac{\ell_2}{2}\right) - 2\pi f_0\right)\Delta t$$

$$\ell_3 = \frac{\pi f_0}{H}\left(P_m - P_{e(0)}^{(2)}\right)\Delta t$$

where $P_{e(0)}^{(2)}$ is obtained from load flow using the phase angle $\delta(0) + \frac{k_2}{2}$. $P_{e(0)}^{(2)}$ is zero as in the case of the calculation of second estimates.

$$k_3 = \left(\left(314.16 + \frac{0.785}{2}\right) - 100\pi\right) \times 0.02 = 0.00785$$

$$\ell_3 = \frac{\pi \times 50}{4}(1 - 0) \times 0.02 = 0.785$$

Step 4
Fourth Estimates Fourth estimates for changes in phase angle and speed are

$$k_4 = ((\omega(0) + \ell_3) - 2\pi f_0) \times 0.02 = 0.01571$$

$$\ell_4 = \frac{\pi f_0}{H}\left(P_m - P_{e(0)}^{(3)}\right)\Delta t = 0.785 \text{ since } P_{e(0)}^{(3)} = 0$$

$P_{e(0)}^{(3)}$ is obtained in general from a load flow using phase angle $\delta(0) + k_3$. Finally the phase angle and speed at $t = 0.02$ are

$$\delta(0.02) = \delta(0) + \frac{1}{6}(k_1 + 2k_2 + 2k_3 + k_4)$$

$$\omega(0.02) = \omega(0) + \frac{1}{6}(\ell_1 + 2\ell_2 + 2\ell_3 + \ell_4).$$

Hence, by substitution, we get

$$\delta(0.02) = 13.365 + \frac{1}{8}(0 + 2\times0.00785 + 2\times0.00785 + 0.01571)\times\frac{180}{\pi} = 13.815^\circ$$

$$\omega(0.02) = 314.16 + \frac{1}{6}(0.785 + 2\times0.785 + 2\times0.785 + 0.785) = 314.945 \text{ rad/s}$$

A load flow is then obtained using phase angle $\delta(0.02)$ in order to compute the voltage and generator power at $t = 0.02$. This power is zero and will remain zero until the fault is removed. This completes the calculations for $t = 0.02$. Computations are carried on until $t = 0.10$, at which point the steps at $t = 0.10^+$ as indicated under the modified Euler's method have to be repeated to compute $V_{p(0.10^+)}$ and $P_{e(0.10^+)}$. These are given in the R-K method as

$$V_{p(0.10^+)} = 1.0289 + j0.14415$$

$$P_{e(0.10^+)} = 1.80189$$

From this point on, the same four steps as discussed earlier, are repeated to obtain the values of δ, ω at $t = 0.12$ s. Computations are carried until $t = 0.2$ s. The Runge–Kutta method requires four load flows to be calculated for each time increment. The, phase angle and speed for various time intervals from $t = 0$ to $t = 0.2$ s are given in Table 8.2. A comparison of the results in graphical form by the modified Euler's method and the Runge–Kutta method is given in Fig. 8.8. There is very little difference between the two methods.

Table 8.2 *Solution of Example 8.1 by Runge-Kutta method*

Time t s	Speed rad/s	Generator values			Bus p	
		Voltage magnitude $\lvert E' \rvert$ p.u.	Phase angle δ degrees	Real power P_e p.u. MW	Voltage magnitude $\lvert V_p \rvert$ p.u.	Phase angle δ_p degrees
0	314.1593	1.2113	13.3657	0	0	0
0.0200	314.9447	1.2113	13.8157	0	0	0
0.0400	315.7301	1.2113	15.1657	0	0	0
0.0600	316.5155	1.2113	17.4157	0	0	0
0.0800	317.3009	1.2113	20.5657	0	0	0
0.1000	318.0863	1.2113	24.6157	0	0	0
0.1200	317.3436	1.2113	28.7107	1.9549	1.0313	–9.2769
0.1400	316.4274	1.2113	31.8592	2.1979	1.0246	–10.2695
0.1600	315.3752	1.2113	33.8713	2.3661	1.0200	–10.8997
0.1800	314.2491	1.2113	34.6263	2.4545	1.0182	–11.1353
0.2000	313.1126	1.2113	34.0758	2.4613	1.0195	–10.9636

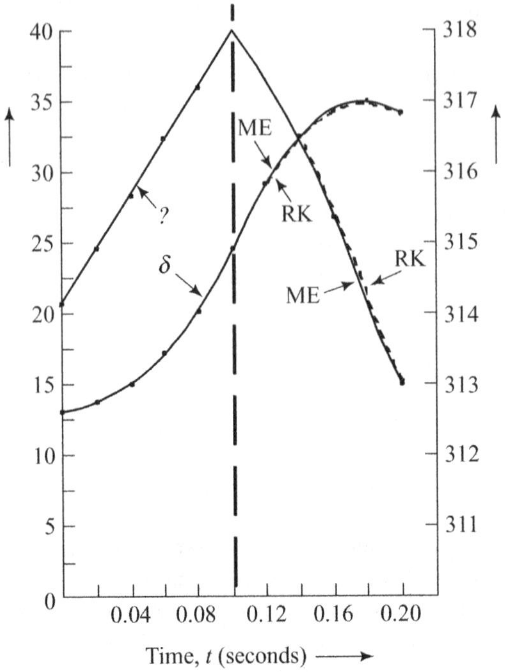

Fig. 8.8 *Swing curves for Example 8.1*

Example 8.2 A sample system is described in Fig. 8.9 for which the machine data and the transmission line data are given in per unit in Tables 8.3 and 8.4, respectively. The load flow data is given in Table 8.5. Loads are to be represented as fixed impedances to the ground. (a) Using the alternate cycle solution method, compute the changes in phase angles of generators for a three-phase fault at bus 2 which clears itself (*self-clearing fault*) at 0.1 s. Take a time step of 0.02 s and the duration of time study as 1.0 s. Take 1 as the slack bus. (b) Repeat the study by the direct method of integrating (using modified Euler's) the swing equations. All values are on 100 MVA base.

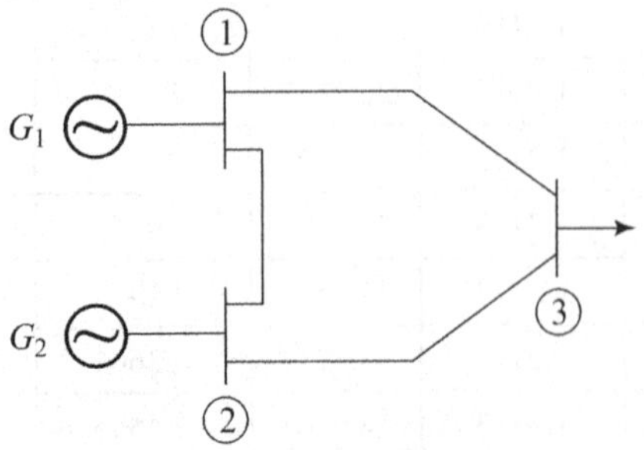

Fig. 8.9 *Two-machine system*

Table 8.3 *Synchronous machine data for sample system*

Bus code p	*Inertia constant H in sec.*	*Direct axis transient reactance x'_d in p.u.*
1	160	0.1
2	3	0.3

Table 8.4 *Line admittances and line charging in p.u. for sample system*

Line no.	*Bus code p-q*	*Admittance Y_{pq}*	*Half line charging $y'_{pq}/2$*
1	1-2	$1.47 - j\,5.88$	$j\,0.15$
2	1-3	$2.94 - j\,11.77$	$j\,0.07$
3	2-3	$2.75 - j\,9.17$	$j\,0.04$

Table 8.5 *Base load flow results for sample system*

Bus code p	*Bus voltages*	*Generation*		*Load*	
		MW	*MVARs*	*MW*	*MVARs*
1	$1.04 \angle 0°$	212.14	93.26	0.0	0.0
2	$1.02 \angle -3.09°$	100.0	70.0	50.0	20.0
3	$0.93 \angle -7.01°$	0.0	0.0	250.0	150.0

Solution:

1. *Conversion of loads into equivalent admittances*—Using Eq. (8.25), the loads are represented as fixed admittances to ground as

$$y_{L2} = 0.48 - j0.19; \quad y_{L3} = 2.91 - j1.75$$

2. *Conversion of scheduled generation to equivalent voltage sources*—Using Eq. (8.27), we obtain the voltages behind the transient reactances for machines 1 and 2 as

$$E_1 = 1.15\angle 10.24°; \quad y_1 = -j/x'_{d1} = -j10.0$$

$$E_2 = 1.26\angle 10.39°; \quad y_2 = -j/x'_{d2} = -j3.33$$

Hence, $\delta_1 = 10.24° = 0.1786$ radians; $\delta_2 = 10.39° = 0.181$ radianst

3. *Partitioned explicit method*— By augmenting the generator representation to the transmission network (including loads represented as fixed impedances), the network equations corresponding to Eq. (8.32) are obtained as

$$\begin{bmatrix} 4.41 - j27.43 & -1.47 + j5.88 & -2.94 + j11.77 \\ -1.47 + j5.88 & 4.70 - j18.39 & -2.75 + j9.17 \\ -2.94 + j11.77 & -2.75 + j9.17 & 8.61 - j22.58 \end{bmatrix} \begin{bmatrix} V_1 \\ V_2 \\ V_3 \end{bmatrix} = \begin{bmatrix} -j10E_1 \\ -j3.3E_2 \\ 0 \end{bmatrix}$$

At $t = 0$, the fault at bus **2** is simulated in the above equations by setting $V_2 = 0$ and iteratively solving for V_1 and V_2 with E_1 and E_2 given by the expressions in step 2 above. The bus voltages thus obtained are $V_1 = 0.53 \angle 0.56°$, $V_2 = 0$, $V_3 = 0.27 \angle -6.27°$. The generator currents and power outputs of the generators are obtained from Eq. (8.33) as

$$I_1 = 1.99 - j6.0; \quad I_2 = 0.76 - j4.13$$

$$P_{el} = 1.0212 \text{ and } P_{e2} = 0.0$$

The differential equations are

$$\frac{d\delta_i}{dt} = \omega_i - \omega_0$$

$$\frac{d\omega_i}{dt} = \frac{\pi f_0}{H_i}(P_{mi} - P_{ei}) \quad i = 1,\ 2.$$

with the initial conditions $\delta_1(0) = 0.1786$ rad, $\delta_2(0) = 0.181$ rad., and $\omega_1(0) = \omega_2(0) = 314.16$ rad/s.

Using the modified Euler's method, the first estimates of the variables are obtained as follows:

$$\delta_{1(0.02)}^{(1)} = 0.1786 \text{ rad};\ \delta_{2(0.02)}^{(1)} = 0.1814 \text{ rad};\ \omega_{1(0.02)}^{(1)} = 314.1816;\ \omega_{2(0.02)}^{(1)} = 315.2072.$$

The voltages behind the transient reactances with these δ_i's are given as per Eq. (8.34) and the network equations solved again for the voltages V_i and hence the powers P_{ci} $(i = 1, 2)$. The second and final estimates of δ's and ω's using the modified Euler's method are obtained as

$$\delta_{1(0.02)} = 0.1789 \text{ rad} = 10.25^\circ;\quad \delta_{2(0.02)} = 0.1918 \text{ rad} = 10.99^\circ;$$

$$\omega_{1(0.02)} = 314.1816;\quad \omega_{2(0.02)} = 315.2072$$

The voltages behind the transient reactances are now given by

$$E_1 = 1.15 \angle 10.25^\circ;\quad E_2 = 1.26 + \angle 10.99^\circ$$

With these voltages, the network equations are again solved to get V_1, V_2, P_{e1}, and P_{e2} at $t = 0.02$ s as

$$V_1 = 0.53\angle 0.57^\circ;\quad V_2 = 0;\quad V_3 = 0.27\angle \square 6.26^\circ;\quad P_{e1} = 1.02;\quad P_{e2} = 0.$$

This completes the calculation for $t + \Delta t = 0.02$ s. Calculation is then done in a similar manner for the next time increment, etc. When the time is advanced to $t = 0.10$ s, however, the network equations are solved without the fault to obtain the post-fault conditions before proceeding with the calculations. Only the magnitudes of the voltages behind transient reactances are held constant. The process is continued until $t = 1.0$ s. The rotor angles of the generators 1 and 2 are plotted in Fig. 8.10. The system is stable for this disturbance.

4. *Summary of numerical computations—modified Euler's method* (Table 8.6)

Initial values: $E_1 = 1.1479 \angle 10.2354^\circ$; $E_2 = 1.2607 \angle 10.4016^\circ$;

The magnitude part remains constant, although in the simulation, the angle, i.e., delta varies.

On application of fault:

Bus voltage values, $V_1 = 0.5291 \angle 0.5551^\circ$; $V_2 = 0$; $V_3 = 0.2655 \angle -6.2768^\circ$. Real power generated, $P_{e1} = 1.0212$; $P_{e2} = 0$.

State variables after the predictor step of the modified Euler's method $\delta_1 = 0.1786$; $\delta_2 = 0.18115$; $\omega_1 = 314.1809$; $\omega_2 = 315.2069$.

State variables after the corrector step, i.e., the final values after the first time step (0.02 s),

$\delta_1 = 0.1789 = 10.2478°$; $\delta_2 = 0.1920 = 11.0018°$; $\omega_1 = 314.1809$; $\omega_2 = 315.2069$; $\omega_2 = 315.2069$.

Bus voltages at this point, $V_1 = 0.5291 \angle 0.5675°$; $V_2 = 0$; $V_3 = 0.2655 \angle -6.2645°$.
Real power generated, $P_{e1} = 1.0212$; $P_{e2} = 0$.

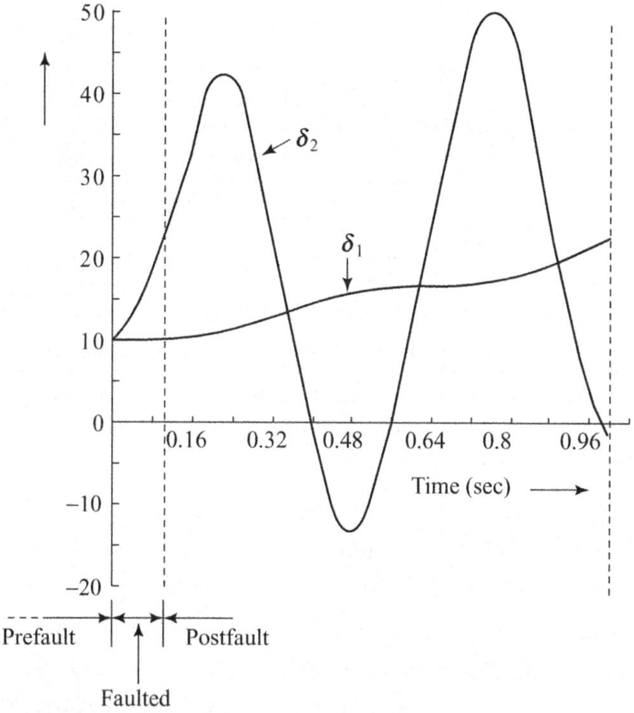

Fig. 8.10 *Swing curves for Example 8.2*

This completes the calculation for $t = 0.02$ s. Proceeding like this, we get the simulations for 1 s (0.1 s faulted and 0.9 s post-fault). In Table 8.6, the values of deltas, omegas, and powers are given in tabular form. Note that the time-steps for both the faulted and the post-fault condition are 0.02 s. However, results are given here at intervals of 0.1 s for the post-fault system to save space.

Table 8.6 *Solution of the 2 m/c system by modified Euler's method*

	Generator 1			*Generator 2*		
Time (in sec)	*Delta (degree)*	*Speed (radian)*	*Power (in p.u.)*	*Delta (degree)*	*Speed (radian)*	*Power (in p.u.)*
0	10.2354	314.1593	1.1215	10.4016	314.1593	1.0004
0.0200	10.2478	314.1809	1.0212	11.0018	315.2069	0
0.0400	10.2849	314.2025	1.0212	12.8026	316.2545	0
0.0600	10.3468	314.2241	1.0212	15.8038	317.3021	0
0.0800	10.4334	314.2457	1.0212	20.0055	318.3498	0

Contd.

Table 8.6 *Contd.*

	Generator 1			Generator 2		
Time (in sec)	*Delta (degree)*	*Speed (radian)*	*Power (in p.u.)*	*Delta (degree)*	*Speed (radian)*	*Power (in p.u.)*
0.1000	10.5449	314.2673	1.0212	25.4077	319.3974	0
0.2000	11.4106	314.3630	0.9991	42.1341	314.2200	2.1465
0.3000	12.8742	314.4564	1.3533	26.3902	309.1705	1.7702
0.4000	14.6102	314.4432	2.6839	–4.0021	309.8184	0.4633
0.5000	15.9377	314.3305	3.3906	–13.9766	315.3984	–0.1709
0.6000	16.6150	314.2400	2.7621	8.2276	319.9207	0.3905
0.7000	17.1014	314.2714	1.4083	39.9316	318.2384	1.7134
0.8000	18.0401	314.3784	0.9829	47.2392	312.4190	2.1644
0.9000	19.5496	314.4504	1.6807	23.5914	308.5492	1.4365
1.0000	21.1361	314.3989	3.0328	–4.3762	311.1501	0.1427

6. *Direct simulation method*—The modified Y_{bus}, which includes the loads represented as shunt impedances to ground, is augmented by the impedances of the machines. The resulting Y_{bus} is

$$\begin{array}{c} \\ \mathbf{1'} \\ \mathbf{2'} \\ \mathbf{1} \\ \mathbf{2} \\ \mathbf{3} \end{array}\begin{array}{c} \begin{array}{ccccc} \mathbf{1'} & \mathbf{2'} & \mathbf{1} & \mathbf{2} & \mathbf{3} \end{array} \\ \begin{bmatrix} -j10.0 & 0 & j10.0 & 0 & 0 \\ 0 & -j3.33 & 0 & +j3.33 & 0 \\ +j10.0 & 0 & 4.41-j27.43 & -1.47+j5.88 & -2.94+j11.77 \\ 0 & +j3.33 & -1.47+j5.88 & 4.7-j18.39 & -2.75+j9.17 \\ 0 & 0 & -2.94+j11.77 & -2.75+j9.17 & 8.61-22.58 \end{bmatrix} \end{array}$$

Since the fault is at bus **2** with zero impedance to the ground, it is simulated by deleting the row and column corresponding to bus **2** in the resulting Y_{bus}. The bus admittance matrix at the internal nodes **1′** and **2′** during faulted condition is obtained by applying the network reduction formula discussed in Chapter 3.

For the faulted system, the reduced bus admittance matrix is

$$\begin{array}{c} \\ \mathbf{1'} \\ \mathbf{2'} \end{array}\begin{array}{c} \begin{array}{cc} \mathbf{1'} & \mathbf{2'} \end{array} \\ \begin{bmatrix} 0.78-j5.46 & 0 \\ 0 & -j3.33 \end{bmatrix} \end{array}$$

The swing equations during the faulted state are

$$\frac{d\delta_1}{dt} = \omega_1 - 314.16\,; \quad \frac{d\delta_2}{dt} = \omega_2 - 314.16$$

$$\frac{d\omega_1}{dt} = \frac{\pi \times 50}{160}(2.1214 - 1.0212) = 1.08\,; \qquad \frac{d\omega_2}{dt} = \frac{\pi \times 50}{3}(1-0) = 52.36$$

The initial conditions at $t = 0^+$ are $\delta_1 = 10.24° = 0.1786$ rad; $\delta_2 = 10.39° = 0.18$ rad;

$\omega_1 = \omega_2 = 314.16$ rad/s $= \omega_2$. Using the modified Euler's method, the equations are integrated until $t = 0.1$ s. The values of δ_1 and ω_i are shown in Table 8.8 for $i = 1, 2$.

For the post-fault state, there is no fault at bus and hence, the bus admittance matrix at the internal nodes of the generators is obtained by network reduction as

$$\begin{array}{cc} & \begin{array}{cc} 1' & 2' \end{array} \\ \begin{array}{c} 1' \\ 2' \end{array} & \begin{bmatrix} 1.21 - j2.92 & 0.37 + j1.67 \\ 0.37 + j1.67 & 0.29 - j2.23 \end{bmatrix} \end{array}$$

The post-fault state system swing equations valid for $t \geq 0.10^+$ are

$$\ddot{\delta}_1 = 0.98\left[2.12 - \left(1.6 + 2.48 \cos\left(77.67° - \delta_{12}^0\right)\right)\right]$$

$$\ddot{\delta}_2 = 52.36\left[1.0 - \left(0.46 + 2.48 \cos\left(77.67° - \delta_{21}^0\right)\right)\right]$$

These equations can be recast in the form of first-order differential equations as in the faulted state. They are then integrated using the values of δ_i, $\omega_i (i = 1,2)$ at $t = 0.1$ s. Thus, we integrate the two systems of equations (faulted and post-fault states) by any of the well known numerical methods. The internal voltage angles of generators 1 and 2 are the same as in Fig. 8.8, which were obtained by the PE method. Table 8.8 gives the values obtained at each time-step using the modified Euler's method. The system is seen to be stable for this disturbance.

Table 8.8 *Rotor angles and speed using direct simulation with modified Euler's method*

	Generator 1			*Generator 2*		
Time (in sec)	*Delta (degree)*	*Speed (radian)*	*Power (in p.u.)*	*Delta (degree)*	*Speed (radian)*	*Power (in p.u.)*
0	10.2354	314.1593	2.1214	10.4016	314.1593	1.0000
0.0200	10.2478	314.1809	1.0204	11.0016	315.2065	0
0.0400	10.2849	314.2025	1.0204	12.8016	316.2537	0
0.0600	10.3469	314.2241	1.0204	15.8016	317.3009	0
0.0800	10.4336	314.2457	1.0204	20.0016	318.3481	0
0.1000	10.5450	314.2674	1.0204	25.4016	319.3953	0
0.2000	11.4371	314.3750	0.8295	42.2614	314.2632	2.1396
0.3000	13.0116	314.4804	1.2830	26.8272	309.2282	1.7692
0.4000	14.8937	314.4709	2.6333	– 3.3578	309.8677	0.4543
0.5000	16.4100	314.3703	3.2421	– 11.9772	315.8728	– 0.2648
0.6000	17.3646	314.2974	2.5767	12.8823	320.1428	0.5163
0.7000	18.2307	314.3468	1.2253	43.5595	317.7422	1.8183
0.8000	19.6565	314.4716	0.8318	47.5035	311.8604	2.1379
0.9000	21.6992	314.5388	1.8137	22.0986	308.5942	1.2905
1.0000	23.7597	314.4797	3.0427	– 3.2065	312.0400	–0.0172

8.6 INCLUSION OF FACTS CONTROLLERS

The FACTS devices described in Chapter 7 are implemented with some suitable controller. The controller modulated the value of some variable/parameter of the system. So, for dynamic simulation, the equations corresponding to the controllers are to be solved to obtain the value of the controlled variable which is then used in the solution of the differential and algebraic equations representing the system. The control structure is chosen depending on the objective of installation of the FACTS device. Hence, for the same FACTS device, the controller equations may be different. Here, the cases of two FACTS devices, one series (TCSC) and one shunt type (STATCOM) are discussed.

8.6.1 Inclusion of TCSC Controller

The structure of a TCSC was discussed in Sec. 2.7.1, and shown pictorially in Fig. 2.24. TCSC is represented as a variable capacitor, whose capacitance value changes with the firing angle, α of the thyristors. Therefore, the controller used for TCSC controls the firing angle depending on the value of some control variable. A typical control structure used for the TCSC is shown in Fig. 8.11. Here, the active power flow (P) through the line (where the TCSC is placed) is taken as the control variable. In an effort to maintain the power flow at a pre-specified reference value (P_{ref}), P and P_{ref} are compared and the error is fed to a proportional-integral (P-I) controller. The output of this controller is taken as the firing angle (α) of the TCSC. A limiter is placed on the value of a and the output of the limiter is used to calculate the net capacitance of the TCSC, C_{tcsc}.

The difference between the P_{ref} and P is denoted by error. The error calculated in each time step is integrated to get INT_{error}. The value of INT_{error} at time t_1 is the integration of the errors of all the previous time steps. This is shown in Eq. (8.52). The value of the firing angle is obtained as in Eq. (8.53).

$$\text{error}(t_1) = P_{ref} - P; \quad INT_{error}(t_1) = INT_{error}(t_0) + \text{error}(t_1); \tag{8.52}$$

$$\alpha(t_1) = K_P.\text{error}(t_1) + K_I.INT_{error}(t_1) \tag{8.53}$$

Generally, the value of α is modulated around some reference value of α, say α_{ref} and hence (8.53) is modified as

$$\alpha(t_1) = \alpha_{ref} + K_P.\text{error}(t_1) + K_I.INT_{error}(t_1) \tag{8.54}$$

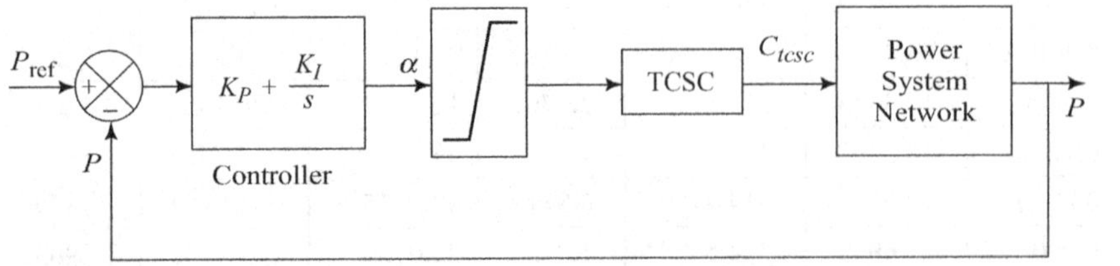

Fig. 8.11 *TCSC controller connected to power system*

Equation (8.52) and (8.54) are solved in every time step and the value of α is used to get the value of C_{tcsc}. This capacitance is included in the bus admittance matrix, Y_{bus} as shown in Eq. (8.55).

$$Y_{\text{bus}}(i, j) = Y_{\text{bus}}(i, j) + 1/(2\pi f\, C_{t\text{csc}}) \tag{8.55}$$

The network equations, Eqs (8.32) are updated with this modified Y_{bus} and thus the effect of the TCSC gets included. The control structure may be different depending on the system requirement. However, method of simulation always involves solution of the controller equations to get the firing angle or the reactance of the TCSC.

Example 8.3 The WSCC 3-machine, 9-bus system is shown in Fig. 8.12 for which the machine data, the transmission line data, and the load flow data are given in Appendix C. The Flux-decay model is considered for the generators. A static exciter is also considered with each generator and the corresponding data are also provided. Loads are to be represented as fixed impedances to ground. If a three-phase fault takes place at bus 6 which clears itself (*self-clearing fault*) after 0.50 s, compute the value of the generator relative rotor angles (taking gen 1 as reference) at the point of fault clearance. Also compute and plot the variation of the generator relative rotor angles for 2 seconds after the fault clearance. Consider a simulation time step of 1/600 s. Now consider a TCSC controller in the line between buses 5 and 7 having a fixed capacitor of reactance equal to half of the reactance of line 5-7 and an inductor of reactance one third that of the fixed capacitor. Re-compute the generator relative rotor angles (taking gen 1 as reference) at the point of fault clearance and also compute and plot the variation of the generator relative rotor angles for 2 seconds after the fault clearance.

Solution: From the load flow data, the initial values for the state variables are obtained as shown below:

The current injected by the generator connected to the ith bus,

$$\bar{I}d_i = (P_i + jQ_i / Vi\angle Q_i)^2 \tag{8.56}$$

where $P_i + jQ_i$ is the power injection by the generator and $V_i\angle\theta_i$ is the bus voltage and its angle.

The generator rotor angle, δ_i is the angle of the voltage

$$\bar{E}_i = V_i \angle Q_i + jx_{qi}\,\bar{I}_i \tag{8.57}$$

The d-axis and the q-axis components of the generator stator currents are obtained from the relation

$$I_{d_i} + jI_{q_i} = \bar{I}_i e^{(\pi/2-\delta_i)} \tag{8.58}$$

So the q-axis component of the voltage behind the transient reactance (E'_q) can be obtained as

$$E'_{q_i} = V_i \cos(\delta_i - \theta_i) + x'_{d_i} I_{d_i} \tag{8.59}$$

The exciter e.m.f. and the bus voltage reference are obtained as

$$E_{fd_i} = E'_{q_i} + \left(x_{d_i} - x'_{d_i}\right) I_{d_i} \tag{8.60}$$

$$V_{\text{ref}_i} = \frac{E_{fd_i}}{K_{A_i}} + V_i \tag{8.61}$$

The pre-fault values of the state variables of the three machines are listed in Table 8.8.

Table 8.8

Machine no	*δ(deg)*	*$\omega_s - \omega_r$ (rad/s)*	*E'_q (p.u.)*	*E_{fd} (p.u.)*
1	3.5862	0	1.0562	1.0819
2	61.0661	0	0.7886	1.7893
3	54.0844	0	0.7687	1.4034

Using Eq. (8.25), the loads (at buses 5, 6 and 8) are represented as fixed admittances to ground.

$$y_{L5} = 1.2605 - j0.5042; \quad y_{L6} = 0.8773 - j0.2924; \; y_{L8} = 0.9684 - j0.3389;$$

The direct method is used to carry out the simulation. The values of the relative rotor angles $(\delta_2 - \delta_1)$ and $(\delta_3 - \delta_1)$ just at the instant of fault clearance are enlisted in Table 8.10. The plot of the relative rotor angle $(\delta_2 - \delta_1)$ for 2 seconds after the fault clearance is shown in Fig. 8.12. Next, the TCSC with specified parameter is included as described in Sec. 8.6.1. The controller constants selected are

$$K_P = 0.8; \quad K_I = 0.12;$$

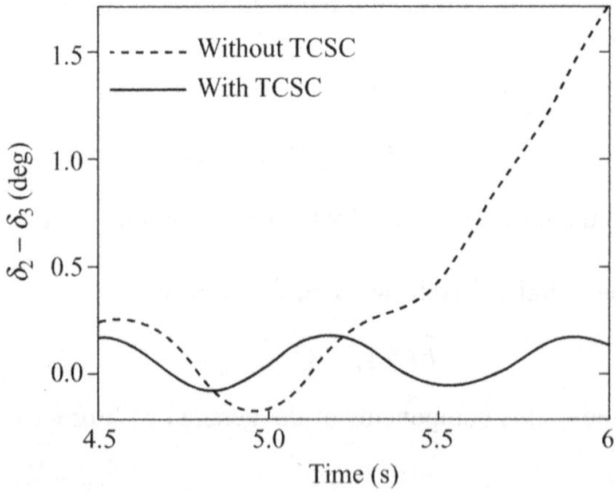

Fig. 8.12 *Change in post-fault excursion of relative rotor angle due to TCSC controller*

Simulation is again carried out and the values of the relative rotor angles $(\delta_2 - \delta_1)$ and $(\delta_3 - \delta_1)$ just at the instant of fault clearance are enlisted in Table 8.9. The plot of the relative rotor angle $(\delta_2 - \delta_1)$ for 2 seconds after the fault clearance is shown in Fig. 8.12 for comparison.

Table 8.9

Relative rotor angle at fault clearance	*System without TCSC*	*System with TCSC*
$(\delta_2 - \delta_1)$	57.4800	55.1850
$(\delta_3 - \delta_1)$	50.4982	49.0341

8.6.2 Inclusion of STATCOM

The structure of the STATCOM was shown in Sec. 2.7.2, (Fig. 2.55). It consists of a voltage sourced converter (VSC), connected to a system bus through some coupling transformer. Usually, the VSC is represented by a controllable voltage source. The STATCOM is mainly used for the reactive power support and to maintain the bus voltage. Therefore the control variable might be either the reactive power injection or the voltage of the bus where the STATCOM is connected. Various control strategies can be used to control the output voltage of the VSC and thus to control the reactive power supplied by it. For simulation, just like the case of TCSC, the controller equations are solved in each time step to obtain the VSC voltage (magnitude and angle). This voltage is then included in the solution of the network.

8.7 CONCLUSION

In this chapter we have explained the fundamental concepts underlying the simulation of multi-machine systems when subjected to large and sudden disturbances. Since these concepts are basic to an understanding when more complex synchronous machine models are involved, the various steps have been carefully explained. Basically, two methods are explained: (1) the partitioned explicit method, and (2) the direct integration method. Method (1) has the advantage that voltage variables at various buses could be monitored during the disturbance. In method (2) that is quicker, this information is missing. The numerical schemes can be either the modified Euler or the fourth-order Runge–Kutta method. However, method (2) is used in transient stability analysis using Energy Function method.

Problems

8.1 A single machine connected to infinite bus has the following parameters: $H = 3.0$ s, $x'_d = 0.25$ p.u., line reactance $x = 0.10$ p.u. Initially the generator bus voltage is 1.40 p.u. and delivers a power of 75 MW. At $t = 0$, there is a three-phase short-circuit at the generator bus which is cleared in 0.1 s. The base MVA is 100 and the infinite bus voltage is $1.0 \angle 0°$.

(a) Write the swing equations for the faulted and the post-fault states with all numerical values computed and assume the damping coefficient $D = 0.1$ p.u.

(b) Integrate the equations using the Euler's, the modified Euler's, and the fourth-order R-K method.

8.2 A synchronous machine having an inertia constant of $H = 5.00$ s and transient reactance of 0.25 p.u. delivers power of 25 MW over two parallel transmission lines each having reactance of 0.20 p.u. The other end of the transmission line may be assumed to be an infinite bus, having voltage of $1 \angle 0°$. The voltage behind the transient reactance of the generator is 1.025 p.u. A three-phase short-circuit occurs at the middle of one of the lines and is cleared in 0.3 s by simultaneous opening of the circuit breakers at both ends of the transmission line. From the swing equations for the faulted and the post-fault state, compute the swing curve using (a) modified Euler's method, and (b) fourth-order R–K method. Use 50 as the base MVA.

8.3 Repeat Problem 8.2 if the fault is at the generator end of the line.

8.4 The following example is adapted from the text by Stagg and El-Abiad (Figure P.8.1).

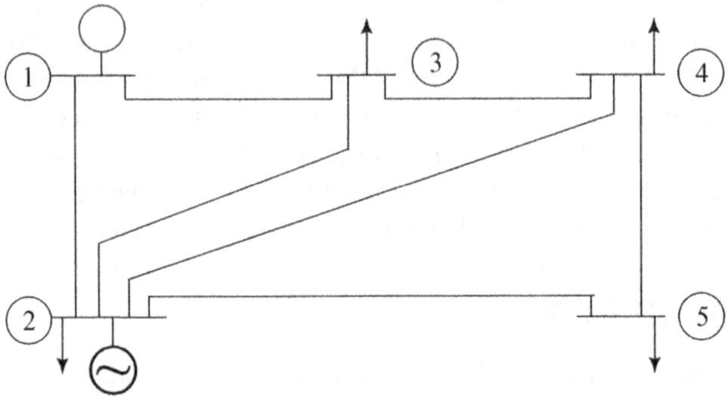

Fig. P.8.1 *A 2-machine, 5-bus system*

The line data are as follows:

Line data bus code	*Impedance p.u.*	*Half-line charging admittance p.u.*
1-2	$0.02 + j\,0.06$	$j\,0.03$
1-3	$0.08 + j\,0.24$	$j\,0.025$
2-3	$0.06 + j\,0.18$	$j\,0.02$
2-4	$0.06 + j\,0.18$	$j\,0.02$
2-5	$0.04 + j\,0.12$	$j\,0.015$
3-4	$0.01 + j\,0.03$	$j\,0.01$
4-5	$0.08 + j\,0.24$	$j\,0.025$

The results of load flow calculation prior to the fault are given below:

Bus number	*Bus voltage*	*Generation*		*Load*	
		MW	*MVAR*	*MW*	*MVAR*
1	$1.06 + j\,0.0$	129.565	−7.48	0	0
2	$1.04621 - j\,0.05128$	40.0	30.0	20	10
3	$1.02032 - j\,0.0892$	0	0	45	15
4	$1.01917 - j\,0.09506$	0	0	40	5
5	$1.01209 - j\,0.10906$	0	0	60	10

Choose 100 as the base MVA.

The synchronous machine data on the common 100 MVA base is as follows:

Machine connected to bus 1 $H = 50$ s $x'_d = 0.25$ p.u.

Machine connected to bus 2 $H = 7.0$ s $x'_d = 1.50$ p.u.

The fault is at bus 2 for a duration of 0.1 s.

(a) Set up the equations for dynamic simulation by the alternate cycle solution method, i.e., Eqs (8.15), (8.16), (8.36), and (8.37).

(b) Choosing a time increment of $\Delta t = 0.02$ s, find out the values of δ_1, δ_2, ω_1, ω_2, and the bus voltages at $t + \Delta t = 0.02$ s using modified Euler's integration method.

(c) Obtain a two-port representation at the internal nodes of the generators connected to buses **1** and **2,** and by the direct method, obtain δ_1, δ_2, ω_1, and ω_2 at $t = 0.02$ and 0.04 s. Use both the modified Euler's method and the fourth-order R–K method.

8.5 Write a computer program for the transient stability evaluation of a single-machine infinite bus system. Assume all the parameters in the swing equation for the faulted and the post-fault state to be given. Use this program to verify the results of Problems 8.1, 8.2, and 8.3.

8.6 For the transient stability of a multi-machine system, write a computer program using both (a) alternate cycle solution method, and (b) direct method. The network reduction needed in the program must form one of the subroutines. Verify the results of Example 8.2 and also work out Problem 8.4 using the program.

8.7 Write a computer program to integrate a system of differential equations $x = f(x)$ by (a) modified Euler's method, (b) fourth-order R–K method, and using the program, integrate the following differential equations:

(a). $\ddot{x} + 4\left(x^2 - 1\right)\dot{x} + x = 0; \quad x(0) = 0.75,\ \dot{x}(0) = 0$

(b). $\ddot{x} + 2x\dot{x} + x = 0; \quad x(0) = 1,\ \dot{x}(0) = 0$

Use time increment $\Delta t = 0.05$ s for (a) and 0.1 s for (b).

9 RENEWABLE SOURCES—WIND ENERGY SYSTEM

9.1 INTRODUCTION

Utilization of the mechanical power available in wind through wind mills to carry out different jobs is not new. Subsequently, small generators were connected to the wind turbines to convert the mechanical power to electrical form. From the initial stages till thirty years back, these generating facilities were isolated and catering to small local loads. Afterwards, large numbers of such wind turbine-generator combinations were installed in the same location, thus forming a wind farm. As the search for renewable sources of energy intensified, wind power emerged as a prominent candidate as the technology was reasonably mature and the cost of generation was very much affordable. So, people started looking for large sites with available wind speed beyond the threshold (4 m/s) and new and larges wind farms were established. The generation of these farms started exceeding 100 MW and is approaching GW levels. Naturally, local loads are not big enough to consume this power and the question of connecting to the grid to transmit the power to other locations has become necessary. Wind speed being variable in nature, the integration of this power with the grid poses a technical challenge. Hence, modeling of the wind turbine and generators for including them in power system studies becomes important.

Early versions of wind turbines used to have fixed speed operation (Danish concept), where wind energy is transformed into electrical energy using a simple squirrel-cage induction generator (SCIG) directly connected to the power grid as shown in Fig. 9.1(a). The turbine and generator rotors are coupled by a fixed-ratio gearbox. At all the operating points, this turbine operates at a constant speed. Usually a fixed capacitor is connected at the terminal of the generator to compensate for the reactive power drawn by the singly-excited induction generator (SEIG). However, new wind power plants use variable speed turbines for more efficient utilization of the energy available in the wind. In this type, turbine speed is adjusted as a function of the wind speed to maximize the output power. A synchronous generator can be used with this turbine. However, this produces three-phase power of variable voltage and frequency. Therefore, back to back converters are required to change the voltage and the frequency level to match the grid frequency (Fig. 9.1(b)). The converters have to handle the rated power of the plant, which makes it costly. Doubly-fed induction generators (DFIG) are used as an alternative, where the back to back converters are needed for the rotor circuit handling comparatively less power (typically 30%). The scheme is shown in Fig. 9.1(c). The reactive power compensation (through shunt connected capacitor) is not required in this case.

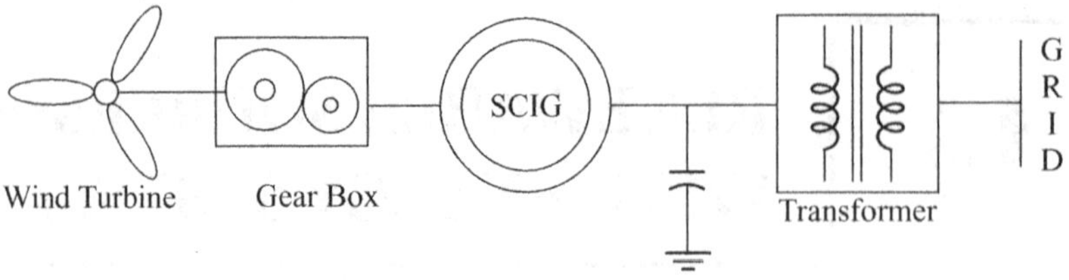

Fig. 9.1(a) *Wind power plant with fixed speed wind turbine and squirrel cage induction generator*

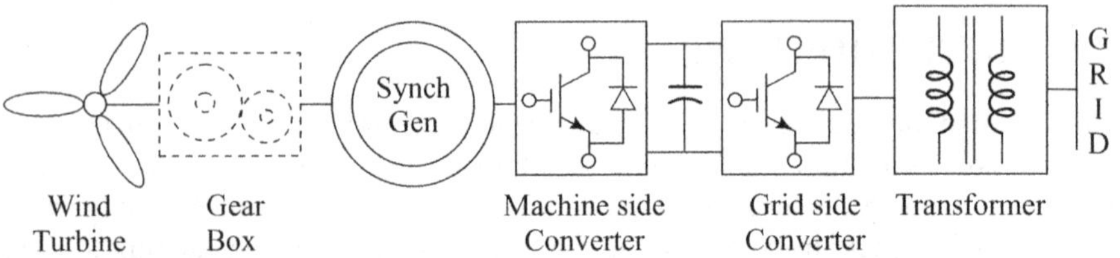

Fig. 9.1(b) *Wind power plant with variable speed wind turbine, synchronous generator and back to back converters*

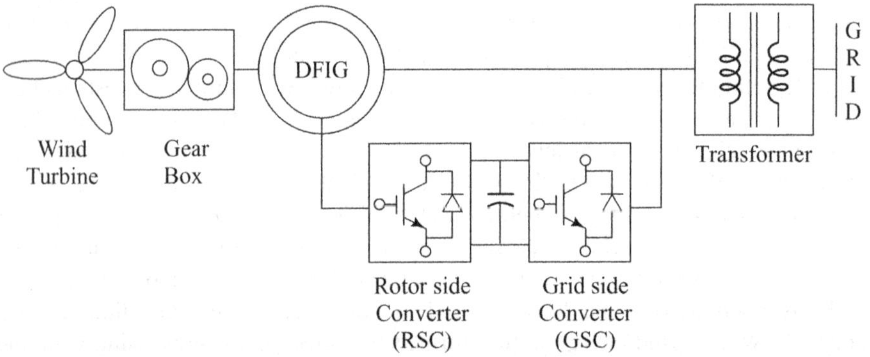

Fig. 9.1(c) *Wind power plant with variable speed wind turbine and doubly-fed induction generator*

DFIG is presently the most popular choice for wind generation. However, DFIG also has few disadvantages like: (a) the presence of slip ring and gear box gives rise to maintenance issues, (b) individual machine usually not available with a rating higher than 5 MW, thus making it necessary to use larger number of machines in a wind farm. These problems can be overcome with the use of permanent magnet synchronous machine (PMSM) though these are very costly at this point of time.

9.2 WIND POWER GENERATION SYSTEM MODEL AND POWER SYSTEM REPRESENTATION

Similar to thermal or hydro generation, wind power generation also has two broad components: (a) the electrical part, i.e. the generator, and (b) the mechanical part which includes mainly the turbine and the generator rotor. The models of these components suitable for inclusion in power system studies are discussed here. But before that, the notations used for different system variables and parameters while describing the models are given below:

v_{ds} and v_{qs} = the stator d-axis and q-axis voltages respectively
v_{dr} and v_{qr} = the rotor d-axis and q-axis voltages respectively
i_{ds} and i_{qs} = the stator d-axis and q-axis currents respectively
i_{qr} and i_{dr} = the rotor d-axis and q-axis currents respectively
e'_d and e'_q = the equivalent d-axis and q-axis voltages behind transient impedance respectively
T_m and T_e = the mechanical torque and the electromagnetic torque respectively
l_m = the mutual inductance, l_{ss} and l_{rr} are the stator and the rotor inductances
r_s and r_r = the stator and the rotor resistance respectively
ω_{eB} = the electrical base speed
ω_r and ω_t = the rotor electrical and mechanical speed respectively
θ_{tw} = the shaft torsional angle (twist angle)
H_t and H_g = the turbine and the generator inertia
k_{sh} = the drive train shaft stiffness
C_{sh} = the drive train damping coefficient
$C_p(\lambda,\beta)$ = the coefficient of performance of the wind turbine
λ = tip speed ratio of the wind turbine
β = pitch angle of the wind turbine
R = the blade length of the wind turbine
ρ = the air density
V_w = the wind velocity

The d-axis and q-axis mentioned above are those of the synchronously rotating reference frame. The convention considered here is that the q-axis leads the d-axis.

9.2.1 Modeling of the Wind Turbine and the Rotating Mass [31, 32]

The wind turbine, the wind generator rotor shaft, and the intermediate gear box can be represented by a 5th order model. However, generally a comparatively reduced order model termed as **two-mass model** is sufficient. The equations representing the two-mass model are given as

$$\frac{d\omega_r}{dt} = \frac{1}{2H_g}\left[k_{sh}\theta_{tw} + C_{sh}\omega_{eB}\left(\omega_t - \omega_r\right) - T_e\right] \tag{9.1}$$

$$\frac{d\theta_{tw}}{dt} = \omega_{eB}\left(\omega_t - \omega_r\right) \tag{9.2}$$

$$\frac{d\omega_t}{dt} = \frac{1}{2H_t}\left[T_m - k_{sh}\theta_{tw} - C_{sh}\omega_{eB}\left(\omega_t - \omega_r\right)\right] \tag{9.3}$$

The mechanical torque T_m is given by

$$T_m = \frac{0.5\rho\pi R^2 C_p(\lambda,\beta)V_w^3}{\omega_t}\left(\frac{1}{T_{mBase}}\right) \tag{9.4}$$

Depending on the nature of study, even a **lumped mass** or **single-mass model** may be used. Here, the mechanical system is represented by only one differential equation as shown below. Since the turbine, the gear box, and the generator rotor is considered as a single-mass, $\omega_r = \omega_t$ and the equivalent inertia of the lumped mass, $H_{eq} = H_g + H_t$.

$$\frac{d\omega_r}{dt} = \frac{1}{2H_{eq}}[T_m - T_e] \tag{9.5}$$

To avoid damages due to very high wind speeds, a provision of controlling the pitch angle (β) of the wind turbine blades is there. Till the generator speed is within the rated value ($\omega_{r_{rated}}$), the pitch angle is maintained at 0° and the torque of the machine is controlled such that maximum available power at the given condition (i.e., wind speed) is extracted. The torque for this maximum power tracking can be obtained using the relation

$$T_e = K_{opt}\omega_r^2 \tag{9.6}$$

where $$K_{opt} = 0.5\rho\pi R^5 C_{p\max}\omega_{t_{rated}}^3 \big/ \lambda_{opt}^3 P_{rated} \tag{9.7}$$

$C_{p_{\max}}$ is the maximum value of C_p when $\beta = 0°$, λ_{opt} is the tip speed ratio when $C_p = C_{p_{\max}}$.

$C_{p_{\max}}$ and λ_{opt} are constants of the wind turbine and can be obtained from the turbine data.

When the generator speed becomes greater than or equal to the rated speed $\omega_{r_{rated}}$, the pitch angle is controlled and the torque is maintained constant at the rated value, $T_{e_{rated}}$.

9.2.2 Wind Generator Modeling [31, 32]

As mentioned in the previous section, the most popularly used machine for the conversion of wind energy to electrical form is the induction machine (IM). Considering that both the stator and the rotor windings of the IM are excited (i.e. doubly-fed induction generator), the stator and the rotor circuits can be represented by the following equations:

$$\frac{di_{ds}}{dt} = \frac{\omega_{eB}}{l_s'}\left[-\left(r_s + \frac{(l_{ss}-l_s')}{T_r}\right)i_{ds} + l_s' i_{qs} - \frac{e_q'}{T_r} + \omega_r e_d' + K_{mrr}v_{dr} - v_{ds}\right] \tag{9.8}$$

$$\frac{di_{qs}}{dt} = \frac{\omega_{eB}}{l_s'}\left[-\left(r_s + \frac{(l_{ss}-l_s')}{T_r}\right)i_{qs} + l_s' i_{ds} - \frac{e_d'}{T_r} + \omega_r e_q' + K_{mrr}v_{qr} - v_{qs}\right] \tag{9.9}$$

$$\frac{de_d'}{dt} = \omega_{eB}\left[\omega_s \frac{(l_{ss}-l_s')}{T_r}i_{qs} - \frac{e_d'}{T_r} + \omega_s(\omega_s - \omega_r)e_q' - \omega_s K_{mrr}v_{qr}\right] \tag{9.10}$$

$$\frac{de'_q}{dt} = \omega_{eB}\left[-\omega_s \frac{(l_{ss} - l'_s)}{T_r} i_{ds} - \frac{e'_q}{T_r} - \omega_s(\omega_s - \omega_r)e'_d - \omega_s K_{mrr} v_{dr}\right] \quad (9.11)$$

where $l' = l_{ss} - (l_m^2/l_{rr})$ (9.12)

$$K_{mrr} = l_m/l_{rr}; \quad (9.13)$$

$$T_r = l_{rr}/r_r \quad (9.14)$$

$$e'_d = \omega_s l_m (i_{qr} + K_{mrr} i_{qs}) \quad (9.15)$$

$$e'_q = -\omega_s l_m (i_{dr} + K_{mrr} i_{ds}) \quad (9.16)$$

Here, e'_d and e'_q are actually voltages proportional to the d-axis and the q-axis flux linkage of the rotor. The rotor winding of the DFIG is supplied by the rotor side converter (RSC), as shown in Fig. 9.1(c). So v_{dr} and v_{qr} are the d-axis and the q-axis components of the output voltage of the RSC.

Generally, the dynamics of the stator and the rotor windings, given by Eqs (9.8) – (9.11), are much faster than the dynamics of the mechanical system and hence, for power system application, usually the stator and rotor dynamics are neglected. This means setting the differential terms in Eqs (9.8) – (9.11) to zero and converting them to algebraic equations as given below.

$$v_{ds} = -r_s i_{ds} + \omega_s l'_s i_{qs} + e'_d = -r_s i_{ds} + \omega_s (l_{ss} i_{qs} + l_m i_{qr}) \quad (9.17)$$

$$v_{qs} = -r_s i_{qs} - \omega_s l'_s i_{ds} + e'_q = -r_s i_{qs} - \omega_s (l_{ss} i_{ds} + l_m i_{dr}) \quad (9.18)$$

$$v_{dr} = -r_r i_{dr} + (\omega_s - \omega_r)(l_{rr} i_{qr} + l_m i_{qs}) \quad (9.19)$$

$$v_{qr} = -r_r i_{qr} - (\omega_s - \omega_r)(l_{rr} i_{dr} + l_m i_{ds}) \quad (9.20)$$

The total active and reactive power delivered to the grid by the DFIG can be expressed as

$$P_{tot} = P_s + P_r = v_{ds} i_{ds} + v_{qs} i_{qs} + v_{dr} i_{dr} + v_{qr} i_{qr} \quad (9.21)$$

$$Q_{tot} = Q_s + Q_{GSC} = v_{qs} i_{ds} - v_{ds} i_{qs} \quad (9.22)$$

The grid side converter (GSC) is operated at unity power factor and hence, $Q_{GSC} = 0$. The electromagnetic torque T_e is given by

$$T_e = (e'_q i_{qs} + e'_d i_{ds})/\omega_s = L_m (i_{ds} i_{qr} - i_{qs} i_{dr}) \quad (9.23)$$

9.3 INCLUSION OF WIND POWER IN LOAD FLOW

The wind power generation usually takes place at a voltage much less than the transmission system voltages. Also, the site is usually distant from the existing system buses. Hence, the wind generator is connected to the power system buses through a transformer and a distribution line. Sometimes, local loads are also connected at the wind generator bus. A typical connection arrangement is shown in Fig. 9.2.

9.3.1 Inclusion of DFIG-based Wind Power Generation

The DFIG has control over both the active and the reactive power output due to the presence of the back to back converters in the rotor circuit (Fig. 9.1(c)). So in a load flow program, the DFIG can be represented to the rest of the power system as a negative load with specified values of P and Q. If the DFIG is connected as shown in Fig. 9.2, then the values of P_{tot} and Q_{tot} are obtained by subtracting the losses in the connecting line and the transformer impedance $(r + jx)$ from P and Q. Thus, solving the load flow for the power system, the voltage (and its angle) of the load buses are

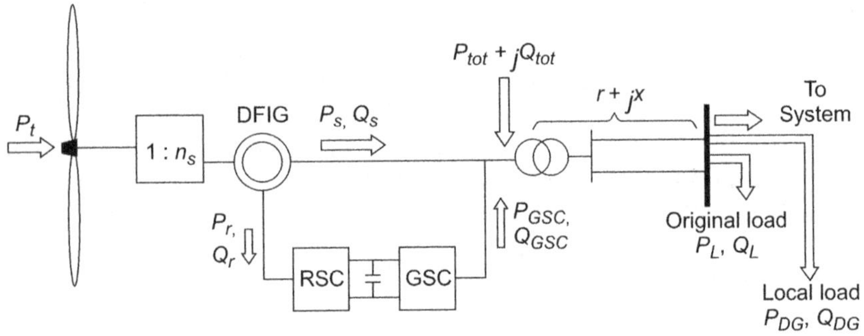

Fig. 9.2 *Connection of DFIG to the system load bus*

obtained. Since, the DFIG is represented as negative load, the voltage of the bus at which the DFIG is connected is also obtained. In the synchronously rotating reference, the values of v_{ds} and v_{qs} are the real and the reactive part of the bus voltage respectively. So, in the six equations, Eqs (9.17) – (9.22), there are 7 unknowns, namely, (i_{ds}, i_{qs}, i_{dr}, i_{qr}, v_{dr}, v_{qr} and ω_r). Hence, one more equation is needed for the solution, which can be obtained using the two expressions of torque given in Eq. (9.6) and Eq. (9.23) as shown below.

$$L_m \left(i_{ds} i_{qr} - i_{qs} i_{dr} \right) = K_{opt} \omega_r^2 \tag{9.24}$$

These seven equations, Eqs (9.17) – (9.22), and Eq. (9.24) are solved using the Newton-Raphson method to get the stator and rotor currents, rotor voltages and the rotor speed. These are used to obtain the value of T_e. Under steady-state, $T_m = T_e$; $\omega_t = \omega_r$; $\theta_{tw} = T_m / k_{sh}$.

Example 9.1 A DFIG based wind farm of rating 50 MW is connected to bus 5 of the power system described in Example 6.2 through a transformer and a distribution line as shown in Fig. 9.3. The combined impedance of the transformer and the distribution line is $(r + jx) = (0.003 + j0.02)$ p.u. At a particular time, the wind farm injects 25 MW active power (and no reactive power) at bus 5. An additional local load of 25 MW is also connected at bus 5. Find the load flow solution of the complete system using Newton–Raphson method. The data for individual DFIG of rating 5 MW is given in Appendix C. The data is on the own base of the machine.

It is to be noted that the wind farm of 50 MW consists of 10 such DFIGs connected in parallel. Since the system base MVA is also taken as 50 MVA (as in Example 6.2), the aggregated DFIG data in the system base will be the same as the individual machine data.

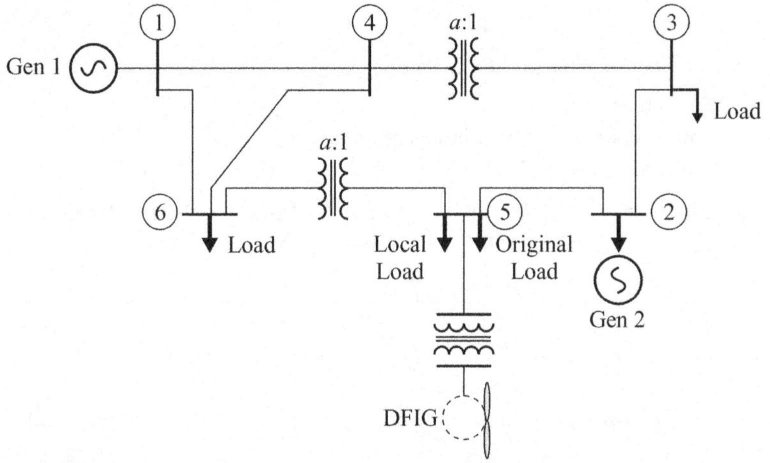

Fig. 9.3 *Power system of Example 6.2 with a DFIG based wind farm connected*

Solution: At first, the DFIG terminal bus is included as the 7th bus and the bus admittance matrix is formed.

Y_{BUS} =

$$\begin{bmatrix}
0.9922 - j4.3925 & 0 & 0 & -0.5583 + j2.5820 & 0 & -0.4339 + j1.8275 & 0 \\
0 & 1.0214 - j1.9545 & -0.4449 + j0.6461 & 0 & -0.5765 + j1.3085 & 0 & 0 \\
0 & -0.4449 + j0.6461 & 0.4449 - j8.1649 & 0 + j6.8353 & 0 & 0 & 0 \\
-0.5583 + j2.5820 & 0 & 0 + j6.8353 & 1.1124 - j11.1062 & 0 & -0.5541 + j2.3249 & 0 \\
0 & -0.5765 + j1.3085 & 0 & 0 & 7.9115 - j53.5416 & 0 + j3.2520 & -7.3350 + j48.8998 \\
-0.4339 + j1.8275 & 0 & 0 & -0.5541 + j2.3249 & 0 + j3.2520 & 0.9880 - j7.3075 & 0 \\
0 & 0 & 0 & 0 & -7.3350 + j48.8998 & 0 & 7.3350 - j48.8998
\end{bmatrix}$$

The power injected at bus 5 = 25 MW + 0 MVAR = (0.5 + j0.0) p.u.

From the converged solution of the system without wind farm (Table 6.7), the voltage of bus 5, $|V_5|\angle\theta_5 = 0.8511\ \angle -13.03°$

Current flowing from the DFIG to bus 5, $|I_{dg}|\ \angle\ \phi_{gd} = (0.5/0.8511\ \angle -13.03°)^* = 0.5875\ \angle -13.03°$

Voltage at bus 7,

$|V_7|\ \angle\ \theta_7 = 0.8511\ \angle -13.03° + j(0.003 + j0.02)\ (0.5875\ \angle -13.03°) = 0.8521\ \angle -12.04°$

Hence, $v_{ds} = 0.832$; $v_{qs} = -0.184$;

Power at bus 7 = $P_{tot} + jQ_{tot} = |V_7|\ \angle\ \theta_7\ (|I_{dg}|\ \angle\ \phi_{dg})^* = (0.8521\ \angle -12.04°)\ (0.5875\ \angle -13.03°)\ x = 0.501 + j0.007$

The initial guess values of the variables are taken as

$i_{ds} = 0.5$; $i_{qs} = 0.5$; $i_{dr} = 0.5$; $i_{qr} = 0.5$; $v_{dr} = 0.5$; $v_{qr} = 0.5$; $\omega_r = 1.0$;

With the values of v_{ds}, v_{qs}, P_{tot} and Q_{tot} being available, and considering the initial guess, the mismatches for the DFIG equations can be found.

$$v_{ds} - [-r_s i_{ds} + \omega_s\ (l_{ss} i_{qs} + l_m i_{qr})] = 0.8321 - 4.0175 = -3.1854$$

$$v_{qs} - [-r_s i_{qs} + \omega_s\ (l_{ss} i_{ds} + l_m i_{dr})] = -0.1835 + 4.0225 = 3.8390$$

$$0 - (-r_r i_{dr} + (\omega_s - \omega_r)\ (l_{rr} i_{qr} + l_m i_{qs}) - v_{dr}) = 0 - (-0.0028 - 0.5) = 0.5028$$

$$0 - (-r_r i_{qr} - (\omega_s - \omega_r)\ (l_{rr} i_{dr} + l_m i_{ds}) - v_{qr}) = 0 - (-0.0028 - 0.5) = 0.5028$$

$$P_{tot} - (v_{ds} i_{ds} + v_{qs} i_{qs} + v_{dr} i_{dr} + v_{qr} i_{qr}) = 0.5010 - 0.8243 = -0.3233$$

$$Q_{tot} - (v_{qs} i_{ds} - v_{ds} i_{qs}) = 0.0069 + 0.5078 = 0.5147$$

$$0 - [L_m(i_{ds} i_{qr} - i_{qs} i_{dr}) - K_{opt}\omega_r^2] = 0 + 0.5787 = 0.5787$$

The Jacobian matrix is obtained by evaluating the derivatives of the seven DFIG equations with respect to the DFIG variables, (i_{ds}, i_{qs}, i_{dr}, i_{qr}, v_{dr}, v_{qr} and ω_r). Thus we get the 7 × 7 matrix as shown below.

$$\begin{bmatrix} -0.005 & 4.04 & 0 & 4.0 & 0 & 0 & 0 \\ -4.04 & -0.005 & -4.0 & 0 & 0 & 0 & 0 \\ 0 & 0 & -0.0055 & 0 & -1.0 & 0 & -4.0301 \\ 0 & 0 & 0 & -0.0055 & 0 & -1.0 & 4.0301 \\ 0.8321 & -0.1835 & 0.5 & 0.5 & 0.5 & 0.5 & 0 \\ -0.1835 & -0.8321 & 0 & 0 & 0 & 0 & 0 \\ 2.0 & -2.0 & -2.0 & 2.0 & 0 & 0 & -1.1574 \end{bmatrix}$$

The updated values of the variables after the first iteration are

$i_{ds} = 1.8036$; $i_{qs} = -0.4061$; $i_{dr} = -1.7753$; $i_{qr} = 0.6204$; $v_{dr} = -30.0477$; $v_{qr} = 30.0541$; $\omega_r = 8.4582$;

Continuing in the same way, convergence (tolerance = 0.0001) is achieved in 8 iterations. The values of the variables in iterations 2-8 are shown in Table 9.1.

Table 9.1 *DFIG variable values at successive iterations*

Iteration	i_{ds}	i_{qs}	i_{dr}	i_{qr}	v_{dr}	v_{qr}	ω_r
2	1.6337	– 0.3686	– 1.6037	0.5823	–2.9773	0.1452	4.3763
3	0.8953	– 0.2058	– 0.8582	0.4170	– 1.1225	0.2777	2.3430
4	0.6164	– 0.1442	– 0.5765	0.3545	– 0.3093	0.0714	1.3710
5	0.5827	– 0.1368	– 0.5425	0.3469	– 0.0036	0.0003	1.0080
6	0.6007	– 0.1408	– 0.5607	0.3510	0.0408	– 0.0076	0.9562
7	0.6020	– 0.1411	– 0.5620	0.3513	0.0412	– 0.0075	0.9558
8	0.6020	– 0.1411	– 0.5620	0.3513	0.0412	– 0.0075	0.9558

9.3.2 Inclusion of Fixed Speed Wind Turbine and Squirrel Cage Induction Generator (SCIG)

In case of SCIG, the stator and the rotor are still described by Eqs (9.17)–(9.20). However, the rotor is short-circuited in itself and hence the voltage applied to the rotor is equal to zero, i.e.,

$$v_{dr} = 0;\ v_{qr} = 0;$$

However, due to the absence of any converter, there is no control over the torque and the reactive power. Also, there is no question of maximum power tracking and hence Eq. (9.6) is not valid. The active power output is usually specified but not the reactive power. Therefore, if the terminal (or stator) voltage and its angle, and the active power output of the SCIG is specified, then Eqs (9.17)–(9.20) (with rotor voltages zero), and Eq. (9.21) are solved to get the values of (i_{ds}, i_{qs}, i_{dr}, i_{qr} and ω_r). Then the reactive power consumed by the SCIG can be obtained using Eq. (9.22).

Example 9.2 A 100 MW wind farm consists of 20 SCIGs connected in parallel. Each SCIG is rated at 5 MW and the detailed relevant data of the machine is considered to be the same as that of the DFIG mentioned in Example 9.1 (Appendix C). Considering a system base of 100 MVA and terminal voltage of $1.0\angle 0°$ (p.u.), find the values of the stator and the rotor currents, the slip and reactive power consumed when the wind farm is supplying 90 MW. Use the Newton–Raphson method.

Solution: It is to be noted that since the system base MVA is the same as the wind farm rating, the aggregated SCIG data in the system base will be the same as the individual machine data.

Given $v_{ds} = 1.0;\ v_{qs} = 0;\ P_{tot} = 0.9$

The initial guess values of the variables are taken as

$$i_{ds} = 0.5;\ i_{qs} = 0.5;\ i_{dr} = 0.5;\quad i_{qr} = 0.5;\ \omega_r = 1.0;$$

In the first iteration, the mismatches for the SCIG equations can be found.

$$v_{ds} - \left[-r_s i_{ds} + \omega_s \left(l_{ss} i_{qs} + l_m i_{qr}\right)\right] = 1.0 - 0.5472 = 0.4528$$

$$v_{qs} - \left[-r_s i_{qs} - \omega_s \left(l_{ss} i_{ds} + l_m i_{dr}\right)\right] = 0 - 0.4478 = -0.4478$$

$$0-\left(-r_r i_{dr}+(\omega_s-\omega_r)\left(l_{rr}i_{qr}+l_m i_{qs}\right)\right)=0-(-0.0033)=0.0033$$

$$0-\left(-r_r i_{qr}-(\omega_s-\omega_r)(l_{rr}i_{dr}+l_m i_{ds})\right)=0-0.0020=-0.0020$$

$$P_{tot}-\left(v_{ds}i_{ds}+v_{qs}i_{qs}+v_{dr}i_{dr}+v_{qr}i_{qr}\right)=0.9-0.5=0.4$$

The Jacobian becomes $\begin{bmatrix} -0.005 & 0.0993 & 1.0 & 0 & 0 \\ -0.0993 & -0.005 & 0 & 1.0 & 0 \\ -0.0053 & 0 & 0 & -0.0014 & 0.5 \\ 0 & 0.0053 & -0.0014 & 0 & -0.5 \\ 1.0 & 0 & 0 & 0 & 0 \end{bmatrix}$

The updated values of the variables after the first iteration are

$$i_{ds}=0.9000;\ i_{qs}=1.1501;\ i_{dr}=0.8903;\ i_{qr}=0.0951;\ \omega_r=1.0099;$$

Convergence is achieved in 4 iterations. The values of the variables in iterations 2-4 are shown in Table 9.2.

Table 9.2 *SCIG variable values at successive iterations*

Iteration	i_{ds}	i_{qs}	i_{dr}	i_{qr}	ω_r
2	0.9000	0.3215	0.9726	0.0910	1.0046
3	0.9000	0.3330	0.0910	0.9714	1.0051
4	0.9000	0.3330	0.9714	0.0910	1.0051

However, problem arises while carrying out the load flow solutions of large power systems with SCIG based wind farm connected at one or more bus. This is because the reactive power exchange by the SCIG is not controlled and hence cannot be specified. Therefore, unlike the DFIG-based wind farm connected system, the method of representing the wind generator in the rest of the system as negative load equal to the specified *P* and *Q* (of the wind farm) cannot be used. The alternate method that can be followed is described below.

Suppose the wind generation system is connected to the *j*th bus of a *n* bus power system as shown in Fig. 9.4.

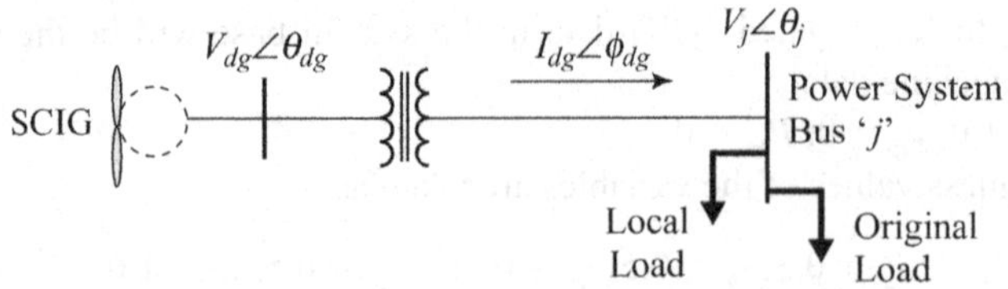

Fig. 9.4 *SCIG based wind farm connected to bus 'j' of large power system*

The power mismatch equations of bus '*j*' get modified as follows:

$$\Delta P_j = P_j^{sp} - P_j$$

$$= P_j^{sp} - \sum_{k=1}^{n}\left(G_{jk}\cos\theta_{jk} + B_{jk}\sin\theta_{jk}\right)|V_j||V_k| - \text{real}\,[(|V_j|\angle\theta_j)(|I_{dg}|\angle\phi_{dg})^*] \quad (9.25)$$

$$\Delta Q_j = Q_j^{sp} - Q_j$$

$$= Q_j^{sp} - \sum_{k=1}^{n}\left(G_{jk}\sin\theta_{jk} - B_{jk}\cos\theta_{jk}\right)|V_j||V_k| - \text{imag}\,[(|V_j|\angle\theta_j)(|I_{dg}|\angle\phi_{dg})^*] \quad (9.26)$$

where $|I_{dg}|\angle\phi_{dg}$ is the current coming from the SCIG, flowing through the distribution line and transformer, and going into bus 'j'. This current is actually the stator current of the SCIG.

$$|I_{dg}|\angle\phi_{dg} = i_{ds} + ji_{qs} \quad (9.27)$$

The terminal voltage (stator voltage) of the SCIG can be expressed as

$$v_{ds} + jv_{qs} = |V_j|\angle\theta_j + j(r + jx)\left(|I_{dg}|\angle\phi_{dg}\right) \quad (9.28)$$

Now, the five equations, Eqs (9.17)–(9.20) (with $v_{dr} = 0$; $v_{qr} = 0$;), and Eq. (9.21) for the SCIG, and the equations of the 'n' bus system (of the form (6.48)–(6.49)) are solved simultaneously using the Newton–Raphson method to obtain the values of the stator and the rotor currents and the rotor speed (or slip) of the SCIG along with the bus voltages and angles of the remaining buses of the system. The equations (9.25)–(9.28) are used to relate the SCIG equations and the 'n' bus system equations while forming the mismatch equations and the Jacobian matrix.

9.4 INCLUSION OF WIND POWER IN STABILITY STUDIES

As described in Chapter 8, the transient stability condition of a power system following a large disturbance like a fault is studied by carrying out time domain simulation (TDS). The evolution of the rotor angles (δ) of the synchronous machines in the post-disturbance condition are obtained, and the maintenance/loss of synchronism is observed to decide about the possible stability condition of the system. The TDS involves the simultaneous solution of one set of differential equations (corresponding to the alternators) and one set of algebraic equations (corresponding to the network). Now, if some wind farm is connected to the power system, then the differential and the algebraic equations of the wind generator, the wind turbine and the rotating mass are also to be solved. Considering the two-mass model of the wind turbine and the rotating mass, they are represented by the differential equations, Eqs (9.1)–(9.3). If the DFIG is used as the wind power generator, it can be represented by the differential equations, Eqs (9.8)–(9.11). However, as mentioned in Sec. 9.2.2, the stator and rotor dynamics of the induction generator is generally neglected and they are represented by the algebraic equations, Eqs (9.17)–(9.20). The other algebraic equations of the induction generator are Eqs (9.21)–(9.22). For the single-mass model, Eq. (9.5) is used instead of Eqs (9.1)–(9.3). If SCIG is used as the wind generator, then Eqs (9.17)–(9.20) are used but with $v_{dr} = 0$; $v_{qr} = 0$.

In the case of DFIG-based wind power generation, controllers are used for the two converters—the rotor side converter (RSC) and the grid side converter (GSC) in the rotor circuit of the machine. The converters are connected back-to-back through a common DC bus as shown in Fig. 9.1(c). Usually, the GSC is controlled such that it allows bi-directional power flow between the grid and the RSC (and subsequently the rotor) via the DC bus as per the demand of the RSC. Also, the GSC is usually operated at unity power factor, exchanging no net reactive power with the grid. Moreover, the control of the GSC is much faster than the electro-mechanical transients. Because of these reasons, generally it is not required to represent the GSC controller dynamics in a transient stability program. Instead it is represented by a current source (I_{GSC}) which accounts for the exchange of power between the RSC and the grid. Hence, the total current exchanged by the DFIG with the grid is the algebraic sum of the stator current and the I_{GSC}. The dynamics of the DC bus capacitor being very fast is also neglected and it is reasonable to assume (for transient stability study) that the capacitor is large enough so that there is no appreciable change in the DC bus voltage during power system disturbances. However, the RSC controls should be incorporated as these have impact on the torque/active power output and the reactive power output of the DFIG.

The RSC being a voltage source converter, supplies the voltage ($v_{dr} + jv_{qr}$) to the rotor winding. The current flowing from the RSC to the rotor depends on this voltage through Eqs (9.10)–(9.11) or Eqs (9.19)–(9.20), and thus the RSC controls the rotor current. Different control structures may be put in place for the RSC, but ultimately the solution of the equations representing those controls will provide the required values of v_{dr} and v_{qr} at every time step of the TDS. These values are put in Eqs (9.19)–(9.20). Also, the value of ω_r obtained from the solution of the differential equations, Eqs (9.1)–(9.3) at every time step, is substituted in Eqs (9.19)–(9.20). After substituting the values of v_{dr}, v_{qr}, and ω_r, Eqs (9.17)–(9.20) turn into linear equations which can be solved to obtain the values of (i_{ds}, i_{qs}, i_{dr}, i_{qr}). While doing this, it should be checked that the values of i_{dr}, i_{qr} do not exceed their respective limits. This is required from a practical point of view as the converter has a current limit. In case the computed value of i_{dr}, i_{qr} exceeds the limit, the limiting values are taken as the value of the current for that particular time step.

To elaborate this, a typical control structure used for DFIG is shown in Fig. 9.5. This is called the field oriented control where the d-axis of the synchronously rotating reference frame is considered to be aligned/oriented with the axis of the DFIG stator voltage, thus giving rise to the relation,

$$v_{ds} = |V_{dg}|;\ v_{qs} = 0 \tag{9.29}$$

This orientation leads to the decoupling of the d-axis control (for active power/ torque) and the q-axis control (for reactive power). Further simplification gives the relations shown in Fig. 9.5. The reference for the rotor voltage is obtained as shown below in Eq. (9.30). The switching pulses of the RSC are obtained by pulse width modulation using this rotor voltage reference.

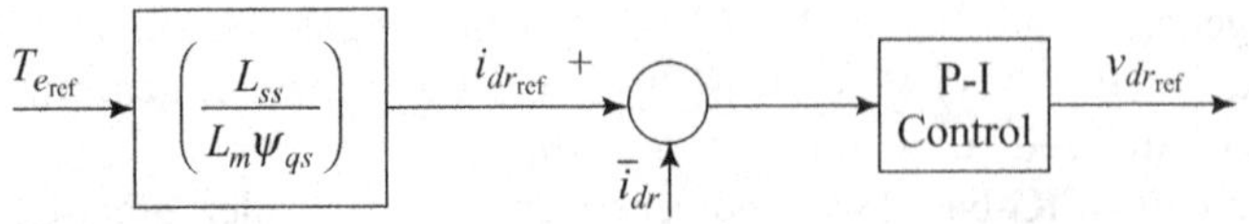

Fig. 9.5(a) *d-axis control for the rotor side converter of the DFIG*

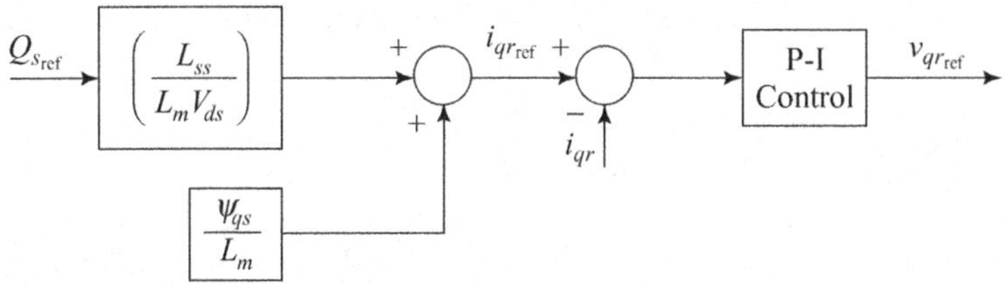

Fig. 9.5(b) *q-axis control for the rotor side converter of the DFIG*

$$\left|V_{r_{ref}}\right| = \sqrt{v_{dr_{ref}}^2 + v_{qr_{ref}}^2}; \quad \theta_r = \tan^{-1}\left(v_{qr_{ref}} / v_{dr_{ref}}\right) \tag{9.30}$$

However, as the dynamics of the rotor is neglected, the relation between the rotor current and the rotor voltage becomes algebraic and the P-I controller can be removed. Thus, the model is simplified further by considering the current references as the converter (rotor) current. That means

$$i_{dr} = i_{dr_{ref}}; \quad i_{qr} = i_{qr_{ref}};$$

This assumption is reasonable since the time constant of the current controllers are usually in the range of 1 ms, which is negligible as compared to the time constant of the electromechanical transients (0.5-2.0 s).

Thus at every time step of the TDS

1. The solution of the synchronous machine differential equations and the network algebraic equations provides the bus voltage magnitudes and angles. The DFIG terminal bus is included in the network and hence voltage at that bus is also obtained. Considering the stator voltage orientation, the values of v_{ds} and v_{qs} are obtained as in Eq. (9.29).
2. The rotor current references are obtained using the values of stator voltage, stator flux, torque, and reactive power references and system parameters following the relationships shown in Fig. 9.5(a) and 9.5(b).
3. These rotor current references are considered to be the rotor currents and used in Eqs (9.19)–(9.20).
4. The differential equations corresponding to the wind turbine and the rotating mass, i.e., Eqs (9.1)–(9.3) are solved using the values of system parameters and the previous time-step values of the variables to obtain the current time-step values of (ω_t, ω_r and θ_{tw}). This value of ω_r is used in Eqs (9.19)–(9.20).
5. Steps 2 and 3 convert Eqs (9.17)–(9.20) to linear algebraic equations which are then solved to get the current time step values of (i_{ds}, i_{qs}, v_{dr}, v_{qr}). If the values of v_{dr} and v_{qr} exceed their pre-defined limit (decided on the basis of the converter voltage rating), then they are considered to be equal to the limiting values.
6. Using the computed values of (i_{ds}, i_{qs}, i_{dr}, i_{qr}, v_{dr}, v_{qr}) and the values of v_{ds} and v_{qs} in Eqs (9.21)–(9.22), the active and reactive power output of the DFIG (P_{tot} and Q_{tot}) are obtained. This P_{tot} and Q_{tot} are then used for the solution of the power system network in the next time step. Similarly, the electromagnetic torque of the DFIG is obtained which is used in solving

Eqs (9.1)–(9.3) in the next time-step. This complete procedure is repeated in every time-step of the TDS.

Example 9.3 A DFIG based wind farm of rating 50 MW is connected to bus 5 of the power system described in Example 6.2 through a transformer and a distribution line as shown in Fig. 9.3. The combined impedance of the transformer and the distribution line is $(r + jx) = (0.003 + j0.02)$ p.u. At a particular time, the wind speed is 13.95 m/s and the wind farm injects 40 MW active power (and no reactive power) at bus 5. An additional local load of 40 MW is also connected at bus 5. The data for individual DFIG of rating 5 MW is given in Appendix C. The data is on the own base of the machine. It is considered that the synchronous machines are represented by the flux decay model with static exciters. The data of the synchronous machines and the exciters are given in Table 9.3. A 3-phase short-circuit fault takes place in bus 4 and gets cleared after 0.20 s. Find and plot the changes in the following variables of the DFIG using Modified Euler's method for the 2 seconds after the fault clearance: (i) Terminal bus voltage magnitude ($|V_{dg}|$), (ii) rotor current (I_r), (iii) Active power output at the terminal (P_{tot}), (iv) Rotor speed (ω_r).

Also plot the relative rotor angle δ_{21} (i.e., $\delta_1 - \delta_2$) for the same period. Consider the rotor angle of the synchronous generator 1 as the reference.

Assume that during the fault condition, the rotor is short-circuited through a resistance of 0.25 p.u. and the machine behaves like a squirrel cage induction machine.

Table 9.3 *Synchronous machine and exciter data for the system of Example 9.3*

M/C No	*Bus No*	x_d *(p.u.)*	x'_d *(p.u.)*	T'_{do} *(sec)*	x_q *(p.u.)*	x'_q *(p.u.)*	T'_{qo} *(sec)*	*H (sec)*	k_D *(p.u.)*	K_A	T_A
1	5	1.3125	0.1813	5.89	1.2578	0.2500	5.600	3.01	1.0	20	0.2
2	6	0.8958	0.1198	6.00	0.8645	0.1969	0.535	6.40	2.5	20	0.2

Solution: Load flow solution of the system is obtained considering the following:

(i) The DFIG terminal bus as an additional bus

(ii) Including the DFIG power (computed as shown in Example 9.1) as negative load at the DFIG terminal bus

(iii) Adding the additional local load at bus 5

The converged solutions are given in Table 9.4. The DFIG terminal bus is considered to be the 7th bus.

Table 9.4 *The load flow solution of the 6 bus system including the DFIG terminal bus*

Bus no.	*Voltage*		*Generation*		*Load*	
	Magnitude (p.u.)	*Angle (deg)*	*MW*	*MVAR*	*MW*	*MVAR*
1	1.05	0.0	48.3	23.97	0	0
2	1.02	–3.2844	25.0	11.22	0	0
3	0.8260	–14.0886	0	0	27.5	6.5
4	0.9299	–9.9014	0	0	0	0

Contd.

Table 9.4 *Contd.*

Bus no.	Voltage		Generation		Load	
	Magnitude (p.u.)	*Angle (deg)*	*MW*	*MVAR*	*MW*	*MVAR*
5	0.8502	–13.2274	0	0	55.0	9.0
6	0.9057	–12.5786	0	0	25.0	2.5
7	0.8533	-11.9636	0	0	– 40.13	– 0.0088

The voltage of the DFIG terminal bus (bus 7), $|V_{dg}| \angle \theta_{dg} = 0.8533 \angle -11.96°$.

With the terminal voltage and power output of the DFIG being known, the initialized values of the DFIG variables (converged solution found using the Newton–Raphson method as shown in Example 9.1) are as given below.

$i_{ds} = 0.8448$; $i_{qs} = -0.0208$; $i_{dr} = -0.8523$; $i_{qr} = 0.2353$; $i_{dr} = -0.0990$; $i_{qr} = -0.0114$; $P_{tot} = 0.8027$; $T_e = 0.7244$; $\omega_r = 1.1189$

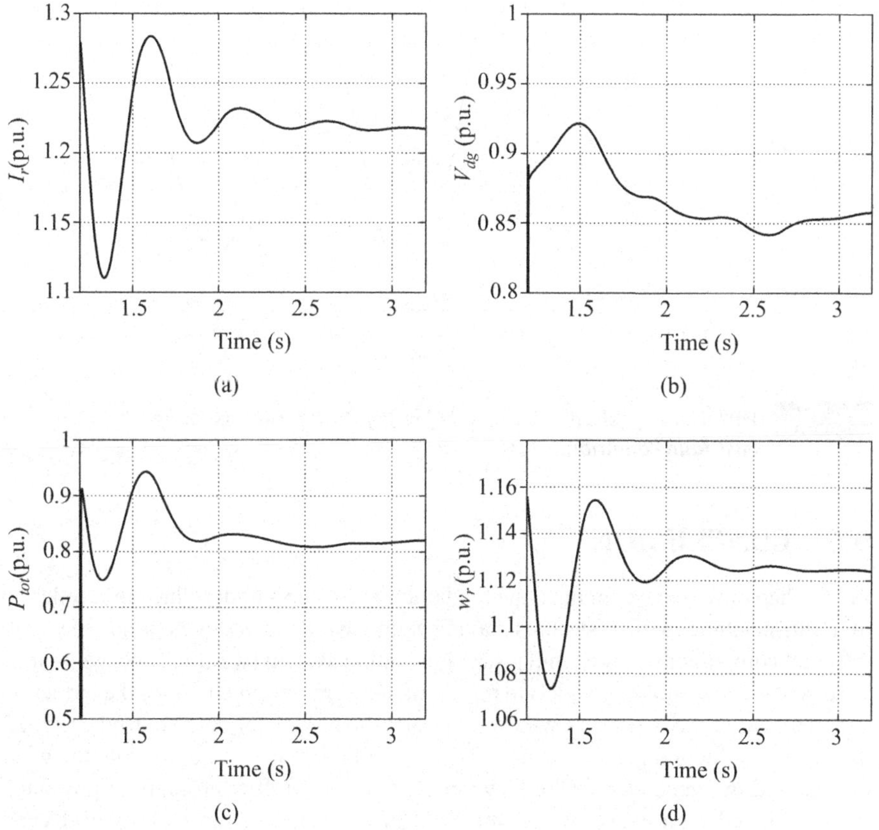

Fig. 9.6 *Variations of different DFIG variables—(a) rotor current (I_r), (b) terminal voltage (V_{dg}), (c) power output at its terminal bus (P_{tot}), and (d) rotor speed in the post-fault condition*

During the fault, the rotor is shorted and hence the converter is not in the circuit, and the machine behaves like a squirrel cage induction machine. At fault clearance, the values of the DFIG variables are as given below:

$i_{ds} = 0.2531;\quad i_{qs} = 0.1216;\quad i_{dr} = -0.2558;\quad i_{qr} = -0.0158;\quad v_{dr} = 0;\quad v_{qr} = 0;$
$P_{tot} = 0.1079;\quad T_e = 0.1083;\quad \omega_r = 1.1549$

The value of the rotor angles of the two synchronous machines at the point of fault clearance are $\delta_1 = 18.82°;\quad \delta_2 = 42.17°$

Time domain simulation is continued with these initial values. The plots of the following variables (i) terminal bus voltage magnitude ($|V_{dg}|$), (ii) rotor current (I_r), (iii) active power output at the terminal (P_{tot}), (iv) rotor speed (ω_r), and (v) relative rotor angle (δ_{21}) of the synchronous generators are shown in Fig. 9.6 and 9.7.

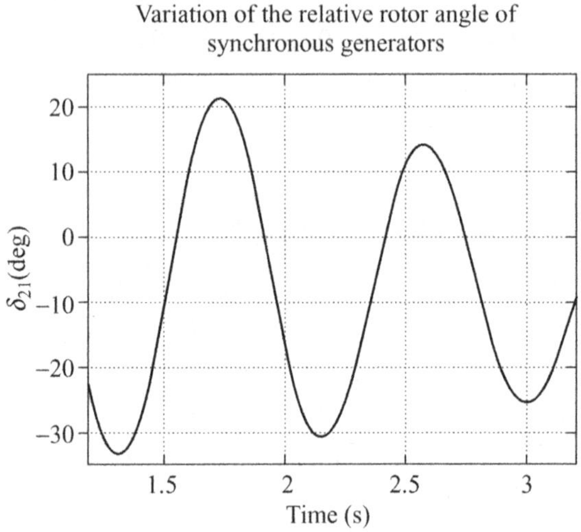

Fig. 9.7 *Variations of relative rotor angle (δ_{21}) of the alternators in the post-fault condition*

9.5 CONCLUSION

In this chapter we have explained the fundamental concepts underlying the simulation of multi-machine power systems having some wind power generation. At first, different components of the wind power generation system (wind turbine, generator) are discussed and suitable models to represent them are presented. Since the induction generators are mostly used as wind generators, the models of both squirrel cage and doubly-fed induction generators are discussed. The methods to incorporate the wind turbine and the generator in the Newton–Raphson load flow program is presented and elaborated with examples. At the end, the inclusion of the DFIG based wind power generation system in the transient stability study of large power systems is discussed. A typical control structure for the DFIG rotor side converter is presented to explain the method of incorporation of the DFIG in the time domain simulation. Possible simplification of the control structure with some reasonable assumptions is also discussed and finally all the steps for the TDS are mentioned.

Problems

9.1 In the system of Example 9.1, the wind farm is rated at 60 MW. Considering a wind power generation (injection at bus 5) of 30 MW and 12 MVAR, and an additional local load exactly equal to the wind generation, find the load flow solution of the system.

9.2 The generator 2 of the power system described in Example 6.2 is replaced by a wind farm of 50 MW. The wind farm generates active power exactly equal to the generation of the synchronous generator 2. Find the load flow solution of the power system.

9.3 For the power system of Example 9.1, the wind farm is considered to be SQIG based with the same rating and system data as the DFIG. Carry out the load flow solution of the system to find the final voltages of all the buses and the generations of the two synchronous machines. Also find the reactive power consumed by the SQIG.

9.4 Repeat the problem of Example 9.3 for the case when the DFIG is connected to bus 4 and supplying 25 MW, there is no local load connected to bus 4 and a 3-phase short-circuit fault takes place at bus 5. Find the critical clearing time for this case.

Appendix A

NUMERICAL SOLUTION OF SYSTEM OF EQUATIONS AND THE USE OF MATRIX ALGEBRA

A.1 INTRODUCTION

In power system problems, such as load flow and transient stability, we encounter both linear and non-linear systems of equations. Since the order of these equations is very high, it is important to have fast, efficient and accurate numerical algorithms. In this appendix we outline a few of these techniques.

We are concerned with solving the following system of n simultaneous equations in n unknowns $x_1, \ldots, x_n$.

$$\begin{aligned} f_1(x_1, x_2, \ldots, x_n) &= 0 \\ f_2(x_1, x_2, \ldots, x_n) &= 0 \\ \vdots \quad & \quad \vdots \\ f_n(x_1, x_2, \ldots, x_n) &= 0 \end{aligned} \tag{A.1}$$

A special case of the above, when $f_1, \ldots, f_n$ are linear in x_i's, results in a system of linear equations of the form

$$\begin{aligned} a_{11}x_1 + a_{12}x_2 + \cdots + a_{1n}x_n &= u_1 \\ a_{21}x_1 + a_{22}x_2 + \cdots + a_{2n}x_n &= u_2 \\ \vdots \quad & \quad \vdots \\ a_{n1}x_1 + a_{n2}x_2 + \cdots + a_{nn}x_n &= u_n \end{aligned} \tag{A.2}$$

More compactly we have

$$\mathbf{AX} = \mathbf{U} \tag{A.3}$$

where A is the matrix of coefficients a_{ij}

$$\mathbf{X} = [x_1, \ldots, x_n]^{\mathrm{T}} \quad \text{and } \mathbf{U} = [u_1, \ldots, u_n]^{\mathrm{T}}$$

Systems of equations of type (A.1) that are non-linear are invariably solved using *iterative methods*. Systems of linear equations of type (A.2) can be solved using either (1) the iterative technique, or (2) employing *direct methods* involving a finite number of operations. We first deal with iterative methods.

A.2 ITERATIVE METHODS FOR SOLUTION of LINEAR SYSTEM OF EQUATIONS

A.2.1 Jacobi Methods

Consider Eq. (A.2). Solving for x_i's

$$x_1 = (u_1 - a_{12}x_2 - a_{13}x_3 - \cdots - a_{1n}x_n)/a_{11}$$

$$x_2 = (u_2 - a_{21}x_1 - a_{23}x_3 - \cdots - a_{2n}x_n)/a_{22} \qquad \text{(A.4)}$$

$$\vdots$$

$$x_n = (u_n - a_{n1}x_1 - a_{n2}x_2 - \cdots - a_{n,n-1}x_{n-1})/a_{11}$$

Equation (A.4) can be written compactly as

$$x_i = \frac{\left(u_i - \sum_{\substack{j=1 \\ j\neq i}}^{n} a_{ij}x_j\right)}{a_{ii}}, \quad i = 1, 2, \ldots, n \qquad \text{(A.5)}$$

In the above formulation, it is assumed that $a_{ii} \neq 0$. Usually, the equations are reordered if necessary so that *diagonal dominance* holds, i.e. each diagonal element a_{ii} is larger in magnitude than the magnitudes of other entries in row i and column i.

For an iterative solution, we assume initial values for x_i's and substitute them in the right-hand side of equations (A.5). The resulting new approximations for the x_i's are resubstituted on the right-hand sides of Eq. (A.5) and the process repeated. After several iterations, if the process has converged, the x_i's computed will show very little change. A criterion for convergence can be set up as follows:

$$\underset{i}{\text{Max}}\left|x_i^{(k+1)} - x_i^{(k)}\right| < \varepsilon \qquad \text{(A.6)}$$

where the superscripts (k) and $(k+1)$ denote the kth and $(k+1)$th iteration, respectively.

A.2.2 Gauss–Seidel Iterative Method

In this method, the linear system of equations are arranged in the form of Eq. (A.5). In the iterations, the new values of x_i's are always used on the right-hand sides as soon as they are computed. In contrast, in the Jacobi method, the new values are not used until iteration for all the n components is completed. Symbolically, values obtained at the end of the first iteration are denoted by $x_i^{(1)}$ $(i = 1, 2, \ldots, n)$ and so on. The values at the end of the kth and $(k + 1)$th iteration are denoted by $x_i^{(k)}$ and $x_i^{(k+1)}$ $(i = 1, 2, \ldots, n)$. The iterates are defined by

$$x_i^{(k+1)} = \left(u_{ii} - a_{i1}x_1^{(k+1)} \ldots - a_{i,i-1}x_{i-1}^{(k+1)} - a_{i,i+1}x_{i+1}^{(k)} \ldots - a_{in}x_n^{(k)}\right) / a_{ii}, i = 1, 2, \ldots, n \qquad \text{(A.7)}$$

Both the Jacobi and Gauss–Seidel methods of solving the system of equations iteratively are sometimes referred to as *relaxation* methods (the errors in the initial starting values of x_i's are decreased or relaxed as computation continues).

A.3 DIRECT METHOD OF SOLVING LINEAR SYSTEM OF EQUATIONS

This method is known as the *Gaussian elimination method* and involves a finite number of operations in arriving at the solutions. Consider the system of three equations:

$$a_{11}x_1 + a_{12}x_2 + a_{13}x_3 = u_1$$

$$a_{21}x_1 + a_{22}x_2 + a_{23}x_3 = u_2$$

$$a_{31}x_1 + a_{32}x_2 + a_{33}x_3 = u_3 \tag{A.8}$$

As a first step, divide the first equation by a_{11}. Next, replace the second equation by adding to it the first equation multiplied by $-a_{21}/a_{11}$. Similarly, replace the third equation by adding to it the first equation multiplied by $-a_{31}/a_{11}$. The result is the system

$$x_1 + \frac{a_{12}}{a_{11}}x_2 + \frac{a_{13}}{a_{11}}x_3 = \frac{u_1}{a_{11}} \tag{A.9}$$

$$(a_{22} - a_{12}a_{21}/a_{11})x_2 + (a_{23} - a_{13}a_{21}/a_{11})x_3 = u_2 - u_1a_{21}/a_{11}$$

$$(a_{32} - a_{12}a_{31}/a_{11})x_2 + (a_{33} - a_{13}a_{31}/a_{11})x_3 = u_3 - u_1a_{31}/a_{11}$$

More compactly, these equations can be written as

$$x_1 + b_{12}x_2 + b_{13}x_3 = v_1$$

$$a_{22}^{(1)}x_2 + a_{23}^{(1)}x_3 = u_2^{(1)} \tag{A.10}$$

$$a_{32}^{(1)}x_2 + a_{33}^{(1)}x_3 = u_3^{(1)}$$

Now divide the second equation in Eq. (A.10) by $a_{22}^{(1)}$ and replace the third equation by adding to it the second equation multiplied by $-a_{32}^{(1)}/a_{22}^{(1)}$. The result is as follows:

$$x_1 + b_{12}x_2 + b_{13}x_3 = v_1$$

$$x_2 + \left(a_{23}^{(1)}/a_{22}^{(1)}\right)x_3 = u_2^{(1)}/a_{22}^{(1)} \tag{A.11}$$

$$\left[a_{33}^{(1)} - \left(a_{23}^{(1)}a_{32}^{(1)}/a_{22}^{(1)}\right)\right]x_3 = \left(u_3^{(1)} - u_2^{(1)}a_{32}^{(1)}/a_{22}^{(1)}\right)$$

More compactly, the above sets of equations can be written as

$$x_1 + b_{12}x_2 + b_{13}x_3 = v_1$$

$$x_2 + b_{23}x_3 = v_2 \tag{A.12}$$

$$a_{33}^{(2)}x_3 = u_3^{(2)}$$

As a final trivial step, divide the third equation by $a_{33}^{(2)}$ and in terms of the new coefficients, the system of equations becomes

$$x_1 + b_{12}x_2 + b_{13}x_3 = v_1$$

$$x_2 + b_{23}x_3 = v_2$$

$$x_3 = v_3 \tag{A.13}$$

where $\qquad v_3 = u_3^{(2)} / a_{33}^{(2)}$.

In matrix notation, Eq. (A.13) becomes

$$\begin{bmatrix} 1 & b_{12} & b_{13} \\ 0 & 1 & b_{23} \\ 0 & 0 & 1 \end{bmatrix} \begin{bmatrix} x_1 \\ x_2 \\ x_3 \end{bmatrix} = \begin{bmatrix} v_1 \\ v_2 \\ v_3 \end{bmatrix} \tag{A.14}$$

The process of computation in arriving at the form (A.13) or (A.14) is termed *forward substitution*. To find x_1, x_2, and x_3, we perform the *backward substitution* procedure starting from the third equation in (A.14) as follows:

$$x_3 = v_3$$

$$x_2 = v_2 - b_{23}v_3$$

$$x_1 = v_1 - b_{12}v_2 - b_{13}x_3 \tag{A.15}$$

The preceding procedure can be generalized to a system of n equations as follows: Define

$$b_{kj} = \frac{a_{kj}^{(k-1)}}{a_{kk}^{(k-1)}} \qquad j \geq k+1 \tag{A.16}$$

$$v_k = u_k^{(k-1)} / a_{kk}^{(k-1)} \tag{A.17}$$

$$a_{ij}^{(k)} = a_{ij}^{(k-1)} - a_{ik}^{(k-1)} b_{kj} \qquad i, j \geq k+1 \tag{A.18}$$

$$u_i^{(k)} = u_i^{(k-1)} - a_{ik}^{(k-1)} v_k \qquad i \geq k+1 \tag{A.19}$$

The final solution is

$$x_i = v_i - \sum_{j=i+1}^{n} b_{ij}x_j, \qquad i = n, n-1, \ldots, 2, 1 \tag{A.20}$$

A.4 SYSTEM OF NONLINEAR EQUATIONS

The system of nonlinear equations of the type (A.1) are invariably solved iteratively using efficient computational schemes. There are basically two competitive algorithms: (1) the Jacobi and Gauss–Seidel method, and (2) the Newton–Raphson method. The relative merits of both these methods will be discussed at the end.

A.4.1 Jacobi and Gauss–Seidel Method

If the n equations (A.1) are expressible in the form

$$\begin{aligned} x_1 &= \Phi_1(x_1, x_2, \ldots, x_n) \\ x_2 &= \Phi_2(x_1, x_2, \ldots, x_n) \\ &\vdots \\ x_n &= \Phi_n(x_1, x_2, \ldots, x_n) \end{aligned} \tag{A.21}$$

then a straightforward iterative technique is possible.
More compactly, let

$$x_i = \Phi_i(x), \quad i = 1, 2, \ldots, n \tag{A.22}$$

Let the initial values be assumed to be $x_i^{(0)}$. Then the successive estimates of the vector $x^{(k)} = \left[x_1^{(k)}, \ldots, x_n^{(k)}\right]^{\mathrm{T}}$ can be obtained by (1) the Jacobi method, or (2) the Gauss–Seidel method.
In Jacobi's method, the iterates are defined by

$$x_i^{(k+1)} = \Phi_i\left(x_1^{(k)}, x_2^{(k)}, \cdots, x_n^{(k)}\right), \quad i = 1, 2, \ldots, n \tag{A.23}$$

In the Gauss–Seidel method, the recently computed values are used in the equations, i.e.

$$x_i^{(k+1)} = \Phi_i\left(x_1^{(k+1)}, x_2^{(k+1)}, \cdots, x_{i-1}^{(k+1)}, x_i^{(k)}, x_{i+1}^{(k)}, \ldots, x_n^{(k)}\right), \quad i = 1, 2, \ldots, n \tag{A.24}$$

The iterations are continued until the maximum difference between consecutive values of $x_i \, (i = 1, 2, \ldots, n)$ is less than a predetermined value ε, i.e.

$$\operatorname*{Max}_i \left| x_i^{(k+1)} - x_i^{(k)} \right| < \varepsilon \tag{A.25}$$

A.4.2 Newton–Raphson Method

The Newton–Raphson method is applied directly to Eq. (A.1). Before explaining the method to the system of the non-linear equations, it is helpful to illustrate it for a non-linear equation in one variable.

Consider the scalar non-linear equation $f(x) = 0$ in the variable x. Let the initial guess be $x^{(0)}$. Expand $f(x)$ around $x^{(0)}$ in a Taylor series. Then the equation becomes

$$f\left(x^{(0)}\right) + \left(x - x^{(0)}\right) f'\left(x^{(0)}\right) + \frac{1}{2!}\left(x - x^{(0)}\right)^2 f''\left(x^{(0)}\right) + \ldots = 0 \tag{A.26}$$

Neglecting second-order terms and higher, we get

$$f\left(x^{(0)}\right)+\left(x-x^{(0)}\right)f'\left(x^{(0)}\right)=0 \tag{A.27}$$

Solving this equation for x, we get an improved estimate as

$$x^{(1)}=x^{(0)}-\frac{f'\left(x^{(0)}\right)}{f\left(x^{(0)}\right)} \tag{A.28}$$

Generalizing for the kth iteration, we get

$$x^{(k+1)}=x^{(k)}-\frac{f'\left(x^{(k)}\right)}{f\left(x^{(k)}\right)} \tag{A.29}$$

Convergence can be tested as before or by an alternative criterion such as $|f|<\varepsilon$. Indeed if the iterations converge, then $f\to 0$. The equation $f(x)=0$ may have several solutions, in which case the converged value is the one closest to the initial guess.

System of Non-linear Equations Consider first a system of two equations

$$f_1(x_1,x_2)=0$$

$$f_2(x_1,x_2)=0 \tag{A.30}$$

Expansion in a Taylor series around $x_1^{(0)}$ and $x_2^{(0)}$, and the retention of only the linear terms yields

$$f_1\left(x_1^{(0)},x_2^{(0)}\right)+\left(x_1-x_1^{(0)}\right)\left.\frac{\partial f_1}{\partial x_1}\right|_0+\left(x_2-x_2^{(0)}\right)\left.\frac{\partial f_1}{\partial x_2}\right|_0=0$$

$$f_2\left(x_1^{(0)},x_2^{(0)}\right)+\left(x_1-x_1^{(0)}\right)\left.\frac{\partial f_2}{\partial x_1}\right|_0+\left(x_2-x_2^{(0)}\right)\left.\frac{\partial f_2}{\partial x_2}\right|_0=0 \tag{A.31}$$

Casting the above equations in matrix form and solving for x_1 and x_2, we obtain

$$\begin{bmatrix}\dfrac{\partial f_1}{\partial x_1} & \dfrac{\partial f_1}{\partial x_2}\\[2ex] \dfrac{\partial f_2}{\partial x_1} & \dfrac{\partial f_2}{\partial x_2}\end{bmatrix}_0\begin{bmatrix}\left(x_1-x_1^{(0)}\right)\\[2ex] \left(x_2-x_2^{(0)}\right)\end{bmatrix}=\begin{bmatrix}-f_1\left(x_1^{(0)},x_2^{(0)}\right)\\[2ex] -f_2\left(x_1^{(0)},x_2^{(0)}\right)\end{bmatrix} \tag{A.32}$$

Hence, the first approximation for x_1 and x_2 yields

$$\begin{bmatrix}x_1^{(1)}\\[2ex] x_2^{(1)}\end{bmatrix}=\begin{bmatrix}x_1^{(0)}\\[2ex] x_2^{(0)}\end{bmatrix}-\begin{bmatrix}\dfrac{\partial f_1}{\partial x_1} & \dfrac{\partial f_1}{\partial x_2}\\[2ex] \dfrac{\partial f_2}{\partial x_1} & \dfrac{\partial f_2}{\partial x_2}\end{bmatrix}_0^{-1}\begin{bmatrix}f_1\left(x_1^{(0)},x_2^{(0)}\right)\\[2ex] f_2\left(x_1^{(0)},x_2^{(0)}\right)\end{bmatrix} \tag{A.33}$$

Using the values of $x_1^{(1)}$ and $x_2^{(1)}$, improved estimates $x_1^{(2)}$ and $x_2^{(2)}$ can be obtained following a similar procedure. The iterations are continued until convergence is obtained. The convergence criterion can be one of the following:

$$1.\ \underset{i}{\text{Max}} \left| x_i^{(k+1)} - x_i^{(k)} \right| < \varepsilon_1, \quad i = 1, 2$$

$$2.\ \underset{i}{\text{Max}} \left| f_i \right| < \varepsilon_2, \qquad i = 1, 2 \tag{A.34}$$

Values of ε_1 and ε_2 are chosen depending on the problem, accuracy desired, etc.

The above procedure for two equations can be easily generalized for a system of n equations of the form (A.1) repeated below for convenience.

$$\begin{aligned} &f_1(x_1, x_2, \ldots, x_n) = 0 \\ &f_2(x_1, x_2, \ldots, x_n) = 0 \\ &\vdots \\ &f_n(x_1, x_2, \ldots, x_n) = 0 \end{aligned} \tag{A.35}$$

In matrix notation, these equations are denoted by $F(X) = 0$. Define the Jacobian matrix denoted by $F'X$ as follows:

$$F'(X) = \left(\frac{\partial f_i}{\partial xj} \right) = \begin{bmatrix} \dfrac{\partial f_1}{\partial x_1} & \cdots & \dfrac{\partial f_1}{\partial x_n} \\ \dfrac{\partial f_2}{\partial x_1} & \cdots & \dfrac{\partial f_2}{\partial x_n} \\ \vdots & & \vdots \\ \dfrac{\partial f_n}{\partial x_1} & \cdots & \dfrac{\partial f_n}{\partial x_n} \end{bmatrix} \tag{A.36}$$

Then the Newton–Raphson iterates are defined by

$$X^{(k+1)} = X^{(k)} - \left[F'(X)^{(k)} \right]^{-1} F(X)^{(k)} \tag{A.37}$$

Convergence can be tested either in terms of x_i or f_i as in Eq. (A.34).

There are no strict guidelines as to which method to choose in a given problem. The Newton–Raphson method has superior convergence properties as compared to the Gauss–Seidel method, i.e. it takes fewer iterations to converge to a solution. This is true provided the initial guess is fairly close to the desired solution. The disadvantage of the Newton–Raphson method is that the first-order derivatives have to be computed. The Gauss–Seidel method is easier to program and no matrix inversion is involved.

In power-system problems $F'(X)$ in Eq. (A.36) is sparse. Special techniques of programming which uses only the non-zero elements together with triangular-factorization obviating the need to take inverse in Eq. (A.37) are used. There are a number of books dealing with sparsity-based techniques. This combined with simplification of the Jacobian, as discussed in Chapter 6, makes this technique a preferred one in power-system problems.

A.5 MATRIX FACTORIZATION

If we have to solve x for a different set of vectors b, then it will be convenient to systematize the procedure in the preceding section. One such method to factorize the matrix A as $A = LU$ or $L'DU$.

A.5.1. LU Factorization

Let

$$\begin{bmatrix} a_{11} & a_{12} & a_{13} \\ a_{21} & a_{22} & a_{23} \\ a_{31} & a_{32} & a_{33} \end{bmatrix} = \begin{bmatrix} \ell_{11} & 0 & 0 \\ \ell_{21} & \ell_{22} & \\ \ell_{31} & \ell_{32} & \ell_{33} \end{bmatrix} \begin{bmatrix} 1 & u_{12} & u_{13} \\ 0 & 1 & u_{23} \\ 0 & 0 & 1 \end{bmatrix}$$

$$= LU \tag{A.38}$$

Multiplying L and U and equating the elements with those of A gives

$$a_{11} = \ell_{11} \qquad a_{12} = \ell_{11}u_{12} \qquad a_{13} = \ell_{11}u_{13}$$

$$a_{21} = \ell_{21} \qquad a_{22} = \ell_{21}u_{12} + \ell_{22} \qquad a_{23} = \ell_{21}u_{13} + \ell_{22}u_{23}$$

$$a_{31} = \ell_{31} \qquad a_{32} = \ell_{31}u_{12} + \ell_{32} \qquad a_{33} = \ell_{31}u_{13} + \ell_{32}u_{23} + \ell_{33} \tag{A.39}$$

Solving for ℓ'_{ij} 's and u'_{ij} 's gives

$$\ell_{11} = a_{11} \qquad u_{12} = \frac{a_{12}}{\ell_{11}} = \frac{a_{12}}{a_{11}} = a_{12}^{(1)} \qquad u_{13} = \frac{a_{13}}{\ell_{11}} = \frac{a_{13}}{a_{11}} = a_{13}^{(1)}$$

$$\ell_{21} = a_{21} \qquad \ell_{22} = a_{22} - \ell_{21}u_{12} \qquad u_{23} = \frac{1}{\ell_{22}}(a_{23} - \ell_{21}u_{13})$$

$$= a_{22} - a_{21}a_{12}^{(1)} \qquad = \frac{1}{a_{22}^{(1)}}\left(a_{23} - a_{21}a_{13}^{(1)}\right)$$

$$= a_{22}^{(1)} \qquad = \frac{a_{23}^{(1)}}{a_{22}^{(1)}} = a_{23}^{(2)} \quad (6.105)$$

$$\ell_{31} = a_{31} \qquad \ell_{32} = a_{32} - \ell_{31}u_{12} \qquad \ell_{33} = a_{33} - \ell_{31}u_{13} - \ell_{32}u_{23}$$

$$= a_{32} - a_{31}a_{12}^{(1)} \qquad = a_{33} - a_{31}a_{13}^{(1)} - a_{32}^{(1)}a_{23}^{(1)}$$

$$= a_{32}^{(1)} \qquad = a_{33}^{(1)} - a_{32}^{(1)}a_{23}^{(2)}$$

$$= a_{33}^{(2)}$$

Thus,

$$\begin{bmatrix} a_{11} & a_{12} & a_{13} \\ a_{21} & a_{22} & a_{23} \\ a_{31} & a_{32} & a_{33} \end{bmatrix} = \begin{bmatrix} a_{11} & 0 & 0 \\ a_{21} & a_{22}^{(1)} & 0 \\ a_{31} & a_{32}^{(1)} & a_{33}^{(2)} \end{bmatrix} \begin{bmatrix} 1 & a_{12}^{(1)} & a_{13}^{(1)} \\ 0 & 1 & a_{23}^{(2)} \\ 0 & 0 & 1 \end{bmatrix} \tag{A.40}$$

i.e. $A = LU$. Note that ℓ_{ij}'s and u_{ij}'s are also obtained during the Gaussian elimination. Thus, the LU factors are derivable from the Gaussian elimination process. We illustrate this by means of an example.

Example A.1 Using Gaussian elimination, write $A = LU$ for the A matrix in Example 6.3.

Solution:

$$[A]=\begin{bmatrix} \underline{2} & 1 & 3 \\ \underline{2} & 3 & 4 \\ \underline{3} & 4 & 7 \end{bmatrix} \quad [b]=\begin{bmatrix} b_1 \\ b_2 \\ b_3 \end{bmatrix}$$

We follow the same steps as in Section 6.5 on Gaussian elimination. The underlined elements are the elements of L and U.

Step 1

$$x_1 + \frac{1}{\underline{2}}x_2 + \frac{3}{\underline{2}}x_3 = \frac{b_1}{2} = b_1^{(1)}$$

$$\underline{2}x_2 + x_3 = b_2 - 2\left(b_1^{(1)}\right) = b_2^{(1)}$$

$$\frac{5}{\underline{2}}x_2 + \frac{5}{2}x_3 = b_3 - 3\left(b_1^{(1)}\right) = b_3^{(1)} \quad \text{(A.41)}$$

Step 2

$$x_1 + \frac{1}{2}x_2 + \frac{3}{2}x_3 = b_1^{(1)}$$

$$x_2 + \frac{1}{\underline{2}}x_3 = \frac{b_2^{(1)}}{2} = b_2^{(2)}$$

$$\frac{5}{\underline{4}}x_3 = b_3^{(1)} - \left(\frac{5}{2}\right)b_2^{(2)}$$

$$= b_3^{(2)} \quad \text{(A.42)}$$

Step 3

$$x_1 + \frac{1}{2}x_2 + \frac{3}{2}x_3 = b_1^{(1)}$$

$$x_2 + \frac{x_3}{2} = b_2^{(2)}$$

$$x_3 = \frac{b_3^{(2)}}{5/4} = b_3^{(3)} \quad \text{(A.43)}$$

The elements of L and U matrices are underlined. Therefore,

$$\mathbf{L} = \begin{bmatrix} 2 & 0 & 0 \\ 2 & 2 & 0 \\ 3 & \frac{5}{2} & \frac{5}{4} \end{bmatrix}, \quad \mathbf{U} = \begin{bmatrix} 1 & \frac{1}{2} & \frac{3}{2} \\ 0 & 1 & \frac{1}{2} \\ 0 & 0 & 1 \end{bmatrix} \tag{A.44}$$

It can be verified that $LU = A$.

L′DU Factorization Suppose L is split up as $L'D$ where

D = diagonal matrix whose entries are diagonal elements of L

L' = lower triangular matrix with diagonal elements = 1

L' is obtained from L by dividing the ith column of L by the diagonal element of that column

For the matrix

$$\mathbf{A} = \begin{bmatrix} a_{11} & a_{12} & a_{13} \\ a_{21} & a_{22} & a_{23} \\ a_{31} & a_{32} & a_{33} \end{bmatrix} \tag{A.45}$$

The $L'DU$ decomposition is

$$\mathbf{L} = \begin{bmatrix} 1 & 0 & 0 \\ \frac{a_{21}}{a_{11}} & 1 & 0 \\ \frac{a_{31}}{a_{11}} & \frac{a_{32}^{(1)}}{a_{22}^{(1)}} & 1 \end{bmatrix} \begin{bmatrix} a_{11} & 0 & 0 \\ 0 & a_{22}^{(1)} & 0 \\ 0 & 0 & a_{33}^{(2)} \end{bmatrix} \begin{bmatrix} 1 & a_{12}^{(1)} & a_{13}^{(1)} \\ 0 & 1 & a_{23}^{(2)} \\ 0 & 0 & 1 \end{bmatrix} \tag{A.46}$$

$$= \begin{bmatrix} 1 & 0 & 0 \\ a_{21}^{(1)} & 1 & 0 \\ a_{31}^{(1)} & a_{32}^{(2)} & 1 \end{bmatrix} \begin{bmatrix} a_{11} & 0 & 0 \\ 0 & a_{22}^{(1)} & 0 \\ 0 & 0 & a_{33}^{(2)} \end{bmatrix} \begin{bmatrix} 1 & a_{12}^{(1)} & a_{13}^{(1)} \\ 0 & 1 & a_{23}^{(2)} \\ 0 & 0 & 1 \end{bmatrix} \tag{A.47}$$

Thus, for the A matrix of Example 6.3,

$$\mathbf{L} = \begin{bmatrix} 2 & 0 & 0 \\ 2 & 2 & 0 \\ 3 & \frac{5}{2} & \frac{5}{4} \end{bmatrix} = \begin{bmatrix} 1 & 0 & 0 \\ 1 & 1 & 0 \\ \frac{3}{2} & \frac{5}{4} & 1 \end{bmatrix} \begin{bmatrix} 2 & 0 & 0 \\ 0 & 2 & 0 \\ 0 & 0 & \frac{5}{4} \end{bmatrix}$$

$$= \boldsymbol{L'D} \tag{A.48}$$

The advantage of this factorization is that if A is symmetric, than both L' and U are transposes of each other. Thus, in the computer, we need store only L' or U. Recall that in the fast decoupled power flow method both B' and B'' are both symmetric if there are no phase shifters. With this factorization, we have

$$\boldsymbol{Ax = b} \tag{A.49}$$

$$\boldsymbol{L'DUx = b} \tag{A.50}$$

which is equivalent to

$$Ux = y \tag{A.51}$$

$$Dy = y' \tag{A.52}$$

$$L'y' = b \tag{A.53}$$

We first solve for y', then y, and finally x.

A.5.2. Partitioning of Matrices

It is sometimes convenient from the point of view of computation, to divide the matrix into a number of lower order matrices called *submatrices*, which can be easily manipulated. It is also helpful in some instances to do such a division to separate one group of elements from another. Thus,

$$A = \begin{bmatrix} A_1 & A_2 \\ A_3 & A_4 \end{bmatrix}$$

Addition and subtraction of these matrices are similar to those for ordinary matrices. The submatrices must be conformable, i.e. they must be of appropriate order.

$$\begin{bmatrix} A_1 & \vdots & A_2 \\ \cdots & \vdots & \cdots \\ A_3 & \vdots & A_4 \end{bmatrix} \pm \begin{bmatrix} B_1 & \vdots & B_2 \\ \cdots & \vdots & \cdots \\ B_3 & \vdots & B_4 \end{bmatrix} = \begin{bmatrix} A_1 \pm B_1 & \vdots & A_2 \pm B_2 \\ \cdots\cdots\cdots & \cdots & \cdots\cdots\cdots \\ A_3 \pm B_3 & \vdots & A_4 \pm B_4 \end{bmatrix}$$

Multiplication

$$\begin{bmatrix} A_1 & \vdots & A_2 \\ \cdots & \vdots & \cdots \\ A_3 & \vdots & A_4 \end{bmatrix} \begin{bmatrix} B_1 & \vdots & B_2 \\ \cdots & \vdots & \cdots \\ B_3 & \vdots & B_4 \end{bmatrix} = \begin{bmatrix} A_1B_1 + A_2B_3 & \vdots & A_1B_2 + A_2B_4 \\ \cdots\cdots\cdots & \vdots & \cdots\cdots\cdots \\ A_3B_1 + A_4B_3 & \vdots & A_3B_2 + A_4B_4 \end{bmatrix}$$

It is evident that in addition to A and B being conformable (i.e. the number of columns of A = the number of rows of B), the number of columns of A to the left of the dotted vertical line must be equal to the number of rows of B above the horizontal dotted line.

Transpose The transpose of a partitioned matrix

$$A = \begin{bmatrix} A_1 & \vdots & A_2 \\ \cdots & \vdots & \cdots \\ A_3 & \vdots & A_4 \end{bmatrix} \text{ is given by } A^{\mathrm{T}} = \begin{bmatrix} A_1^{\mathrm{T}} & \vdots & A_3^{\mathrm{T}} \\ \cdots & \vdots & \cdots \\ A_2^{\mathrm{T}} & \vdots & A_4^{\mathrm{T}} \end{bmatrix}$$

Inverse of a Partitioned Matrix This is useful in the computation of the inverse of a matrix A in terms of inverses of matrices of lower order. Let A be of order $n \times n$. Partition A into four submatrices A_{11}, A_{12}, A_{21}, and A_{22} of dimension $r \times r$, $r \times s$, $s \times r$, and $s \times s$, respectively, such that $r + s = n$.

$$A = \begin{bmatrix} A_{11} & \vdots & A_{12} \\ \cdots & \vdots & \cdots \\ A_{21} & \vdots & A_{22} \end{bmatrix}$$

Inverse of A may be similarly partitioned

$$A^{-1} = \begin{bmatrix} D_{11} & D_{12} \\ D_{21} & D_{22} \end{bmatrix}$$

We wish to find D_{ij} in terms of A_{ij}. Since $AA^{-1} = U$, we have

$$A_{11}D_{11} + A_{12}D_{21} = U$$

$$A_{11}D_{12} + A_{12}D_{22} = 0$$

$$A_{21}D_{11} + A_{22}D_{21} = 0$$

$$A_{21}D_{12} + A_{22}D_{22} = U$$

These are solved for D_{11}, D_{12}, D_{21}, and D_{22}. The best formulation involves a minimum number of matrix inversions. The minimum is 2 and the solution yields the following:

1. Compute A_{11}^{-1}
2. Compute $D_{22} = \left(A_{22} - A_{21}A_{11}^{-1}A_{12}\right)^{-1}$

$$D_{21} = -D_{22}A_{21}A_{11}^{-1}$$

$$D_{12} = -A_{11}^{-1}A_{12}D_{22}$$

$$D_{11} = -A_{11}^{-1} + A_{11}^{-1}A_{12}D_{22}A_{21}A_{11}^{-1}$$

Partitioning does not save time but is useful in inverting matrices that are too large to store in the main memory of the computer.

Appendix B NUMERICAL SOLUTIONS OF ORDINARY DIFFERENTIAL EQUATIONS

B.1 INTRODUCTION

The transient behaviour of a power system under faulted conditions is described mathematically by a set of ordinary differential equations (d.e.) for each machine. The order of these equations for each machine ranges from two, for the simplest classical model of the synchronous machine, to as many as ten or more when governor, excitation effects as well as detailed modelling of the synchronous machine are taken into account. We shall now discuss the solution of such equations from a numerical point of view. The algorithms are first explained for a first-order d.e. Extension to higher order d.e.'s is straightforward when they are cast into a set of first-order d.e.'s.

B.2 EULER'S METHOD

Although this method has limitations from the point of view of accuracy, it is simple to understand and amenable to a geometric interpretation. Consider the equation

$$\frac{dy}{dx} = f(x, y),\ y(x_0) = y_0 \tag{B.1}$$

where x and y are the independent and dependent variables, respectively. If Eq. (B.1) can be integrated exactly, then the solution is of the form

$$y = \varphi(x) \tag{B.2}$$

Equation (B.2) will be a smooth curve in the x-y plane. However, we shall try to approximate this curve by a series of line segments as follows. Divide the x-axis into intervals of length h such that the discrete points are $x_0, x_1 = x_0 + h,\quad x_2 = x_0 + 2h, \ldots,$ and so on. In the interval from x_0 to $x_0 + h$, $\frac{dy}{dx}$ is assumed constant at the value evaluated at x_0. Hence,

$$y_1 = y(x_0 + h) = y(x_0) + h\left.\frac{dy}{dx}\right|_0 = y_0 + hf(x_0, y_0) \tag{B.3}$$

Similarly, over the interval $x_0 + h$ to $x_0 + 2h$, $\frac{dy}{dx}$ is obtained by evaluating it at (x_1, y_1) and thus

$$y_2 = y(x_0 + 2h) = y_1 + h\left.\frac{dy}{dx}\right|_1 = y_1 + hf(x_1, y_1)$$

This process is repeated so that

$$y_{i+1} = y_i + hf(x_i, y_i) \tag{B.4}$$

A geometric interpretation of the method is given in Fig. B.1.

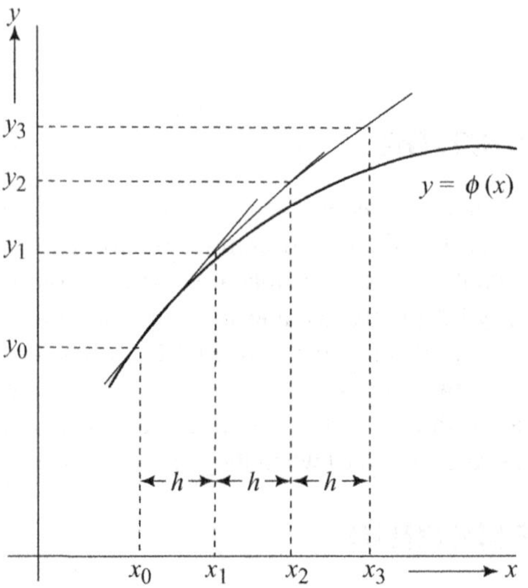

Fig. B.1 *Graph of approximate solution by Euler's method*

Modified Euler's Method Euler's method has the following drawbacks:

1. If the step size h is large, the method can lead to erroneous results.
2. The error increases as we take increasing values of x.

These drawbacks can be partially overcome by the use of modified Euler's method. The steps in this method are the predictor and corrector steps.

Step 1 (Predictor) Obtain an approximate value for y_1 denoted by $y_1^{(1)}$ as before

$$y_1^{(1)} = y_0 + h\left.\frac{dy}{dx}\right|_{x=x_0}$$

$$= y_0 + hf(x_0, y_0) \tag{B.5}$$

Step 2 (Corrector) With the value of $y_1^{(1)}$, obtain an approximate value of

$$\frac{dy}{dx} \text{ at } x_1 = x_0 + h$$

i.e.,
$$\left.\frac{dy}{dx}\right|_{x=x_1}^{(1)} = f\left(x_1, y_1^{(1)}\right) \tag{B.6}$$

Take the arithmetic mean of the values of $\frac{dy}{dx}$ at $x = x_0$ and at $x = x_1$. With this average value of $\frac{dy}{dx}$, obtain an improved estimate for y_1 as

$$y_1^{(2)} = y_0 + \frac{h}{2}\left[\left.\frac{dy}{dx}\right|_{x=x_0} + \left.\frac{dy}{dx}\right|_{x=x_1}^{(1)}\right]$$

$$= y_0 + \frac{h}{2}\left[f(x_0, y_0) + f\left(x_1, y_1^{(1)}\right)\right] \tag{B.7}$$

The above two steps can be repeated to obtain a better estimate as follows.

Step 3 Substitute $y_1^{(2)}$ to obtain the next approximation for

$$\frac{dy}{dx} \text{ at } x = x_1,$$

i.e.,
$$\left.\frac{dy}{dx}\right|_{x=x_1}^{(2)} = f\left(x_1, y_1^{(2)}\right) \tag{B.8}$$

Step 4 An even better approximation for y_1 is obtained as

$$y_1^{(3)} = y_0 + \frac{h}{2}\left[\left.\frac{dy}{dx}\right|_{x=x_0} + \left.\frac{dy}{dx}\right|_{x=x_1}^{(2)}\right]$$

$$= y_0 + \frac{h}{2}\left[f(x_0, y_0) + f\left(x_1, y_1^{(2)}\right)\right] \tag{B.9}$$

Step 5 Repeat steps 3 and 4 until two successive estimates of y_1 are within specified tolerance.

Step 6 Steps 1 through 5 are repeated for the next time interval, and so on.

B.3 RUNGE–KUTTA METHOD

Like Euler's method, the Runge–Kutta (R–K) method is self-starting but generally more accurate than the Euler's method. Consider the d.e.

$$\frac{dy}{dx} = f(x, y),\ y(x_0) = y_0$$

From the Taylor series expansion of $y(x)$ around x_0,

$$y(x_0 + h) = y(x_0) + hy'(x_0) + \frac{h^2}{2!}y''(x_0) + \cdots \tag{B.10}$$

The method consists in approximating the higher-order derivatives on the right-hand sides. We develop the method as follows:

First define

$$y' = \frac{dy}{dx} = f(x, y)$$

$$y'' = \frac{d}{dx}[f(x, y)] = \frac{\partial f}{\partial x} + \frac{\partial f}{\partial y}\frac{dy}{dx} = f_x + ff_y \tag{B.11}$$

$$y''' = f_{xx} + f_{xy}y' + f\left(f_{yx} + f_{yy}y'\right) + f_y\left(f_x + f_y f\right)$$

$$= f_{xx} + 2ff_{xy} + f^2 f_{yy} + f_y\left(f_x + f_y f\right) \tag{B.12}$$

assuming $f_{xy} = f_{yx}$.

Similarly, $$y'''' = f_{xxx} + f_{xxy}f + 2\left(f_x + ff_y\right)f_{xy} + 2f\left(f_{xyx} + f_{xyy}f\right) + \cdots \tag{B.13}$$

Expansion of $y(x_0 + h)$ in a Taylor series around $x = x_0$ gives

$$y(x_0 + h) = y(x_0) + hy'(x_0) + \frac{h^2}{2!}y''(x_0) + \frac{h^3}{3!}y'''(x_0) + \cdots$$

$$= y(x_0) + hf_0 + \frac{h^2}{2!}\left(f_x + ff_y\right)\Big|_0$$

$$+ \frac{h^3}{3!}\left[f_{xx} + 2ff_{xy} + f^2 f_{yy} + f_y\left(f_x + ff_y\right)\right]\Big|_0 + \cdots \tag{B.14}$$

Second-Order R–K Method Define

$$k_1 = hf(x_0, y_0) \tag{B.15}$$

$$k_2 = hf(x_0 + ph, y_0 + qk_1) \tag{B.16}$$

Let $$y(x_0 + h) = y(x_0) + ak_1 + bk_2 \tag{B.17}$$

where p, q, a, and b are constants. These constants are selected in such a manner that Eqs. (B.14) and (B.17) agree up to and including the terms in h^2. From Eq. (B.16),

$$k_2 = h\left[f(x_0, y_0) + \left(ph\frac{\partial f}{\partial x} + qk_1\frac{\partial f}{\partial y}\right)\right]\Bigg|_0 + \cdots$$

$$= hf_0 + ph^2 f_x\big|_0 + h^2 qf_0 f_y\big|_0$$

$$+ \text{terms of order greater than } h^2. \tag{B.18}$$

Substituting Eqs. (B.15) and (B.18) in Eq. (B.17) and combining terms

$$y(x_0 + h) = y(x_0) + (ah + bh)f_0 + h^2\left(bpf_x\big|_0 + bqf_0 f_y\big|_0\right) + \cdots \tag{B.19}$$

Comparing the right-hand sides of Eqs. (B.14) and (B.19), we get

$$\left.\begin{aligned} a+b&=1\\ bp&=\tfrac{1}{2}\\ bq&=\tfrac{1}{2}\end{aligned}\right\} \Rightarrow \begin{aligned} p&=q\\ a+b&=1\\ bp&=\tfrac{1}{2}\end{aligned} \tag{B.20}$$

Equation (B.20) consists of three equations in four unknowns. One of the unknowns is fixed arbitrarily which then determines the rest uniquely. Selecting an arbitrary value for $a = \frac{1}{2}$, then $b = \frac{1}{2}$, $p = 1$, and $q = 1$. We get the algorithm as

$$y_1 = y_0 + \frac{1}{2}(k_1 + k_2) \tag{B.21}$$

where,
$$k_1 = hf(x_0, y_0)$$

$$k_2 = hf(x_0 + h, y_0 + k_1) \tag{B.22}$$

Generalizing, we have

$$y_{i+1} = y_i + \frac{h}{2}[f(x_i, y_i) + f(x_{i+1}, \bar{y}_{i+1})] \tag{B.23}$$

where,
$$\bar{y}_{i+1} = y_i + hf(x_i, y_i) \tag{B.24}$$

The application of the R–K method requires computation of k_1 and k_2, and the error involved is of the order of h^3, since terms after h^2 were ignored.

Fourth-Order R–K Method The general fourth-order R–K method consists in assuming

$$y = y_0 + ak_1 + bk_2 + ck_3 + dk_4 \tag{B.25}$$

where,
$$k_1 = hf(x_0, y_0)$$

$$k_2 = hf(x_0 + ph, y_0 + qk_1)$$

$$k_3 = hf(x_0 + rh, y_0 + sk_2)$$

$$k_4 = hf(x_0 + th, y_0 + uk_2)$$

Following the same procedure as for the second-order R–K method and neglecting terms of order greater than h^4 in the approximation, several fourth-order algorithms are possible depending on the choice of the arbitrary constant. The widely used fourth-order R–K approximation is

$$y_{i+1} = y_i + \frac{h}{6}(k_1 + 2k_2 + 2k_3 + k_4) \tag{B.26}$$

where
$$k_1 = hf(x_i, y_i)$$

$$k_2 = hf\left(x_i + \frac{h}{2}, y_i + \frac{k_1}{2}\right)$$

$$k_3 = hf\left(x_i + \frac{h}{2}, y_i + \frac{k_2}{2}\right)$$

$$k_4 = hf(x_i + h, y_i + k_3)$$

The error involved in this approximation is of the order of h^5.

B.4 HIGHER-ORDER DIFFERENTIAL EQUATIONS

The methods discussed so far are easily extended to higher-order differential equations by writing them as a set of first-order d.e.'s. Thus, consider an nth order d.e. of the form which can be expressed as

$$\frac{d^n y}{dx^n} = F\left(y, \frac{dy}{dx}, \cdots, \frac{d^{n-1} y}{dx^{n-1}}\right) \tag{B.27}$$

Then define a set of new variables as

$$\begin{aligned} y_1 &= y \\ y_2 &= y' \\ &\vdots \\ y_n &= y^{(n-1)} \end{aligned} \tag{B.28}$$

Then differentiating Eq. (B.28) and making use of Eq. (B.27), we have

$$\begin{aligned} \frac{dy_1}{dx} &= y_2 \\ \frac{dy_2}{dx} &= y_3 \\ &\vdots \\ \frac{dy_n}{dx} &= F(x, y_1, y_2, \cdots, y_n) \end{aligned} \tag{B.29}$$

By choosing a set of variables other than those in Eq. (B.28), we can get a representation different than from Eq. (B.29). In general, therefore, we can reduce Eq. (B.27) to a set of simultaneous first-order d.e.'s of the form,

$$\begin{aligned} \frac{dy_1}{dx} &= f_1(x, y_1, \cdots, y_n) \\ \frac{dy_2}{dx} &= f_2(x, y_1, \cdots, y_n) \\ &\vdots \\ \frac{dy_n}{dx} &= f_n(x, y_1, \cdots, y_n) \end{aligned} \tag{B.30}$$

Initial conditions are specified as

$$y_1(x_0) = y_{1,0}, y_2(x_0) = y_{2,0}, \cdots, y_n(x_0) = y_{n,0}$$

The numerical solution of a set of equations of the form (B.30) is, at least in principle, no more difficult than the solution of a first-order d.e. The algorithm selected is applied to each of the n equations in parallel at each step.

B.5 CONCLUSION

The three basic methods of numerical integration for the continuous time domain, namely Euler, modified Euler, and the Runge–Kutta method, have been discussed in this appendix. The method for numerical integration in the discrete domain is also discussed. Generally, for power-system problems, the modified Euler method or the fourth-order R–K method gives results within acceptable accuracy. Another method, known as the simultaneous implicit method with trapezoidal integration, is widely used in industry. It is particularly useful when the system contains elements with widely varying time constants or when we have to integrate a system of Differential Algebraic Equations (DAE). This is often the case with multi-machine systems with detailed machine models. A brief discussion of this method is given in Section 7.4.2 in the power system context.

Appendix C

TEST SYSTEM DATA AND SOFTWARE FOR POWER SYSTEM ANALYSIS

In this appendix, we present system data and load-flow results for IEEE 14-bus, 30-bus, and 57-bus test systems. The data is presented in sections C.1–C.3 in the following format.

1. Bus data and load-flow results
2. Line data
3. Transformer data
4. Shunt capacitor data
5. Regulated bus data (P-V buses)

Data for a 50 Hz, 5 MVA, doubly fed induction generator and a 5 MW variable speed wind turbine are presented in C.4.

C.1 IEEE 14-BUS TEST SYSTEM [38]

Table C.1.1 *Bus data and load-flow results*

Bus No.	*Bus Voltage*		*Generation*		*Load*	
	Magnitude Per Unit	*Phase Angle Degrees*	*Real MW*	*Reactive MVAR*	*Real MW*	*Reactive MVAR*
1	1.060	0.0	232.4	–16.9	0.0	0.0
2	1.045	–4.98	40.0	42.4	21.7	12.7
3	1.010	–12.72	0.0	23.4	94.2	19.0
4	1.019	–10.33	0.0	0.0	47.8	–3.9
5	1.020	–8.78	0.0	0.0	7.6	1.6
6	1.070	–14.22	0.0	12.2	11.2	7.5
7	1.062	–13.37	0.0	0.0	0.0	0.0
8	1.090	–13.36	0.0	17.4	0.0	0.0
9	1.056	–14.94	0.0	0.0	29.5	16.6
10	1.051	–15.10	0.0	0.0	9.0	5.8

Contd.

Table C.1.1 *Contd.*

11	1.057	–14.79	0.0	0.0	3.5	1.8
12	1.055	–15.07	0.0	0.0	6.1	1.6
13	1.050	–15.16	0.0	0.0	13.5	5.8
14	1.036	–16.04	0.0	0.0	14.9	5.0

Table C.1.2 *Line data*

Line No.	*Between Buses*	*Line Impedance*		*Half-Line Charging Susceptance Per Unit*
		R Per Unit	*X Per Unit*	
1	1-2	0.01938	0.05917	0.02640
2	2-3	0.04699	0.19797	0.02190
3	2-4	0.05811	0.17632	0.01870
4	1-5	0.05403	0.22304	0.02460
5	2-5	0.05695	0.17388	0.01700
6	3-4	0.06701	0.17103	0.01730
7	4-5	0.01335	0.04211	0.0064
8	5-6	0.0	0.25202	0.0
9	4-7	0.0	0.20912	0.0
10	7-8	0.0	0.17615	0.0
11	4-9	0.0	0.55618	0.0
12	7-9	0.0	0.11001	0.0
13	9-10	0.03181	0.08450	0.0
14	6-11	0.09498	0.19890	0.0
15	6-12	0.12291	0.25581	0.0
16	6-13	0.06615	0.13027	0.0
17	9-14	0.12711	0.27038	0.0
18	10-11	0.08205	0.19207	0.0
19	12-13	0.22092	0.19988	0.0
20	13-14	0.17093	0.34802	0.0

Table C.1.3 *Transformer data*

Transformer	*Between Buses*	*Tap Setting*
1	4-7	0.978
2	4-9	0.969
3	5-6	0.932

Table C.1.4 *Shunt capacitor data*

Bus Number	*Susceptance Per Unit*
9	0.190

Table C.1.5 *Regulated bus data (P-V buses)*

Bus Number	Voltage Magnitude Per Unit	Reactive Power Limits	
		Minimum MVAR	Maximum MVAR
2	1.045	–40.0	50.0
3	1.010	0.0	40.0
6	1.070	–6.0	24.0
8	1.090	–6.0	24.0

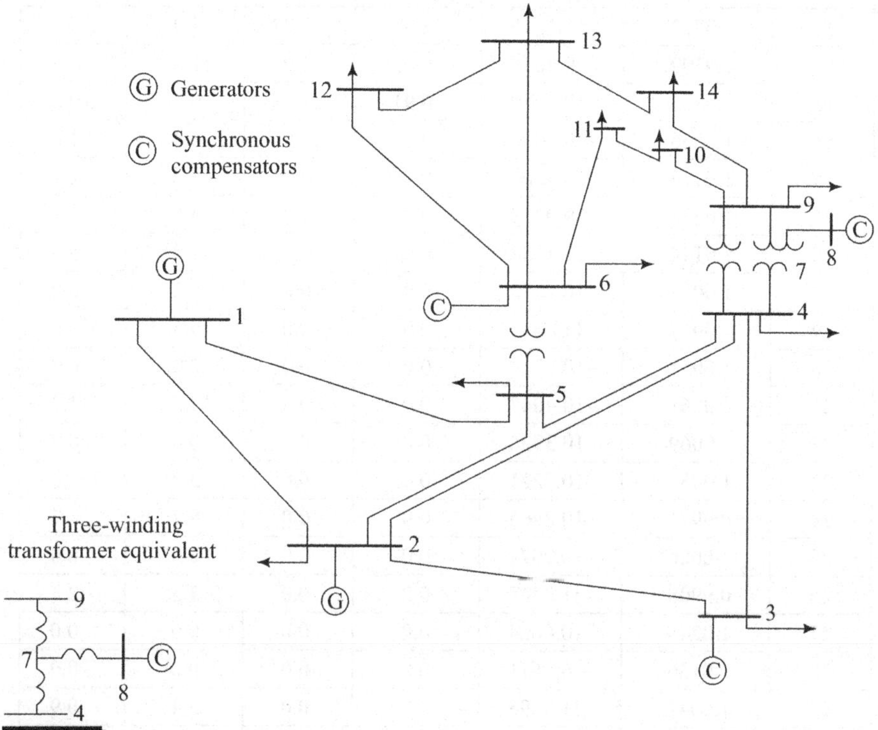

Fig. C.1 *IEEE 14-bus test system*

C.2 IEEE 30-BUS TEST DATA [38]

Table C.2.1 *Bus data and load-flow results*

Bus No.	Bus Voltage		Generation		Load	
	Magnitude Per Unit	Phase Angle Degrees	Real MW	Reactive MVAR	Real MW	Reactive MVAR
1	1.05	0.0	138.48	–2.79	0.0	0.0
2	1.0338	–2.7339	57.56	2.47	21.7	12.7
3	1.0313	–4.6815	0.0	0.0	2.4	1.2
4	1.0263	–5.6077	0.0	0.0	7.6	1.6

Contd.

Table C.2.1 *Contd.*

5	1.0058	–8.993	24.56	22.57	94.2	19.0
6	1.0208	–6.4547	0.0	0.0	0.0	0.0
7	1.0069	–8.0244	0.0	0.0	22.8	10.9
8	1.0230	–6.4733	35.0	34.84	30.0	30.0
9	1.0332	–8.03	0.0	0.0	0.0	0.0
10	1.0183	–9.9268	0.0	0.0	5.8	2.0
11	1.0913	–6.1345	17.93	30.78	0.0	0.0
12	1.0399	–9.4036	0.0	0.0	11.2	7.5
13	1.0883	–8.2049	16.91	37.83	0.0	0.0
14	1.0236	–10.3086	0.0	0.0	6.2	1.6
15	1.0179	–10.3600	0.0	0.0	8.2	2.5
16	1.0235	–9.9028	0.0	0.0	3.5	1.8
17	1.0144	–10.1356	0.0	0.0	9.0	5.8
18	1.0057	–10.9253	0.0	0.0	3.2	0.9
19	1.0017	–11.0615	0.0	0.0	9.5	3.4
20	1.0051	–10.831	0.0	0.0	2.2	0.7
21	1.0061	–10.4047	0.0	0.0	17.5	11.2
22	1.0069	–10.3936	0.0	0.0	0.0	0.0
23	1.0053	–10.7221	0.0	0.0	3.2	1.6
24	0.9971	–10.8465	0.0	0.0	8.7	6.7
25	1.0086	–10.9074	0.0	0.0	0.0	0.0
26	0.9908	–11.3345	0.0	0.0	3.5	2.3
27	1.0245	–10.6624	0.0	0.0	0.0	0.0
28	1.0156	–6.8671	0.0	0.0	0.0	0.0
29	1.0047	–11.8893	0.0	0.0	2.4	0.9
30	0.9932	–12.7699	0.0	0.0	10.6	1.9

Table C.2.2 *Line data*

Line No.	*Between Buses*	*Line Impedance*		*Half-Line Charging Susceptance Per Unit*
		R Per Unit	*X Per Unit*	
1	1–2	0.0192	0.0575	0.0264
2	1–3	0.0452	0.1852	0.0204
3	2–4	0.0570	0.1737	0.0184
4	3–4	0.0132	0.0379	0.0042
5	2–5	0.0472	0.1983	0.0209
6	2–6	0.0581	0.1763	0.0187

Contd.

Table C.2.2 *Contd.*

7	4–6	0.0119	0.0414	0.0045
8	5–7	0.0460	0.1160	0.0102
9	6–7	0.0267	0.0820	0.0085
10	6–8	0.0120	0.0420	0.0045
11	6–9	0.0	0.2080	0.0
12	6–10	0.0	0.5560	0.0
13	9–11	0.0	0.2080	0.0
14	9–10	0.0	0.1100	0.0
15	4–12	0.0	0.2560	0.0
16	12–13	0.0	0.1400	0.0
17	12–14	0.1231	0.2559	0.0
18	12–15	0.0662	0.1304	0.0
19	12–16	0.0945	0.1987	0.0
20	14–15	0.2210	0.1997	0.0
21	16–17	0.0824	0.1932	0.0
22	15–18	0.1070	0.2185	0.0
23	18–19	0.0639	0.1292	0.0
24	19–20	0.0340	0.0680	0.0
25	10–20	0.0936	0.2090	0.0
26	10–17	0.0324	0.0845	0.0
27	10–21	0.0348	0.0749	0.0
28	10–22	0.0727	0.1499	0.0
29	21–22	0.0116	0.0236	0.0
30	15–23	0.1000	0.2020	0.0
31	22–24	0.1150	0.1790	0.0
32	23–24	0.1320	0.2700	0.0
33	24–25	0.1885	0.3292	0.0
34	25–26	0.2544	0.3800	0.0
35	25–27	0.1093	0.2087	0.0
36	27–28	0.0	0.3960	0.0
37	27–29	0.2198	0.4153	0.0
38	27–30	0.3202	0.6027	0.0
39	29–30	0.2399	0.4533	0.0
40	8–28	0.0636	0.2000	0.0214
41	6–28	0.0169	0.0599	0.0065

Table C.2.3 *Transformer data*

Transformer No.	*Between Buses*	*Tap Setting*
1	6–9	1.0155
2	6–10	0.9629
3	4–12	1.0129
4	28–27	0.9581

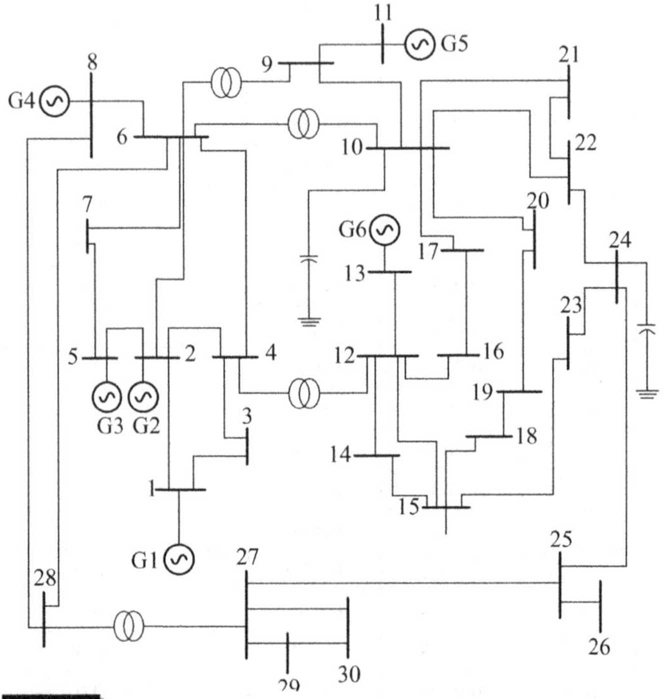

Fig. C.2 *IEEE 30-bus test system*

Table C.2.4 *Shunt capacitor data*

Bus No.	*Susceptance Per Unit*
10	0.19
24	0.04

Table C.2.5 *Regulated bus data*

Bus No.	*Voltage Magnitude Per Unit*	*Reactive Power Limits*	
		Minimum MVAR	*Maximum MVAR*
2	1.0338	–20.0	60.0
5	1.0058	–15.0	62.5
8	1.0230	–15.0	50.0
11	1.0913	–10.0	40.0
13	1.0883	–15.0	45.0

C.3 DATA FOR IEEE 57-BUS SYSTEM [38]

Table C.3.1 *Bus data and load-flow results*

Bus No.	Bus Voltage		Generation		Load	
	Magnitude Per Unit	Phase Angle Degrees	Real MW	Reactive MVAR	Real MW	Reactive MVAR
1	1.040	0.0	478.0	128.9	55.0	17.0
2	1.010	–1.18	0.0	–0.8	3.0	88.0
3	0.985	–5.97	40.0	–1.0	41.0	21.0
4	0.981	–7.32	0.0	0.0	0.0	0.0
5	0.976	–8.52	0.0	0.0	13.0	4.0
6	0.980	–8.65	0.0	0.8	75.0	2.0
7	0.984	–7.58	0.0	0.0	0.0	0.0
8	1.005	–4.45	450.0	62.1	150.0	22.0
9	0.980	–9.56	0.0	2.2	121.0	26.0
10	0.986	–11.43	0.0	0.0	5.0	2.0
11	0.974	–10.17	0.0	0.0	0.0	0.0
12	1.015	–10.46	310.0	128.5	377.0	24.0
13	0.979	–9.79	0.0	0.0	18.0	2.3
14	0.970	–9.33	0.0	0.0	10.5	5.3
15	0.988	–7.18	0.0	0.0	22.0	5.0
16	1.013	–8.85	0.0	0.0	43.0	3.0
17	1.017	–5.39	0.0	0.0	42.0	8.0
18	1.001	–11.71	0.0	0.0	27.2	9.8
19	0.970	–13.20	0.0	0.0	3.3	0.6
20	0.964	–13.41	0.0	0.0	2.3	1.0
21	1.008	–12.89	0.0	0.0	0.0	0.0
22	1.010	–12.84	0.0	0.0	0.0	0.0
23	1.008	–12.91	0.0	0.0	6.3	2.1
24	0.999	–13.25	0.0	0.0	0.0	0.0
25	0.982	–18.13	0.0	0.0	6.3	3.2
26	0.959	–12.95	0.0	0.0	0.0	0.0
27	0.982	–11.48	0.0	0.0	9.3	0.5
28	0.997	–10.45	0.0	0.0	4.6	2.3
29	1.010	–9.75	0.0	0.0	17.0	2.6
30	0.962	–18.68	0.0	0.0	3.6	1.8
31	0.936	–19.34	0.0	0.0	5.8	2.9
32	0.949	–18.46	0.0	0.0	1.6	0.8

Contd.

Table C.3.1 *Contd.*

Bus No.	Bus Voltage		Generation		Load	
	Magnitude Per Unit	Phase Angle Degrees	Real MW	Reactive MVAR	Real MW	Reactive MVAR
33	0.947	–18.50	0.0	0.0	3.8	1.9
34	0.959	–14.10	0.0	0.0	0.0	0.0
35	0.966	–13.86	0.0	0.0	6.0	3.0
36	0.976	–13.59	0.0	0.0	0.0	0.0
37	0.985	–13.41	0.0	0.0	0.0	0.0
38	1.013	–12.71	0.0	0.0	14.0	7.0
39	0.983	–13.46	0.0	0.0	0.0	0.0
40	0.973	–13.62	0.0	0.0	0.0	0.0
41	0.996	–14.05	0.0	0.0	6.3	3.0
42	0.966	–15.50	0.0	0.0	7.1	4.0
43	1.010	–11.33	0.0	0.0	2.0	1.0
44	1.017	–11.83	0.0	0.0	12.0	1.8
45	1.036	–9.25	0.0	0.0	0.0	0.0
46	1.060	–11.09	0.0	0.0	0.0	0.0
47	1.033	–12.49	0.0	0.0	29.7	11.6
48	1.027	–12.57	0.0	0.0	0.0	0.0
49	1.036	–12.92	0.0	0.0	18.0	8.5
50	1.023	–13.39	0.0	0.0	21.0	10.5
51	1.052	–12.52	0.0	0.0	18.0	5.3
52	0.980	–11.47	0.0	0.0	4.9	2.2
53	0.971	–12.23	0.0	0.0	20.0	10.0
54	0.996	–11.69	0.0	0.0	4.1	1.4
55	1.031	–10.78	0.0	0.0	6.8	3.4
56	0.968	–16.04	0.0	0.0	7.6	2.2
57	0.965	–16.56	0.0	0.0	6.7	2.0

Table C.3.2 *Line data*

Line No.	Between Buses	Line Impedance		Half-Line Charging Susceptance Per Unit
		R Per Unit	X Per Unit	
1	1-2	0.0083	0.0280	0.0645
2	2-3	0.0298	0.0850	0.0409
3	3-4	0.0112	0.0366	0.0190
4	4-5	0.0625	0.1320	0.0129
5	4-6	0.0430	0.1480	0.0174
6	6-7	0.0200	0.1020	0.0138

Contd.

Table C.3.2 *Contd.*

Line No.	*Between Buses*	*Line Impedance*		*Half-Line Charging Susceptance Per Unit*
		R Per Unit	*X Per Unit*	
7	6-8	0.0339	0.1730	0.0235
8	8-9	0.0099	0.0505	0.0274
9	9-10	0.0369	0.1679	0.0220
10	9-11	0.0258	0.0848	0.0109
11	9-12	0.0648	0.2950	0.0386
12	9-13	0.0481	0.1580	0.0203
13	13-14	0.0132	0.0434	0.0055
14	13-15	0.0269	0.0869	0.0115
15	1-15	0.0178	0.0910	0.0494
16	1-16	0.0454	0.2060	0.0273
17	1-17	0.0238	0.1080	0.0143
18	3-15	0.0162	0.053	0.0272
19	4-18	0.0	0.555	0.0
20	4-18	0.0	0.430	0.0
21	5-6	0.0302	0.0641	0.0062
22	7-8	0.0139	0.0712	0.0097
23	10-12	0.0277	0.1262	0.0164
24	11-13	0.0223	0.0732	0.0094
25	12-13	0.0178	0.0580	0.0302
26	12-16	0.0180	0.0813	0.0108
27	12-17	0.0397	0.1790	0.0238
28	14-15	0.0171	0.0547	0.0074
29	18-19	0.4610	0.6850	0.0
30	19-20	0.2830	0.4340	0.0
31	20-21	0.0	0.7767	0.0
32	21-22	0.0736	0.1170	0.0
33	22-23	0.0099	0.0152	0.0
34	23-24	0.1660	0.2560	0.0042
35	24-25	0.0	1.182	0.0
36	24-25	0.0	1.230	0.0
37	24-26	0.0	0.0473	0.0
38	26-27	0.165	0.2540	0.0

Contd.

Table C.3.2 *Contd.*

Line No.	Between Buses	Line Impedance		Half-Line Charging Susceptance Per Unit
		R Per Unit	X Per Unit	
39	27-28	0.0618	0.0954	0.0
40	28-29	0.0418	0.0587	0.0
41	7-29	0.0	0.0648	0.0
42	25-30	0.135	0.202	0.0
43	30-31	0.326	0.497	0.0
44	31-32	0.507	0.755	0.0
45	32-33	0.0392	0.0360	0.0
46	32-34	0.0	0.953	0.0
47	34-35	0.052	0.0780	0.0016
48	35-36	0.043	0.0537	0.0008
49	36-37	0.0290	0.0366	0.0
50	37-38	0.0651	0.1009	0.0010
51	37-39	0.0239	0.0379	0.0
52	36-40	0.0300	0.0466	0.0
53	22-38	0.0192	0.0295	0.0
54	11-41	0.0	0.7490	0.0
55	41-42	0.207	0.3520	0.0
56	41-43	0.0	0.412	0.0
57	38-44	0.0289	0.0585	0.0010
58	15-45	0.0	0.1042	0.0
59	14-46	0.0	0.0735	0.0
60	46-47	0.023	0.068	0.0016
61	47-48	0.0182	0.0233	0.0
62	48-49	0.0834	0.1290	0.0024
63	49-50	0.0801	0.1280	0.0
64	50-51	0.1386	0.2200	0.0
65	10-51	0.0	0.0712	0.0
66	13-49	0.0	0.1910	0.0
67	29-52	0.1442	0.1870	0.0
68	52-53	0.0762	0.0984	0.0
69	53-54	0.1878	0.2320	0.0
70	54-55	0.1732	0.2265	0.0

Contd.

Table C.3.2 *Contd.*

Line No.	*Between Buses*	*Line Impedance*		*Half-Line Charging Susceptance Per Unit*
		R Per Unit	*X Per Unit*	
71	11-43	0.0	0.1530	0.0
72	44-45	0.0624	0.1242	0.0020
73	40-56	0.0	1.1950	0.0
74	56-41	0.5530	0.5490	0.0
75	56-42	0.2125	0.3540	0.0
76	39-57	0.0	1.3550	0.0
77	57-56	0.1740	0.2600	0.0
78	38-49	0.1150	0.1770	0.0030
79	38-48	0.0312	0.0482	0.0
80	9-55	0.0	0.1205	0.0

Table C.3.3 *Transformer data*

Transformer No.	*Between Buses*	*Tap Setting*
1	4-18	0.97
2	4-18	0.978
3	7-29	0.967
4	9-55	0.94
5	10-51	0.93
6	11-41	0.955
7	11-43	0.958
8	13-49	0.895
9	14-46	0.9
10	15-45	0.955
11	21-20	1.043
12	24-25	1.000
13	24-25	1.000
14	24-26	1.043
15	34-32	0.975
16	39-57	0.98
17	40-56	0.958

Table C.3.4 *Shunt capacitor data*

Bus Number	*Susceptance Per Unit*
18	0.100
25	0.059
53	0.063

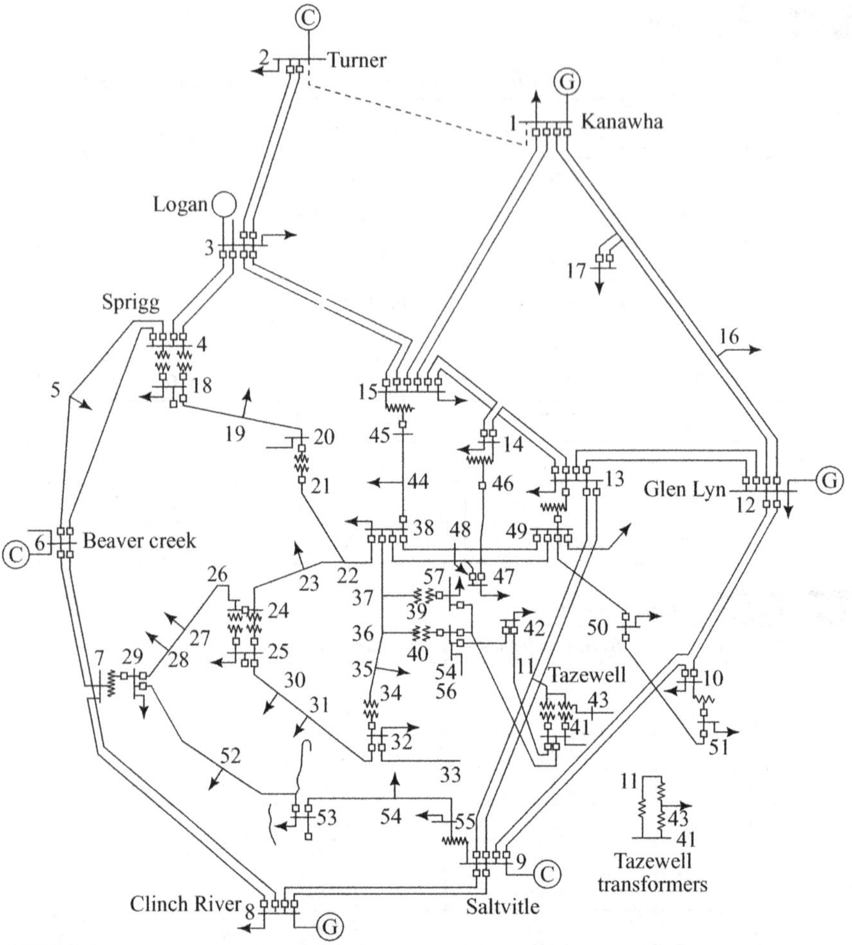

Fig. C.3 *IEEE 57-bus test system*

Table C.3.5 *Regulated bus data*

Bus No.	*Voltage Magnitude Per Unit*	*Reactive Power Limits*	
		Minimum MVAR	*Maximum MVAR*
2	1.01	–17.0	50.0
3	0.985	–10.0	60.0
6	0.98	–8.0	25.0
8	1.005	–140.0	200.0
9	0.98	–3.0	9.0
12	1.015	–50.0	155.0

C.4 DFIG AND WIND TURBINE DATA [31, 32]

5 MW Wind Turbine-Generator Data

Base frequency, $f_B = 50$ [Hz]; Electrical base speed, $\omega_{eB} = 2\pi f_B$ [rad/s]; Base power, $S_B = 5$ MVA

DFIG parameters in p.u. (on machine base):

Mutual inductance, $l_m = 4$ p.u.; Stator inductance, $l_{ss} = 1.01\ l_m$; Rotor inductance, $l_{rr} = 1.005\ l_{ss}$
Stator resistance, $r_s = 0.005$ p.u.; rotor resistance, $r_t = 1.1\ r_s$
Turbine inertia, $H_t = 4$ sec; Generator inertia, $H_g = 0.1\ H_t$
Drive train shaft stiffness, $k_{sh} = 0.3$ p.u./el. rad; Damping coefficient, $C_{sh} = 0.01$ p.u.s/el.rad
Turbine rated power (P_{rated}) = 5 MW
Rated wind speed ($V_{w\ rated}$) = 15 m/s; Blade length (R) = 40.05 m
Generator rated speed ($\omega_{r\ rated}$) = 1.2 p.u. Generator rated speed (ω_{trated}) in rad/s = $V_{w\ rated} \times \lambda_{opt} / R$
Maximum performance co-efficient ($C_{p\ max}$) = 0.48
Tip speed ratio, $\lambda = (\omega_t\ [rad{=}s] \times R[m]) / (V_w\ [m/s])$; Optimum tip speed ratio (λ_{opt}) = 8.1

3.6 MW Wind Turbine-Generator Data [35]

Base frequency, $f_B = 60$ [Hz]; Electrical base speed, $\omega_{eB} = 2\pi f_B$ [rad/s]; Base Power, $S_B = 3.6$ MVA

DFIG parameters in p.u. (on machine base):

Mutual inductance, $l_m = 4.4$ p.u.; Stator inductance, $l_{ss} = 0.07937 + l_m$; Rotor inductance, $l_{rr} = 0.4 + l_m$;
Stator resistance, $r_s = 0.0079$ p.u.; rotor resistance, $r_t = 0.025$;
Turbine inertia, $H_t = 4.29$ secands; Generator inertia, $H_g = 0.9$;
Damping coefficient of the flexible coupling (shaft) between the two masses, $D_{tg} = 1.5$ p.u;
Shaft stiffness, $K_{tg} = 296.7$ p.u.;
Wind turbine: Rated capacity = 3.6 MW; Number of blades = 3; Rotor diameter = 104 m;
Swept area = 8495 m^2; Rotor speed (variable) = 8.5–15.3 rpm; Cut-in wind speed = 3.5 m/s; Cut-out wind speed = 27 m/s.

C.5 FREE SOFTWARE FOR POWER SYSTEM ANALYSIS AND LINKS FOR DOWNLOAD

MATPOWER: http://www.pserc.cornell.edu//matpower/index.html
POWERWORLD: http://www.powerworld.com/gloversarmaoverbye
Power System Toolbox: http://www.ecse.rpi.edu/homepages/chowj

D Appendix

SUPPLEMENTARY PROBLEMS

In this appendix, additional problems for each chapter are given. Some of them are qualitative in nature to test understanding of the concepts discussed in the chapters.

D.1 CHAPTER 2: POWER SYSTEM COMPONENTS

1. A 300 km transmission line, three-phase, 50 Hz, has the following parameters: $r = 0.15$, ohms/km/phase, $x = 0.5$ ohms/km/phase and $y = 3.50 \times 10^{-6}$ ohms/km/phase. Find
   (i) Characteristic impedance, z_c.
   (ii) Propagation constant.
   (iii) Parameters of the nominal π circuit.
2. Using the results of Problem l, find the sending end voltage and current if the receiving end is open circuited and has a voltage of 100 kV line to line.
3. A 50 Hz three-phase transmission line is 250 km long. It has a total series impedance of $35 + j40$ ohms per phase and a shunt admittance of 1000×10^{-6} ohms. If the power delivered is 50 MW at 220 kV at 0.9 p.f. lagging, compute the sending end voltage using
   (i) Short line approximation.
   (ii) Nominal π circuit.
4. An off-nominal turns ratio transformer with a reactance of 0.08 p.u. referred to the tap side and $t = 1.05$ is connected in parallel with a transformer of reactance of 0.01 p.u. If the unity side of the transformer has a voltage of $1\angle 0°$, find the circulating current.
5. A per-phase equivalent of a 115/23/23 kV three-phase three-winding transformer has the following data from short-circuit tests: $Z_{ps} = 0.07$, $Z_{pt} = 0.09$ and $Z_{st} = 0.08$ all in p.u. and referred to 115 kV side.
   (i) Find the equivalent circuit.
   (ii) If a load of 10 MW at 0.8 p.f. lagging is connected to the 23 kV secondary side at rated voltage and the tertiary winding has an impedance of $j0.1$ p.u. connected to the ground, find the current and terminal voltage on the primary side. Take 10 as base MVA.

D.2 CHAPTER 3: TOPOLOGICAL ANALYSIS OF POWER NETWORKS

1. Given a connected graph, define
   (i) Tree
   (ii) Co-tree
   (iii) Fundamental loop
   (iv) Fundmental cutset
2. From the bus incidence matrix given below construct the oriented graph.

$$\begin{array}{c} \\ (A) \\ (B) \\ (C) \\ (D) \end{array} \begin{array}{c} \begin{array}{cccccccc} 1 & 2 & 3 & 4 & 5 & 6 & 7 & 8 \end{array} \\ \begin{bmatrix} -1 & 0 & 0 & 0 & 1 & 0 & 1 & 0 \\ 0 & 0 & 0 & 0 & -1 & 1 & 0 & 1 \\ 0 & 0 & -1 & 1 & 0 & -1 & 0 & 0 \\ 0 & -1 & 0 & -1 & 0 & 0 & -1 & -1 \end{bmatrix} \end{array}$$

3. For the graph in Fig. D.1 and the tree $T(1, 5, 6, 7)$, construct
   (i) Fundamental loop matrix
   (ii) Fundamental cutset matrix
   (iii) Verify the relation $\boldsymbol{BC}^T = \boldsymbol{0}$
4. For the graph in Fig. D.2, write
   (i) The incidence matrix with 4 as the datum node.
   (ii) For the tree $T(1, 2, 3)$ construct the $\boldsymbol{B}$ and $\boldsymbol{C}$ matrices using the incidence matrix $\boldsymbol{A}$.

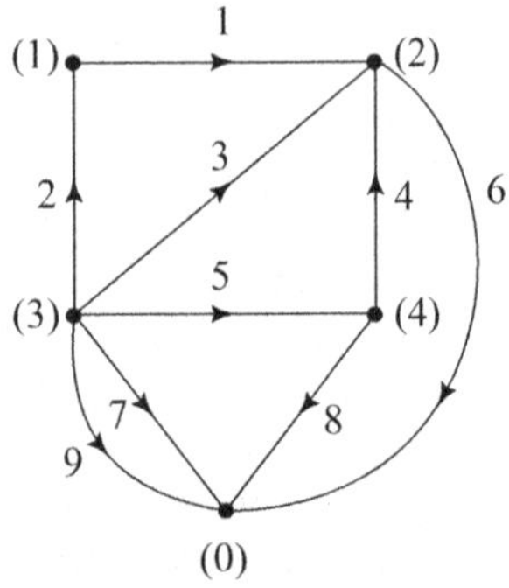

Fig. D.1 Connected graph

5. For the network in Fig. D.3, compute the $\boldsymbol{Y}_{\text{bus}}$ and $\boldsymbol{Z}_{\text{bus}}$ using topological concepts. Take **4** as a reference bus. The impedances are: line 1 = line 2 = line 3 = $j0.2$ p.u., line 4 = line 5 = $j0.1$ p.u.
6. If a line 6 in parallel to line 2 exists between buses **2** and **3**, with a self-impedance of $j0.25$ p.u. and mutual impedance of $j0.01$ p.u., repeat the computation of $\boldsymbol{Y}_{\text{bus}}$ and $\boldsymbol{Z}_{\text{bus}}$.

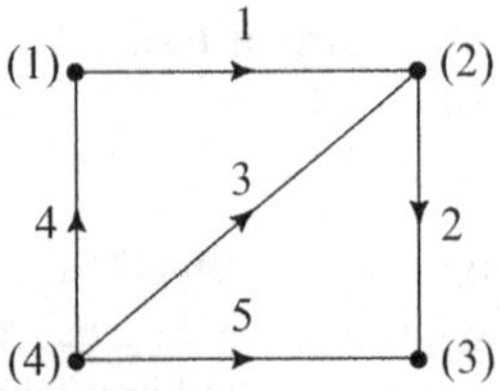

Fig. D.2 Connected graph

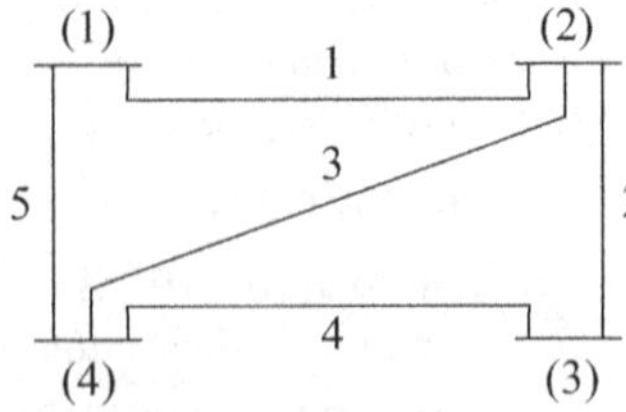

Fig. D.3 Four-bus system

7. In Fig. D. 1, the elements 1–5 have admittances of $-j10$ p.u. and the elements 6–9 have admittances $j0.25$ p.u. Construct $\boldsymbol{Y}_{\text{bus}}$ by inspection. **0** is the ground bus.
8. With a neat sketch show the following.
   (i) An oriented graph having 5 nodes and 8 elements.
   (ii) A tree for the graph.
   (iii) The co-tree.
   (iv) The corresponding $\boldsymbol{B}$ and $\boldsymbol{C}$ matrices.
   (v) The incidence matrix $\boldsymbol{A}$.

D.3 CHAPTER 4: BUS IMPEDANCE ALGORITHM

1. For the network in Fig. D.3 construct $\boldsymbol{Z}_{\text{bus}}$ using the building algorithm.
2. Modify $\boldsymbol{Z}_{\text{bus}}$ of Problem 1 (this section) if an element 6 having self-impedance of $j0.25$ p.u. is added in parallel with line 1.
3. Suppose line 3 is removed form Fig. D.3, what is the resulting $\boldsymbol{Z}_{\text{bus}}$.
4. Repeat Problem 2 (this section) if the added element has mutual impedance of $j0.01$ p.u. with line 1.

D.4 CHAPTER 5: SHORT-CIRCUIT STUDIES

1. Figure D.4 represents a 4 bus, 2-generator system with generators connected at buses **1** and **3** having a transient reactance of 0.2 p.u. and 0.15 p.u., respectively. Assume prefault voltages at generators as $1\angle 0°$. For a symmetrical three-phase-to-ground fault at bus **4**, find the fault current and currents in all the lines as well as currents supplied by the generators.

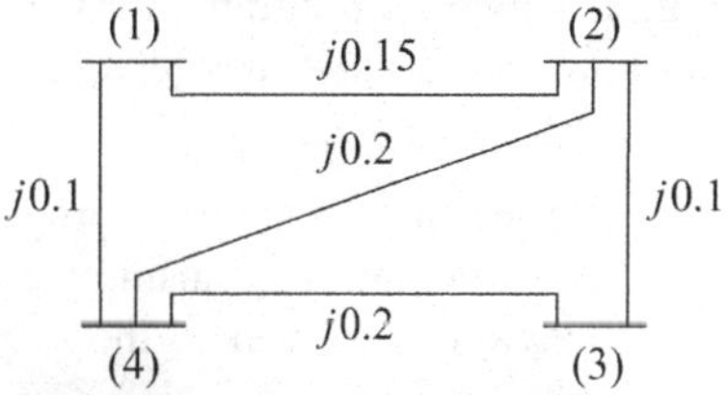

Fig. D.4 A 4-bus system

2. Repeat Problem 1 (this section) with line connecting buses **2** and **4** removed.
3. All lines in Fig. D.5 have a positive sequence impedance of $j0.2$ p.u. Generators with transient reactances 0.05 p.u. are connected at buses **1** and **2**. Assuming prefault voltage as $1\angle 0°$, for a three-phase to ground fault bus **3**, find fault current, fault voltages at buses and currents in all the lines.

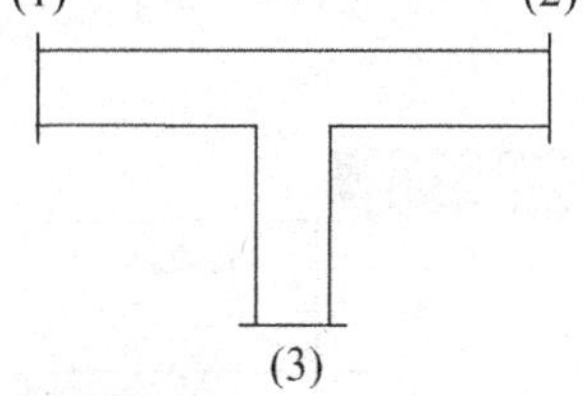

Fig. D.5 A 3-bus system

4. In Fig. D.5, each line has a negative sequence impedance of $j0.15$ p.u. and zero sequence impedance of $j0.1$ p.u. The generators have negative and zero sequence impedances as $j0.1$ p.u. and $j0.05$ p.u., respectively. For a fault at bus **3**, find the fault current.
5. Repeat Problem **4** (this section) for a line-line-ground fault at bus **3**.

6. Write a note on the steps to be followed during Z_{bus} building algorithm for
   (i) removal of a given uncoupled element; and
   (ii) changing the impedance value of an uncoupled element.
7. Form the Z_{bus} matrix for the network in Fig. D.6 by adding the elements in the sequence 1–2, 1–3, 1–4, 4–6, 2–5. The p.u. values of reactances are shown in the Fig. D.6. Select **1** as reference bus.

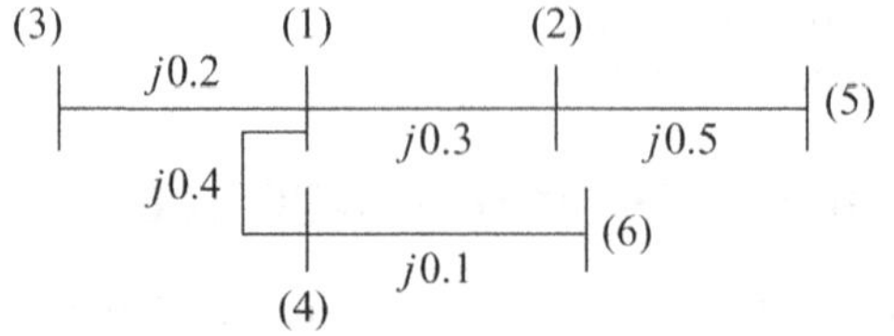

Fig. D.6 Six-bus network

8. In Problem 7 (this section), if the impedance value of line 2–5 is changed from $j0.5$ p.u. to $j0.25$ p.u. find the new Z_{bus}.

D.5 CHAPTER 6: POWER FLOW STUDIES

Students are strongly urged to write computer programs. For a good selection of sample computer programs for load flow and transient stability, see Appendix G of the book by Kothari and Nagrath, *Modern Power System Analysis* [1].

1. Write a computer program for constructing Y_{bus} using topological relations, given the primitive admittance matrix.
2. A three-bus system is shown in Fig. D.7 with the line impedances in p.u. The bus data in p.u. is shown in Table D.1.

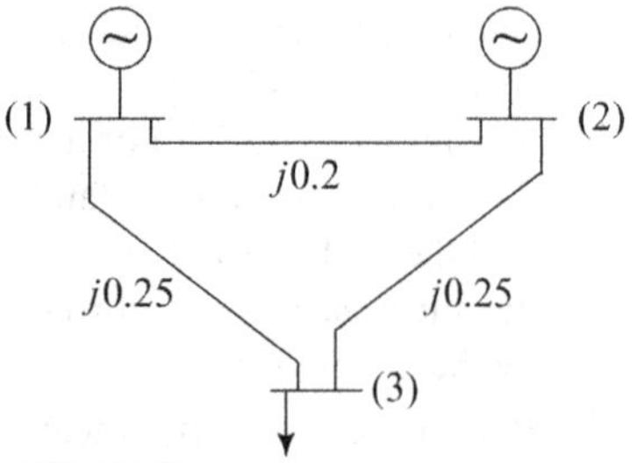

Fig. D.7 Three-bus system

Table D.1

Bus	*Voltage*	*Generation*		*Load*	
		P_g	Q_g	P_L	Q_L
1	$1.2\angle 0°$	–	–	–	–
2	–	0.7	0.5	0.3	0.2
3	–	0	0	0.6	0.4

Note that bus **2** is a P–Q bus too. Bus **1** is the slack bus. Using Y_{bus} and Gauss–Seidel method, compute at the end of first iteration,

(i) voltages at buses **2** and **3** and
(ii) complex power at the slack bus.

3. In Problem 2 (this section), assume Q_g at bus **2** is not specified but has limits of $Q_{g,\,min} = 0.3$ and $Q_{g,\,max} = 0.8$. $|V_2| = 1.1$ p.u. Do two iterations using the Gauss–Seidel method.
4. In Fig. D.8, the net injections are shown at all except the slack bus **1**. Bus **2** is a *P–V* bus.

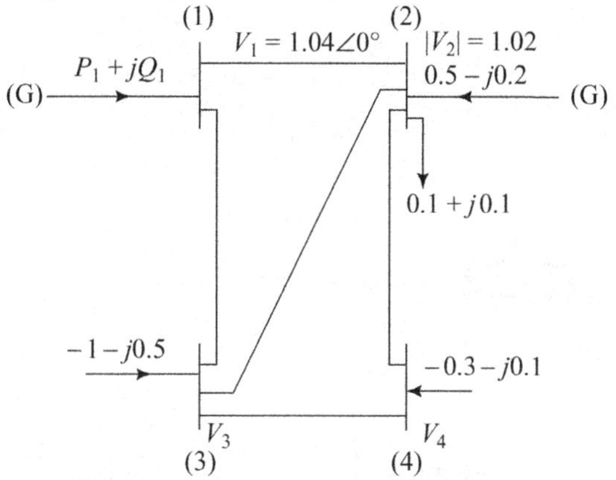

Fig. D.8 A 4-bus network

(i) Write the real and reactive power equations at buses **2** and **4**.
(ii) Do one iteration of Gauss–Seidel power flow and compute the complex power at slack bus **1**.
(iii) Solve the power flow until convergence using a computer program as discussed at the beginning of this section.

$$Y_{bus} = \begin{array}{c} \\ \mathbf{1} \\ \mathbf{2} \\ \mathbf{3} \\ \mathbf{4} \end{array}\begin{array}{c} \begin{array}{cccc} \mathbf{1} & \mathbf{2} & \mathbf{3} & \mathbf{4} \end{array} \\ \begin{bmatrix} 3-j9 & -2+j6 & -1+j3 & 0 \\ -2+j6 & 3.666-j11 & -0.666+j2 & -1+j3 \\ -1+j3 & -0.666+j2 & 3.666-j11 & -2+j6 \\ 0 & -1+j3 & -2+j6 & 3-j9 \end{bmatrix} \end{array}$$

5. Write the Jacobian matrices **H**, **N**, **M**, **L** symbolically using Newton–Raphson method for Problem 4 (this section).
6. Compute the matrices in Problem 5 (this section)–numerically, assumming a flat start of $1\angle 0°$ at buses **3** and **4**.
7. From Problem 6 (this section), show the matrices.for
   (i) decoupled load flow,
   (ii) fast decoupled load flow; and
   (iii) dc load flow.
8. For Problem 4 (this section) do a power flow using a computer program using the exact Newton–Raphson method and each of the methods mentioned in Problem 7 (this section). Discuss the results.

9. In Fig. D.9, bus **3** is the slack bus with $V_3 = 1\angle 0°$. Base MVA = 100.
   (i) Using dc power flow solve for θ_1 and θ_3.
   (ii) Compute the line flows in lines 1–2, 1–3 and 2–3.

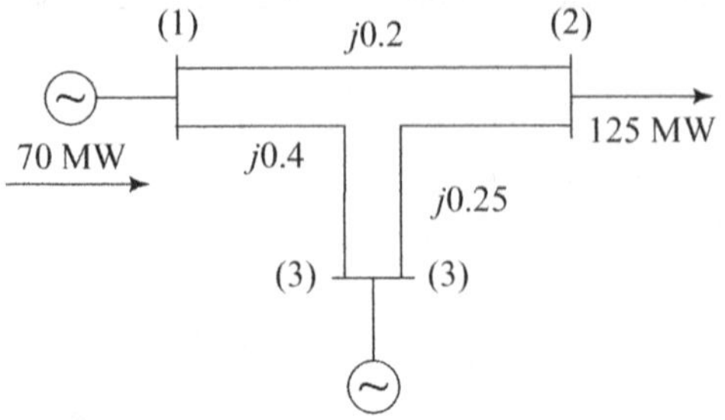

Fig. D.9 A 3-bus system

10. Given $\boldsymbol{Ax} = \boldsymbol{b}$

$$\text{where } \boldsymbol{A} = \begin{bmatrix} 2 & 3 & -1 & 0 \\ -6 & -5 & 0 & 2 \\ 2 & -5 & 6 & -6 \\ 4 & 2 & 2 & -3 \end{bmatrix}$$

   (i) Find the LU factors.

   (ii) With $\boldsymbol{b} = \begin{bmatrix} 20 \\ -45 \\ -3 \\ 30 \end{bmatrix}$ solve for $\boldsymbol{x}$ using the LU factors.

11. Using the computer, invert the $\boldsymbol{A}$ matrix in Problem 10 (this section and compare the results for x.
12. Find LU factors of

$$A = \begin{bmatrix} 5 & -2 & -3 \\ -3 & 7 & -2 \\ -3 & -3 & 8 \end{bmatrix}$$

and using the LU factors, solve $\boldsymbol{Ax} = \boldsymbol{b}$ where $\boldsymbol{b} = \begin{bmatrix} 4 \\ -10 \\ 6 \end{bmatrix}$

D.6 CHAPTER 8: STABILITY STUDIES

1. In Fig. D.10, bus **2** is an infinite bus ($H = \infty$). The complex power delivered by the generator at bus **1** is $1.15 + j0.75$ p.u. All the reactances in Fig. D.10 are in p.u.

(i) Find the voltage behind transient reactance of the generator.

(ii) Assuming $H = 3.0$ sec and $D = 0.01$ p.u., write the swing equations.

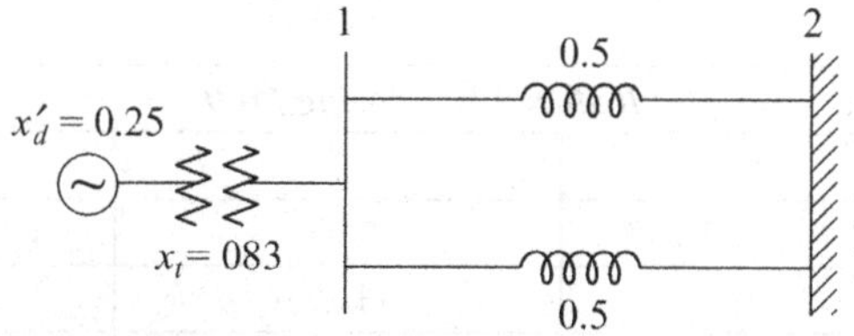

Fig. D.10 SMIB system

2. In Problem 1 (this section), if a self-clearing fault at bus **1** were to be cleared in 0.08 sec, find δ and ω as a function of time.
3. In Problem 2 (this section), change the clearing time in steps of 0.01 s and find the critical clearing time.
4. (i) Repeat Problem 2 (this section) if the fault is at the bus 1 end of one of the transmission lines and the line is cleared at both ends in 0.08.

   (ii) Find the critical clearing time.
5. (i) For the system in Fig. D.11, form the $Y_{\text{bus}}^{\text{aug}}$ as in Eq. (8.45).

   (ii) Form the reduced Y_{bus} at the internal nodes **4** and **5** by reduction method, that is eliminate buses **1**, **2** and **3**.

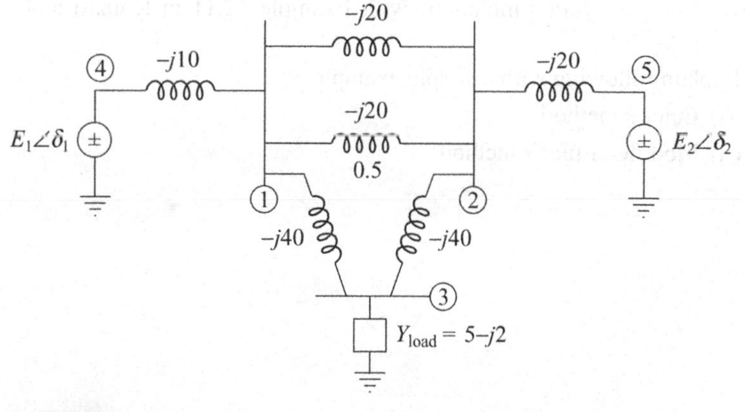

Fig. D.11 Multimachine system (all values in admittances)

6. The prefault power angle curve of a SMIB system is $P_{el} = 2.10 \sin \delta$ and the mechanical power is $P_m = 1.0$ p.u. A three-phase bolted short-circuit fault occurs at the terminals of the machine so that P_{e2} during fault $= 0$ and is cleared with no line clearing, that is P_e after clearing $= 2.10 \sin \delta$. Assume $H = 5.0$.

   (i) Find δ_{cr} and t_{cr} by repeated simulation.

   (ii) Repeat (i) if $P_{e2} = 1.5 \sin \delta$.

7. A three-machine system has the following reduced Y_{bus} for fault postfault conditions (Table D.2).

Table D.2

	Reduced Y_{bus} during fault		
Bus	1	2	3
1	$5.7986 - j35.6301$	0	$-0.0681 + j5.1661$
2	0	$-j11.236$	0
3	$-0.0681 + j5.1661$	0	$0.1362 - j6.2737$
	Reduced Y_{bus} postfault		
1	$1.3932 - j13.8731$	$-0.2214 + j7.6289$	$-0.0901 + j6.0975$
2	$-0.2214 + j7.6289$	$0.5 - j7.7898$	0
3	$-0.0901 + j6.0975$	0	$0.1591 - j6.1168$

The inertias are $H_2 = 12$ s and $H_3 = 9$ s. The internal voltages as **1**, **2** and **3** are $E_1 = 1.0\angle 0°$, $E_2 = 1.096\angle 19.35°$ and $E_3 = 1.071\angle 18.25°$. Bus **1** is an infinite bus and hence $\delta_1 = 0$. Hence, there is no swing equation at bus **1**. Assume damping as zero. $P_{m2} = 3.25$ $P_{m3} = 2.1$ p.u.

(i) Write the swing equations.

(ii) Compute the swing curves if the fault clearing time is 0.275s.

(iii) Repeat (ii) for $t_{cl} - 0.08$ s.

(iv) Compare your simulations with Example 12.11 in Kothari and Nagrath [1].

8. Explain following with a simple example.

(i) Euler's method.

(ii) Modified Euler's method.

REFERENCES

1. G. W. Stagg and A. H. El-Abiad, *Computer Methods in Power Systems*, McGraw-Hill, New York, 1968.
2. D. P. Kothari and I. J. Nagrath, *Modern Power System Analysis*, 3rd ed. New Delhi: Tata McGraw Hill, 2003.
3. I. J. Nagrath and D. P. Kothari, *Power System Engineering*, Tata McGraw Hill, 1994.
4. C. L. Wadhwa, *Electrical Power Systems*, 3rd ed. New Delhi: New Age International, 2000.
5. O. L. Elgerd, *Electric Energy Systems Theory*, New Delhi: Tata McGraw Hill, 2002.
6. H. Saadat, *Power System Analysis*, New Delhi: Tata McGraw Hill, 2001.
7. S. A. Nasar, *Schaum's Outline of Theory and Problems of Electric Power Systems*, New Delhi: Tata McGraw Hill, 2001.
8. G. L. Kusic, *Computer-Aided Power System Analysis*, Prentice-Hall of India.
9. D. P. Kothari and J. S. Dhillon, *Power System Optimization*, Prentice Hall of India.
10. A. R. Bergen and V. Vittal, *Power System Analysis*, India: Pearson Education, 1999.
11. K. R. Padiyar, *Power System Dynamics, Stability and Control*, BS Publications, Hyderabad, 2002.
12. P. W. Sauer and M. A. Pai, *Power System Dynamics and Stability*, Singapore: Pearson Education, 2003.
13. M.A. Pai, D.P. Sengupta, K.R. Padiyar, *Small Signal Analysis of Power Systems*, Narosa Publishing House, 2004.
14. M. A. Pai, *Power Circuits and Electromechanics*, Narosa Publishing House, 2009.
15. P. Kundur, *Power System Stability and Control*, McGraw-Hill, New York, 1994.
16. S. A. Soman, S. A. Khaparde, and S. Pandit, *Computational Methods for Large Sparse Power System Analysis*, Boston, MA: Springer, 2002.
17. John J. Grainger, William D. Stevenson, Jr., *Power System Analysis*, McGraw-Hill Inc, 1994.
18. L. P. Singh, *Advanced Power System Analysis and Dynamics*, New Age International, New Delhi, 2003.
19. Mariesa Crow, *Computational Methods for Electric Power Systems*, CRC Press, Boca-Raton, Florida, 2003.

20. J. Arrillaga, *High Voltage Direct Current Transmission*, The Institution of Electrical Engineers, London, UK, 1998.
21. E. W. Kimbark, *Direct Current Transmission*, Wiley-Intersciece, 1971.
22. K. R. Padiyar, *HVDC Power Transmission Systems: Technology and System Interactions,* New Age International (P) Limited and Publishers.
23. N. G. Hingorani, L. Gyugyi, *Understanding FACTS*, IEEE Press, 2001.
24. N. Christl, R. Hedin, P. E. Krause and S. M. McKenna, "Advanced series compensation (ASC) with thyristor controlled impedance," *CIGRE General Session*, Paris, 1992.
25. Enrique Acha, *FACTS Modelling and Simulation in Power Networks*, Wiley India Pvt. Ltd., 2012.
26. J. Arrillaga, N.R. Watson, *Computer Modelling of Electrical Power Systems*, John Wiley & Sons (Asia) Pvt. Ltd., 2005.
27. Allen J. Wood, Bruce F. Wollenberg, *Power Generation, Operation and Control*, John Wiley & Sons, Inc., 2013.
28. L. K. Kirchmayer, G. W. Stagg, *Analysis of Total And Incremental Losses in Transmission Systems*, Transactions of the American Institute of Electrical Engineers, Vol. 70, Issue 2, 1951.
29. H. W. Dommel and W.F. Tinney, "Optimal Power Flow Solutions" *IEEE Transactions on Power Apparatus and Systems*, Vol. PAS-87, No. 10, October 1968.
30. T. B. Nguyen, and M. A., Pai, 2003, Dynamic security-constrained rescheduling of power systems using trajectory sensitivities, *IEEE Transactions on Power Systems*, **18**(2) pp. 848-854
31. F. Mei, *Small Signal Modeling and Analysis of Doubly Fed Induction Generators in Wind Power applications*, Ph.D. Thesis 2007, Imperial College London, University of London.
32. F. Mei and B. Pal, *Modal Analysis of Grid-Connected Doubly Fed Induction Generators*, IEEE Trans. on Energy Conversion, Vol. 22, No. 3, (2005), pp. 728-736.
33. Thomas Ackermann, *Wind Power in Power Systems*, John Wiley & Sons, 2005.
34. S. N. Bhadra, D. Kastha, S. Banerjee, *Wind Electrical Systems*, Oxford University Press India, 2005.
35. Wei Qiao, Ronald G. Harley, "Effect of Grid-Connected DFIG Wind Turbines on Power System Transient Stability", IEEE Power and Energy Society General Meeting—Conversion and Delivery of Electrical Energy in the 21st Century, 2008.
36. Website of Ministry of Power, Government of India, http://powermin.nic.in/
37. Website of Centre for Wind Energy Technology, Government of India, http://www.cwet.tn.nic.in/
38. Archive of University of Washington, Electrical Engineering http://www.ee.washington.edu/research/pstca/

INDEX

U

V

W

Made in the USA
Monee, IL
05 November 2026

67807958R00174